March of the Pigments
Color History, Science and Impact

March of the Pigments
Color History, Science and Impact

By

Mary Virginia Orna
College of New Rochelle, USA
Email: maryvirginiaorna@gmail.com

Print ISBN: 978-1-83916-315-9

EPUB ISBN: 978-1-83916-326-5

A catalogue record for this book is available from the British Library

The Royal Society of Chemistry is a charity, registered in England and Wales, Number 207890, and a company incorporated in England by Royal Charter (Registered No. RC000524), registered office: Burlington House, Piccadilly, London W1J 0BA, UK, Telephone: +44 (0) 20 7437 8656.

Visit our website at www.rsc.org/books

Printed in the United Kingdom by CPI Group (UK) Ltd, Croydon, CR0 4YY, UK

Dedication

This volume is dedicated to my Ursuline sisters for their support and encouragement and to all of my ACS colleagues who have read, suggested, debated, reviewed and corrected my bumbling efforts. Whatever errors remain are solely my own.

I would be remiss not to mention that this book was written during the most trying months of the COVID-19 pandemic. I salute with great gratitude those dedicated health professionals who risked their lives to keep us safe and healthy and the scientists who worked around the clock to develop vaccines in record time.

March of the Pigments: Color History, Science and Impact
By Mary Virginia Orna
© Mary Virginia Orna 2022
Published by the Royal Society of Chemistry, www.rsc.org

Preface

Pigments speak of passion – passion in their creation and in their application. They have graced our world with joy, delight, symbolism, protection, identity and meaning. They pervade every aspect of human life from the food we eat, the clothes we wear and the buildings we build. They color our bodies inside and out. Optical pigments shapeshift so that we can see form and color; dermal pigments protect our bodies from the harmful effects of ultraviolet light; pigments in our food give delight to the eye or warning of toxicity. Animals depend on pigments for camouflage, caution and connection. They are also the engines that govern growth, climate and the food we eat. They have given form to our artistic expression from the dawn of civilization. They are the products of Nature's wisdom, the forge of Vulcan, Egyptian furnaces, alchemical paint pots and slick modern laboratories. Their march through human history and the effect they have had on that history is the subject of this volume. This journey begins 45 000 years ago in Paleolithic caves around the world. It ends on the threshold of a future that will transform our very definition of a pigment.

This march could never have progressed, nor even begun, without an intimate partnership with chemistry. Beginning with the common hearth, once *Homo sapiens* introduced the reality of chemical change into routine activities, these changes found their way increasingly into the cultural expressions we

March of the Pigments: Color History, Science and Impact
By Mary Virginia Orna
© Mary Virginia Orna 2022
Published by the Royal Society of Chemistry, www.rsc.org

call art. Little by little, the color palette expanded, aided by the trial-and-error of alchemists and early chemical practitioners. Archaeological expeditions and excavations have unearthed priceless signposts to lost cultures that we are still discovering today. Accompanied by occasional quantum leaps, pigment discovery and usage has advanced to the steady drumbeat of technological and theoretical development. Pigments have also been the bellwether of economic progress and prosperity: both the desire and production of colored products have accompanied the advance of civilization.

It is hoped that this comprehensive chronicle of how pigments have impacted humanity from the dawn of history will convey an appreciation of both Mother Nature and the civilizing influence of the chemical enterprise. But most of all, we hope that this entry into the world of pigments opens the reader to the enchantment and beauty that is chemistry.[1]

Mary Virginia Orna

REFERENCE

1. P. Ball, *The Beauty of Chemistry: Art, Wonder and Science*, MIT Press, Cambridge, 2021.

Acknowledgments

I'm grateful to many colleagues and friends for their help in writing this book. First of all to my fine Royal Society of Chemistry editor, Michelle Carey, and next to the gracious interview provided by Dr Narayan Khandekar on the Forbes Pigment Collection at Harvard University.

For help with images I thank Melina Avery, Giacomo Chiari, Judith Dartt, Kathleen Feeney, Jeff Fieberg, Steven Fine, Felix Karlsson, Petr Kostrhun, Michelle LaBerge, MAN Museum Madrid, Jim and Jenny Marshall, Maureen McCarthy, Daniel Meyer, Bob Pribush, Dan Rabinovich and Greg Smith. Special thanks to my advisors on the cover image: three Ursuline artists, Martha Counihan, Anne Therese Dillen, and Terry Eppridge.

For their compelling poetry: Ann Naito Haney, Jim Powell and Alberto Rios.

For help with references: Luis Avila, Hilary Becker, Simon Cotton, Paul Dillon, Marco Fontani, Catherine Giles, Mihaela Leonida, David Lewis, Patrick McGovern, Maria João Melo, Jilleen Nadolny, Bob Pribush, Greg Purcell, Dan Rabinovich, Jonathan Rees, David Reese, Ron Tempest and Tony Travis.

For their thorough, meticulous, expert reviews, and with gratitude for their many needed corrections:

Chapter 1: Jim Marshall, Jerry Sarquis, Jeff Seeman;
Chapter 2: George Fisher, David Hart, Gary Patterson;
Chapter 3: Millie Delgado, Terry Klazer, Nick Tsarevsky;

March of the Pigments: Color History, Science and Impact
By Mary Virginia Orna
© Mary Virginia Orna 2022
Published by the Royal Society of Chemistry, www.rsc.org

Chapter 4: Amina El-Ashmawy, Natalie Foster, Vera Mainz;
Chapter 5: Peter Corfield, Nancy LeMaster, Larry Westmoreland;
Chapter 6: Wayne Haag, Karen Hart, Jan Kochansky;
Chapter 7: Hilary Becker, Marco Fontani, Chris Vyhnal;
Chapter 8: Larry Krannich, Zaida Morales-Martinez, Mickey Sarquis;
Chapter 9: Mark Benvenuto, Carolyn Brockland, Betsy Jamieson;
Chapter 10: Lucy Eubanks, Joe Jeffers, Ken Latham;
Chapter 11: Ruth Beeston, Margaret Comaskey, Mamie Moy;
Chapter 12: Roxie Allen, David Lewis, Dan Rabinovich;
Chapter 13: Sara Hubbard, Val Kuck, Bob Pribush;
Chapter 14: Sharon Haynie, Mary Kochansky, Ron Tempest;
Chapter 15: Terry Eyrich, Zvi Koren, Pat Smith;
Chapter 16: Paul Dillon, Art Greenberg, Seth Rasmussen;
Glossary: Mark Benvenuto.

Any shortcomings and errors that remain I acknowledge as my very own.

Author Biography

Mary Virginia Orna, PhD, is Professor Emerita of Chemistry, College of New Rochelle, New York. Her academic specialties are in the areas of color chemistry and archaeological chemistry. Her more recent books include *The Chemical History of Color* (2013), *Science History: A Traveler's Guide* (2014), *The Lost Elements: The Periodic Table's Shadow Side* (2015), *Sputnik to Smartphones: A Half-Century of Chemistry Education* (2015), *Carl Auer von Welsbach: Chemist, Inventor, Entrepreneur* (2017), *Chemistry's Role in Food Production and Sustainability: Past and Present* (2019) and *Archaeological Chemistry: A Multidisciplinary Analysis of the Past* (2020). She has thirteen other authored, co-authored or edited books on chemical education and history of chemistry to her credit. She is also the recipient of numerous chemical education and service awards, the latest being the American Chemical Society 2021 HIST Award "for her original research in the area of color and pigment chemistry." In 1989, she was designated the New York State Professor of the Year, and in 1994 she served as a Fulbright Fellow in Israel. Her hobby is constructing crossword puzzles; she has contributed many of these to the *New York Times*. She is a religious of the Ursulines of the Roman Union.

March of the Pigments: Color History, Science and Impact
By Mary Virginia Orna
© Mary Virginia Orna 2022
Published by the Royal Society of Chemistry, www.rsc.org

Contents

CHAPTER 1

Dissecting Daylight:
How We See Color[†]

Colour is the place where our brain and the universe meet.
Paul Cézanne.[1]

1.1 INTRODUCTION

The history and usage of pigments are intimately related to the development of technology, physics and chemistry. This chapter introduces some of these technical details, including the nature of light and how it interacts with the human eye and colored substances, and how pigments as colored substances fit in. You can easily skip the technical details of this chapter without detracting from the joy of the pigment journey you are about to embark upon.

1.2 E.T. REVISITED?

If you looked up into Earth's big blue sky on 19 October 2017, you may have sighted a potential visitor from far beyond the solar system. The object, shaped like a cigar and dubbed 1I/2017 U1, was sighted just passing through our astronomical neighborhood.

[†]Glossary entries and chapter cross-references are in boldface.

March of the Pigments: Color History, Science and Impact
By Mary Virginia Orna
© Mary Virginia Orna 2022
Published by the Royal Society of Chemistry, www.rsc.org

Harvard astrophysicist Avi Loeb lent credence to this possible "social call" by aliens in a blog post for *Scientific American*, later transformed into a book.[2,3] If the inhabitants of 1I/2017 U1 had decided to stop off to say "hello," and if their point of origin were a planet tied to the closest red giant, Gamma Crucis, they would have stumbled out of their spaceship functionally blind.

1.3   THE NATURE OF LIGHT

Why is this so? Why couldn't these aliens see non-red objects? We need light to see and there are different forms of light – the light coming from our sun is different from the light emitted by Gamma Crucis. What is the nature of this difference? To find out, we need to back up a bit, examine the nature of light, and then its relationship to color.

All light is accompanied by heat, so we infer that they share the same nature, called energy, *i.e.*, anything that causes a change in motion. Energy exhibits many forms, such as heat, light, and electricity; all forms are interconvertible by interaction with matter. The ancient Greeks were the first to theorize regarding the nature of light and color. Aristotle (384–322 BCE) made the first important contribution to what is now the modern theory of selective absorption (Figure 1.4).[4] Then Seneca (4 BCE – 65 CE), a Roman stoic philosopher, first noted that a prism reproduces the colors of the rainbow. Leonardo da Vinci (1452–1519) noticed that when light struck a water glass that it "spread out" as a colored image, but it was Isaac Newton (1643–1727) in 1666 who formulated modern color theory by experiment.

Newton passed a ray of sunlight through a prism and noted that it dispersed into an array of colors on the opposite wall. When he placed another prism in the path of the dispersed colors, they recombined to form sunlight again, what today we call "white light." Newton had succeeded in dissecting sunlight into what we term the "prismatic" colors because they can be generated by a prism. He initially called them red, yellow, green, blue and violet[‡] in that order.[5,6] Newton was astounded by this revelation,

---

[‡]Newton later included "orange" and "indigo" largely to bring the number of spectral colors up to seven, matching the musical scale or his version of harmony and aesthetics. In his first lecture on the topic, he admitted that the colors were continuous and that specific lines of demarcation were difficult to discern.

a seeming paradox that prompted controversy for the following 300 years.

First Thomas Young (1773–1829) in 1801, and then James Clerk Maxwell (1831–1879) in 1865, theorized that light is a special form of energy that can travel through empty space at great speed§ generating electric and magnetic fields as it goes; these fields oscillate with a certain frequency (symbol v) which defines the energy of the light beam. The distance traveled by a single oscillation is called its **wavelength** (symbol λ), measured in a unit called a nanometer (nm), a billionth of a meter. This is a handy measure because our sunlight has a maximum intensity in the wavelength range of about 400 to 700 nanometers. Wavelength ranges above 700 nm, which defines the limit of what we call the color red, are in the infrared (IR) region; wavelength ranges below 400 nm, which defines the limit of what we call the color violet, are in the ultraviolet (UV) region. All of these energies are present in the sun's rays, which we can now call the solar spectral irradiance curve, shown in Figure 1.1, left.[7] The entire range of wavelengths, called the electromagnetic spectrum, highlighting the visible spectrum and its colors, is shown in Figure 1.1, right.

1.4 HOW WE SEE COLOR

Now, getting back to our visitors from Gamma Crucis, a star that emits a range of energy that maximizes in the IR, invisible to human eyes. But the evolutionary pattern of the light receptor cells in our visitors' eyes would have necessarily maxed out in that range (because that's where all the light is on their planet) to find food or avoid becoming someone else's dinner. Our human eyes, for the same reason – by dint of biological evolution – are sensitive to the energy range of the sun's rays of maximal intensity that actually manages to penetrate our atmosphere – namely the red, orange, yellow, green, blue and violet radiation.[8,9] Hence we call this range, or spectrum of energies, the visible region (for us!). If we traveled to our Gamma Crucis visitors' planet, it would be our turn to be blind – because their solar spectral irradiance curve would maximize in the infrared region.[10]

---

§Light travels with a velocity of 1 86 000 miles s^{-1} (or 3 00 000 km s^{-1}) in a vacuum. In other media, such as oil or water, it slows down, which affects the appearances of paints.

All light sources are not the same, as we have seen. It depends upon the star you were born under, but also the type of lighting in your living room. Incandescent light seems "warm" to us because incandescent lamps emit a greater intensity of red light than the other colors; fluorescent lighting seems "cool" to us because it is poor in red light, but rich in yellow, green and blue light. Each type of light source has its own spectral irradiance curve that is quite different from ordinary daylight or sunlight.

Figure 1.2 (left) illustrates how human retinal photoreceptors, consisting of rods and cones, also theorized by Young,[11] respond to light stimulus.[12,13] The tri-chromatic system involves retinal structures called cones that contain photopigment proteins called photopsins that are sensitive to short (blue), medium (green) and long (red) wavelengths of light respectively. Light

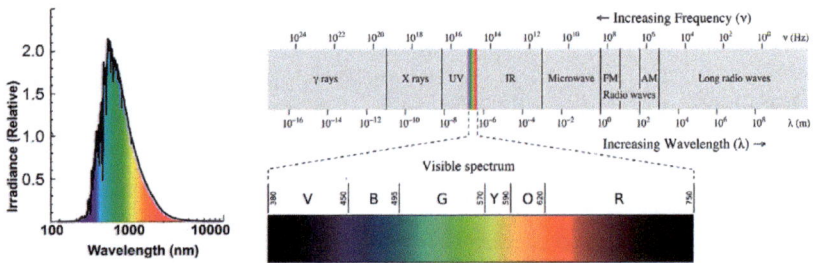

Figure 1.1 Left: Solar spectral irradiance curve. The height of the curve (*y*-axis) shows the intensity of the radiation plotted against the wavelength in nm. Note that maximum intensity is in the green region, falling off on both sides toward the red and violet regions. Reproduced from ref. 7.¶ Right: The electromagnetic spectrum showing all the regions; offset is the visible region, the only part of the spectrum that we can see. Note that the other regions of the spectrum are quite important in modern life as well. Reproduced from https://commons.wikimedia.org/w/index.php?title=File:EM_spectrumrevised.png&oldid=468982208, under the terms of the CC BY-SA 3.0 license, https://creativecommons.org/licenses/by-sa/3.0/deed.en.

¶The Solar Spectral Irradiance (SSI) CDR used in this study was acquired from the NOAA National Centers for Environmental Information (formerly NCDC) (http://www.ncdc.noaa.gov). This CDR was developed by Judith Lean at the Naval Research Laboratory (NRL) and Odele Coddington, and Doug Lindholm at the University of Colorado Boulder's Laboratory for Atmospheric and Space Physics (LASP), in collaboration with Peter Pilewskie, and Marty Snow (also at LASP), through support from NOAA's CDR Program using the NRLSSI2 model.

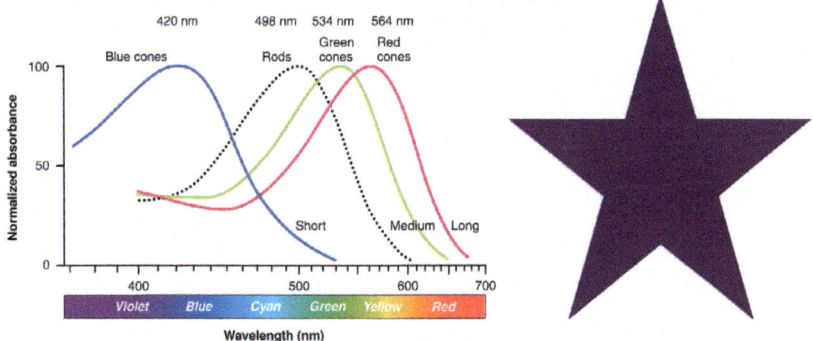

Figure 1.2   Left: Sensitivity curves for the rods and cones in the human retina. Reproduced from ref. 12, https://openstax.org/books/anatomy-and-physiology/pages/1-introduction, with permission from OpenStax under the terms of the CC BY 4.0 license, https://creativecommons.org/licenses/by/4.0/. Right: Magenta star.

stimulus causes these proteins to undergo a structural change that activates their ability to send an electrochemical signal to the brain. The rods, which are far more numerous and sensitive than cones, have only one photopigment, rhodopsin. Rods operate in low light to give gray images.[13] Prolonged gazing at a single color can exhaust the sensitivity of the respective cones, producing an afterimage of the complementary color. If you gaze at the magenta star (a color combination of red and blue) in Figure 1.2 (right) for several minutes and then shift your gaze elsewhere, you should see a faint green afterimage – green is the complementary color of magenta.

1.5   ADDITIVE AND SUBTRACTIVE COLOR MIXING

The color mixing circles shown in Figure 1.3 provide the rationale for complementary colors. Figure 1.3 (left) depicts additive color mixing, which is the mixing of lights.[14] Each of the three major regions of the visible spectrum can be roughly divided into the red (600–700 nm), the green (500–600 nm) and the blue (400–500 nm) regions. If all three colored lights are added together, they will produce white light, as seen in the center of the three overlapping circles (which is why sunlight appears white: it contains all of the spectral colors). We call red, green

Figure 1.3   Left: Additive primary color circles; Reproduced from ref. 14. Right: Subtractive color primary circles. Reproduced from https://commons.wikimedia.org/w/index.php?title=File:Subtractive_color_mixing.jpg&oldid=464911059 under the terms of the CC BY-SA 3.0 license, https://creativecommons.org/licenses/by-sa/3.0/deed.en.

and blue the additive primary colors. When neighboring circles overlap, they produce the complementary colors: green + blue = cyan (a turquoise color); blue + red = magenta (sometimes called purple); red + green = yellow. So the complementary colors are yellow (to blue), magenta (to green) and cyan (to red). Now can you see why staring at a magenta star overwhelmed your red and blue sensitive cones for a while, and you could only see green for a time. In Figure 1.3 (right), we see the subtractive color system. Here, the three subtractive primary colors are yellow, magenta and cyan; their respective complements, shown in the overlapping neighboring circles are, respectively, blue (magenta + cyan), green (cyan + yellow) and red (yellow + magenta). A combination of all three subtractive primaries yields black. This is the system used in four-color printing.[15] Check the colors of your laser printer cartridges – they correspond to the subtractive primaries plus black. We use this system when mixing pigments in a work of art: when complementary colors are placed adjacent to one another in a painting, they enhance one another producing a brightening effect; adjacent non-complementary colors, on the contrary, produce a duller effect (**Chapter 13.4**). Additive color mixing is good for adding lights such as theatre lights and the pixels on your television screen; it is also the principle behind photonic and structural colors (**Chapter 16.3**).

The seemingly contradictory perceptual results obtained from mixing light stimuli and mixing colored materials such as pigments has been a source of confusion for at least 2000 years. The system we describe here is based on color mixing. Other systems based on color perception and on color matching are very useful in other applications, particularly in the fashion world and computer color measurement techniques. The various color order systems that artists and color technologists have dealt with over the years to measure and explain these perceptions have great consequences in areas such as color matching, retouching, and art conservation.[16,17]

1.6 HOW LIGHT INTERACTS WITH MATTER

When light strikes a surface, many things happen to it simultaneously depending upon the nature of the surface. It can be reflected; that is how you can see the surface. Some light can be absorbed. If all the wavelengths of visible light are absorbed, you will see the color black; if almost all of the wavelengths of visible light are reflected, then you will see white. However, if only some of the wavelengths are absorbed, then you will see only the wavelengths that are reflected to your eye. For example, if all of the blue and green wavelengths (400–600 nm) are absorbed, you will see only the wavelengths not absorbed, 600–700 nm, corresponding to red light; you perceive that object as red. The blue–green combination, cyan, is the complementary color to red because when added back into red, it completes the visible spectrum. This effect is called selective absorption, a very important pigment property (Figure 1.4). The degree of selective absorption depends on the intrinsic chemical structure of the pigment: electrons undergo transitions which correspond to discrete energy levels from 400 to 700 nm in the pigment molecule.

The color you see also depends upon the source of the light. The sun Is our universal light source. Other light sources such as incandescent and fluorescent lighting do not have the same spectral irradiance curves as the sun, so objects do not appear the same when lighted with different light sources. Some sources are very rich in the red end of the spectrum, so their light will give

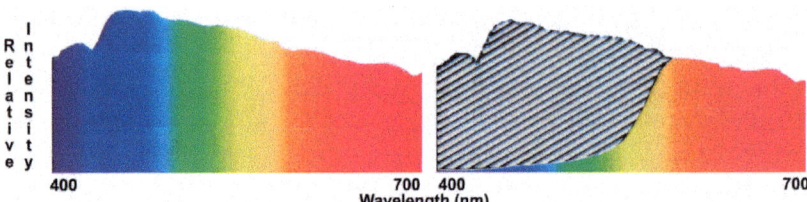

Figure 1.4   Illustration of selective absorption. Left: Solar spectral irradiance
curve for normal daylight over the range of the visible spectrum,
appearing white; Right: Selectively absorbed light (gray lines)
by an object that reflects the remainder of the visible spectrum,
appearing reddish-orange.

a warmish glow; others contain very little red light, so will seem
colder to us. In comparing colors of objects, it is very important to have the same light source, otherwise they will appear
different.[18]

1.6.1   Light Scattering

Light can also be scattered as it travels. This phenomenon
accounts for the fact that the sky is blue and that sunlight, even
though we call it "white" light, has a slightly yellow tinge. The
molecules in our atmosphere, mostly nitrogen and oxygen, act
as scattering centers. As sunlight enters the atmosphere, the air
molecules divert the various rays, and violet and blue rays are
more strongly scattered than the other wavelengths. In fact, the
sky is more violet than blue, but our eyes are not as sensitive to
violet. Physicist Lord Rayleigh (1842–1919) found that the shorter
the wavelength of light, the greater the scattering.[19]

   Scattering also affects the appearance of a pigment's color. Early
artists knew that a pigment's particle size affected the depth of its
color. Finely ground pigments are paler than coarsely ground ones
because scattering power increases as particle size decreases. As
the particle size decreases, the amount of light scattered increases
relative to the amount of light absorbed, so the smaller particles
appear lighter in color. In fact, some pigments can only be ground
very coarsely because too much grinding causes their colors to
fade so much that they become useless.[20] Figure 1.5[21] shows an
example of coarsely and finely ground malachite crystals; not only
does the color depth change but also the hue.

Figure 1.5 Malachite crystals. Left: coarsely ground pigment; Right: finely ground pigment. Reproduced from ref. 21 with permission from Natural Pigments (naturalpigments.com).

Other things can happen if the object receiving the light is transparent. Light striking a clear glass window is transmitted, *i.e.*, all the wavelengths pass right through allowing you to view what is on the other side. If the glass is red, you will see only the wavelengths of the red light. When light travels from one medium into another, *e.g.*, from air into water, both the velocity and the wavelength change, but the frequency always remains the same. Put simply, light slows down or speeds up when it moves from one medium to another. The degree of slowing down compared to travel in a vacuum is measured by the **refractive index** (RI) of the medium. Each frequency of light has its own RI, which accounts for the different behaviors of light traveling through a prism and, indeed, through the atmosphere to produce rainbows and blue skies. This property is also extremely important with respect to paints; it determines whether a paint layer is transparent or opaque when formulated with different pigments and binders because both pigment particles and binders are transparent and each has its own refractive index. If they are similar, the paint will be transparent; if they are quite different, the light will be diverted many times in traveling through the particles, and not much will be reflected, so the paint will be opaque (Figure 1.6).

Early humans, and even late modern humans, were ignorant of the properties of light discussed here. But, by trial and error,

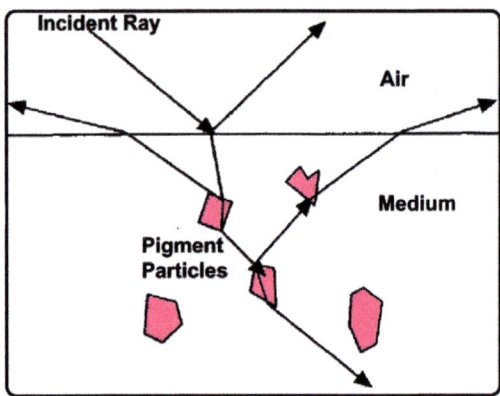

Figure 1.6   An incident ray of light travels from air into a paint layer. When it strikes the paint medium, it is refracted (changes direction) because the RI of the medium is different from the RI of air. It undergoes multiple reflections and refractions in the transparent pigment particles. It may be reflected back into the air in the process. If the ray does not reach the ground, it cannot reflect the ground back to the viewer; in this case the paint is opaque. Reproduced from ref. 18 with permission from Springer Nature, Copyright 2013.

they developed skills in transferring pigments, the concrete substances that selectively absorbed visible light, to any desired surface. Now that we know the principles, we have to see what pigments themselves are all about.

1.7   WHAT IS A PIGMENT?

Broadly defined, a pigment is any substance that imparts color. All materials used for coloring other objects may be classified as either pigments or dyes, depending upon the method of application. Typically, dyes are dissolved in a medium into which the object to be dyed is then immersed. A physical or chemical interaction between the dye and a substrate is necessary for the dye to become "anchored" in place. Pigments, on the other hand, do not react with the substrate and must be applied by first being suspended as insoluble particles in a medium, called the vehicle or binder, which allows the pigment to adhere to the surface. The more general term for both dyes and pigments is "colorant."

The colorants you will encounter in this book are drawn from all three of the natural "kingdoms," animal, vegetable and mineral. We call mineral pigments those derived from either naturally occurring rocks or other geological substances, such as ores, found in the earth, or synthesized from similar starting materials. Chemists term them "inorganic," an adjective that applies to materials made from any element except carbon. Since none of these substances is soluble in ordinary solvents, all of them require a binder. Colorants derived from plants and animals are carbon-based or, in chemical terms, organic compounds. Many are soluble in water or in another liquid and can be used as dyes; many are also often used as pigments, but some require special treatment to render them opaque enough to be used in paints.

1.7.1   Pigment Characteristics

Just as we need three dimensions, length, width and height, to describe an object, so color, and therefore pigments, need three dimensions as well: hue, chroma and value.

- Hue is the actual color of the pigment in terms of where it falls in the visible spectrum. For example, a green-hued pigment will reflect wavelengths largely in the green region, about 500 to 600 nm.
- Chroma, or amount of color, is sometimes called intensity or saturation. It is not related to either the hue or the lightness or darkness of the pigment, but it is an index of how pure or intense the color is.
- Value is a measure of the lightness or darkness of a pigment, usually measured from 1 to 10 on a grayscale, where 1 is more white and 10 is more black. All three characteristics are affected by the choice of a binder and how well it is bound to the pigment.[22]

1.7.2   Pigment Qualities

In addition to the refractive index (RI), described earlier in this chapter, several other pigment qualities are important.

- Compatibility. Pigments can be compatible or incompatible with other pigments or with some binders and varnishes.
- Undertone and Masstone. These qualities describe a pigment's difference of appearance when applied as a transparent or an opaque film.

- Lightfastness. Fastness is a pigment's ability to resist fading when exposed to ultraviolet or visible light.
- Tinting Strength. This quality refers to a pigment's ability to color a mixture.
- Covering Power. A pigment's degree of opacity.
- Toxicity. All inorganic pigments made with **heavy metals** (such as mercury and lead) have varying degrees of toxicity. Some organic pigments may be carcinogenic, another form of toxicity.[23]

1.7.3 Pigments Classified by Color

Color, at least on the visual level, is the most obvious property an object can have. Hardly a day goes by that one does not find it necessary to name a color, and we must presume that this need goes back to the first societies and the first use of human language. You will find as you read the rest of this volume that many pigments were first named according to their colors, as were many of the chemical elements. As the need developed in art, commerce, and in many other areas, color names continued to be invented so that today, there are almost 10 000 color names. The need to translate among color vocabularies so that their meaning is clear[24] is a daunting task since some of them are completely unintelligible to persons working in a different field.

It would take many pages to list the entire body of colorants used as artists' pigments. Furthermore, even if they were all listed by name, ambiguities would creep in because some very different chemical substances have historically been given the same pigment name. If there was confusion and ambiguity about colorant nomenclature in general, the situation was exacerbated many-fold by the developments in the dye and color industry. Almost every day a new colorant was developed in the patent races following the 1856 discovery of the first commercially produced dye. Many of these compounds had uncertain identities, complex structures and arbitrarily bestowed names.[25] *The Colour Index* (CI) dispenses with this uncertainty by listing identification numbers and names given to individual pigments by the Society of Dyers and Colourists in the UK and the American Association of Textile Chemists and Colorists in the United States.[26] Pigments are identified by both their common names and their *Colour Index* Constitution Number when available. *The Colour Index* currently lists over 13 000 generic colorant names under more than 45 000 commercial names, although very few of them are of interest to artists. *The Color of Art Pigment* Database[27] has sliced out the latter and provides full CI information for each pigment. Another reference, equally monumental in its scope, is *The Pigment Compendium*.[28] It contains information and as-complete-as-possible literature sources (by "plundering" specialist surveys and "hunting down" original sources, as the compilers put it) on over 1000 separate compounds listed with a consistent naming system in almost 2600 entries, 80% of them made of inorganic materials, 16% from natural organic sources, and 4% synthetic organic pigments.

1.8 CONCLUSION

In Chapter 16, we will see that the traditional definitions and properties of pigments described here become more fluid. Their colors can be manipulated by applying various forms of energy, and their very substance can be generated by some exotic creatures. In some instances, pigments will have no substance of their own at all but will depend upon assemblies of precisely arranged structures. Meanwhile, our pigments will march on through history, responsive to the drums of cultures and civilizations.

In the next chapter, 21st century technology will give way to the cutting edge skills of humans who lived and worked 45 000 years ago and who still speak to us "through shimmering patterns [thrown] across slabs of solid rock."[29] Their voices remind us of our own vitality and mortality, echoing over a time span nine times greater than our own recorded history.

REFERENCES

1. J. Gasquet and C. Pemberton (Translator), *Joachim Gasquet's Cézanne: a Memoir with Conversations, (1897–1906)*, Thames and Hudson, London, 1991, p. 153.
2. E. Kolbert, *The New Yorker*, 25 January 2021, 60–64.
3. A. Loeb, *Extraterrestrial: The First Sign of Intelligent Life beyond Earth*, Houghton Mifflin Harcourt, New York, 2021.
4. Aristotle, *De Sensu*, The Clarendon Press, Oxford, 1908, ch. III, pp. 439–440.

5. I. Newton, *Opticks*, Dover Publications, New York, 1952, p. 50.
6. D. Topper, *Stud. Hist. Philos. Sci.*, 1990, **21**(2), 269–279.
7. O. Coddington, J. L. Lean, D. Lindholm, P. Pilewskie, M. Snow and NOAA CDR Program, *NOAA Climate Data Record (CDR) of Solar Spectral Irradiance (SSI), NRLSSI Version 2*, NOAA National Centers for Environmental Information, 2015, DOI: 10.7289/V51J97P6, https://www.ncdc.noaa.gov/cdr/atmospheric/solar-spectral-irradiance, accessed June 2021.
8. R. Finlay, *J. World Hist.*, 2007, **18**(4), 383–431.
9. C. Köberl, Our colourful world, in *Colours of Hallstatt: Textiles Connecting Science and Art Exhibition*, 1 February 2012 – 6 January 2013, Catalogue, Natural History Museum, Vienna, 2012, p. 6.
10. A. Lançon, P. H. Hauschildt, D. Ladjal and M. Mouhcine, *Astron. Astrophys.*, 2007, **468**, 205–220.
11. T. Young, *Philos. Trans. R. Soc.*, 1802, 9212–9248.
12. J. Gordon Betts, K. A. Young, J. A. Wise, E. Johnson, B. Poe, D. H. Kruse, O. Korol, J. E. Johnson, M. Womble and P. DeSaix, *Anatomy and Physiology*, OpenStax, Houston, Texas, 25 April 2013, https://openstax.org/books/anatomy-and-physiology/pages/14-1-sensory-perception (access for free), accessed August 2021.
13. H. Zollinger, *Color: a Multidisciplinary Approach*, Wiley VCH, New York, 1999, pp. 79–121.
14. additivecolor.svg, https://commons.wikimedia.org/w/index.php?title=File:Additive_color.svg&oldid=487482422, accessed June 2021.
15. R. M. Christie, *Colour Chemistry*, Royal Society of Chemistry, Cambridge, 2nd edn, 2015, pp. 23–26.
16. R. G. Kuehni and A. Schwarz, *Color Ordered*, Oxford University Press, New York, 2008.
17. R. S. Berns, *Billmeyer and Saltzman's Principles of Color Technology*, Wiley, New York, 4th edn, 2019, pp. 37–84.
18. M. V. Orna, *The Chemical History of Color*, Springer, Heidelberg, 2013, pp. 11–28.
19. M. Minnaert, *The Nature of Light and Color in the Open Air*, Dover Publications, New York, 1954, pp. 238–239.

20. M. V. Orna, Historic mineral pigments: colorful bench-marks of ancient civilizations, in *Chemical Technology in Antiquity*, ed. S. C. Rasmussen, American Chemical Society, Washington, DC, 2015, pp. 17–69, at 58–59.
21. G. O'Hanlon, *Traditional Oil Painting: The Revival of Historical Artists' Materials*, Figure 8, https://www.naturalpigments.com/artist-materials/traditional-oil-painting-revival/, accessed June 2021.
22. B. Baade, The use of dry pigments in inpainting, in *Painting Conservation Catalog, vol. 3, Inpainting*, ed. American Institute for Conservation Paintings Specialty Group, AIC, Washington, DC, 2010.
23. M. D. Gottsegen, *The Painter's Handbook*, Watson-Guptill, New York, 2006.
24. K. L. Kelly, *Color Universal Language and Dictionary of Names*, National Bureau of Standards SP440, Washington, DC, 1976.
25. M. P. Crosland, *Historical Studies in the Language of Chemistry*, Harvard University Press, Cambridge, 1962, p. 293.
26. The colour index, www.colour-index.com, accessed March 2021.
27. The Color of Art Pigment Database, http://www.artiscreation.com/Color_index_names.html#.YDF3qPlOmM9, accessed February 2021.
28. N. Eastaugh, V. Walsh, T. Chaplin and R. Siddall, *The Pigment Compendium: A Dictionary of Historical Pigments*, Elsevier, New York, 2004.
29. S. Kriss, *Pocket Worthy*, https://getpocket.com/explore/item/what-the-caves-are-trying-to-tell-us?utm_source=pocket-newtab, accessed February 2021.

Dark Unfathom'd Caves: The Earliest Cultural Use of Color[†]

Full many a gem of purest ray serene, The dark, unfathom'd caves of ocean bear.[1]

Thomas Gray

2.1 INTRODUCTION

Messages from our distant forebears decorate ancient caves and rock shelters. Their images are astoundingly modern, seeming to have anticipated the works of Kandinsky and Picasso. Though their discovery mesmerized the world, their meaning remains one of the greatest of archaeological mysteries.

2.1.1 Discovery of Paleolithic Art

"¡Mira, papá, bueyes!" "Look, papa, oxen!" The most famous three words in paleontology, breathed by an 8-year-old little girl, ushered in the study of the painted caves of Europe. But not without some rocky starts. In 1878, Spanish amateur archaeologist,

---

[†]Glossary entries and chapter cross-references are in boldface.

March of the Pigments: Color History, Science and Impact
By Mary Virginia Orna
© Mary Virginia Orna 2022
Published by the Royal Society of Chemistry, www.rsc.org

Don Marcelino Sanz de Sautuola (1831–1888), carried back visions of the Stone Age from the Universal Exhibition in Paris. Fascinated by what he saw, he started to dig about his property in Santander, Spain. Accompanied by his 8-year-old daughter, María, he cleared away some brush and overgrowth from the almost-forgotten Altamira cave entrance, and they both clambered down into a low-ceilinged chamber. Don Marcelino began to search the cave floor for tools and artifacts when María suddenly made her famous discovery.[2] While Papá was looking down, María was looking up. Wide-eyed, she spied an ox, over seven feet long, and so much more: the now-famous polychrome ceiling paintings hidden from human eyes for over 35 000 years.

Paleolithic cave art, the first known cultural use of color, brings us face to face with a unique communication from our distant forbears. It opens a window on a long lost world that we will explore by visiting three of these caves in southern France: Lascaux, Chauvet and Rouffignac. We will thrill to their discovery and wonder at the sophistication of an art that seems thoroughly "modern." Each cave has its own unique story to tell: its discovery, its chronology, its subsequent treatment and fate at the hands of modern humans. We will encounter controversy and differences of opinion. We will wonder at the possible reasons for the art and the artists' choice of materials. We will meet animal species that no longer exist documented in detail on the cave walls and ceilings. We will marvel at the degree of chemical know-how these people perfected with only simple tools and the family hearth. Finally, we will learn how modern chemical instrumentation can unveil the timeline, methods, materials, accomplishments and possible modes of preservation of this priceless heritage.

2.1.2 Early Controversy and Validation

Don Sautuola speedily enlisted the help of University of Madrid archaeologist Juan Vilanova y Piera to verify Altamira's antiquity, and in 1880, he published a small volume documenting his find.[3] In it he remarked: "I was surprised to find and contemplate a large number of painted animals on the ceiling of the cave. They were painted with red and black ochre and were very large, mostly figures looking like bison due to the curvature of their hump. I saw two complete bison in profile, others missing their heads, and some in incomprehensible positions, and others made with just a few strokes because the colors with which they had been made had more or less disappeared. There was also the figure of an entire roe deer, very well made, and a head that looked like a horse. In all, there were 23 figures, not counting several others of which only traces remain. One figure was 125 cm (4 ft) high, one 155 cm (5 ft) long, and the roe deer was 220 cm (over 7 ft) long and 140 cm (4.5 ft) high. On examination, the painter was very well practiced and in making them had a firm hand because there was no hesitation but, on the contrary, every figure was done at once in one clear swoop, remarkable given that the ceiling surface was so uneven. But I must ask the question as to who painted these figures and have come to the conclusion that they had to have been done in ages past since none of the animals in the drawings existed at present in Europe, particularly elephants and bison."

Perhaps it was the low-key presentation. Maybe it was his lack of standing in the scientific community. Whatever it was, Sautuola experienced the biggest letdown of an aspiring archaeologist: general dismissal. His assertion that the paintings were Paleolithic in origin even raised accusations of fraud and naiveté. One of Sautuola's most vociferous attackers was Émile Cartailhac (1845–1921), a recognized and prestigious prehistory expert. It was only in 1902, after findings in southern France vindicated Sautuola's work, that Cartailhac acknowledged his own errors by publishing a public admission[4] and by apologizing to María in person. It was, sadly, too late for her father, who had died in 1888.

Once the Altamira cave's antiquity was verified, dated to the Aurignacian period, 35 000 to 23 000 before present (BP) (Table 2.1),[5] our perception of early human beings immediately

Table 2.1 Chronology of the Upper Paleolithic Era (50 000–15 000 BP). Reproduced from ref. 4 with permission from American Chemical Society, Copyright 2015.

Culture	Time Period (BP)	Span (years)
Aurignacian	35 000–23 000	12 000
Gravettian	28 000–22 000	6000
Solutrean	21 000–16 000	5000
Magdalenian	18 000–10 000	8000

emancipated "cavemen" from their commonly held image of lumbering doltish brutes. In the face of some of the most remarkable art ever conceived, we had to admit that the Stone Age artists were creative, sensitive, abstract thinkers with incredible technological and aesthetic know-how engaged in collaborative efforts over intergenerational eons. Pablo Picasso (1881–1973), after viewing the Lascaux paintings in 1940, is said to have remarked "They've invented everything."[6] And taking "them" as his teachers, he honed his skills at stripping away the unnecessary details that encumbered his own artistic breakthrough. His famous bull studies are a case in point.[7]

2.1.3   The "What?" of Cave Art: First Chemical Analysis

Henri Moissan (1852–1907), 1906 Nobel laureate in chemistry, was among the first to carry out analyses on the paints found in the caves of Font-de-Gaume and La Mouthe in the Dordogne valley.[8] He reported, after examination with a microscope, the presence of manganese oxide mixed together with calcite and small grains of quartz. He also remarked that the irregular red hematite (iron oxide, Fe_2O_3) deposit was completely different from what "had been fixed to the wall by the hand of man as a primitive fresco." A few years later, Cartailhac and the Abbé Henri Breuil (1877–1961), acknowledged doyen of French prehistory, described, without a shred of experimental justification, a black pigment from the Salon Noir of Grotte de Niaux as carbon, manganese oxide, and animal fat.[9] Chemical analysis in those days consisted more of opinion and comparison than of actual hard data. It would take another 65 years before chemists could pin down the chemistry of the caves. In 1977, instrumental investigations began where it all started: at Altamira.[10,11]

2.1.4   The "When?" of Cave Art: Chronology

This "thoroughly modern" cave art (according to Picasso) coincided with what archaeologists dub the Upper Paleolithic transition, marked by a paradigm shift in artistic expression by way of sculpture, painting, and even body painting. Archaeologists have found what seem to be tattoo kits among the artifacts uncovered in some excavations, a topic we will explore in the next chapter.

Box 2.1   U-Series Dating

Uranium-series dating uses the **radioactive decay** of uranium relative to one or more of its daughter products to calculate the age of an object. Natural uranium is present in extremely tiny amounts in all rocks on earth. Since its principal salts are soluble in water, it is carried into caves where the slow evaporation of the water forms stalagmites, or other types of shapes, over eons of time. Some of these formations take the form of calcite (calcium carbonate) films over cave paintings, so even though the painting itself cannot be directly dated by this method, some calcite lying very near to the painting can. One of the major daughters of uranium used in this method, thorium, has insoluble salts, so it will not be carried into the cave by the water. So at time zero, when the cave uranium starts decaying, no thorium is present. By measuring the buildup of thorium and knowing the rate of decay of the uranium to form the thorium, we can calculate how long ago the buildup began. The date will give us the minimum age of the painting underneath the calcite, assuming that the calcite began to accrue soon after the painting was executed, assuming no movement of the thorium or uranium as well. This method, applied to rocks or cave formations, can measure time periods of up to 5 00 000 years.

However, one could not conceive of just how far back this transition occurred until uranium–thorium (U-series) dating techniques (see Box 2.1) showed that Paleolithic paintings flowered over a 25 000-year period,[12] time spans that have never been achieved by any other culture. From this point of view, our own historical time periods, ancient, medieval, and modern, seem like a mere blink of an eye.

2.1.5   More Discoveries

In the early 20th century, many more caves were discovered. Some of them, including Altamira and Lascaux, were open to the public for decades but are now closed due to the over-tourism that led to an alarming rate of visitor-induced deterioration, threatening their very integrity.[13] The constant temperatures, atmospheres and infrequent human interventions that once prevailed stabilized the caves; it became quite clear that the impact of human presence upset this balance by introducing changes to the internal climate.[14] Happily, the Grotte Chauvet, discovered in 1994 after 23 000 years of inaccessibility and concealment, is a classic example of what can and should be done to preserve this cultural heritage. It was never subjected to the erroneous policies that

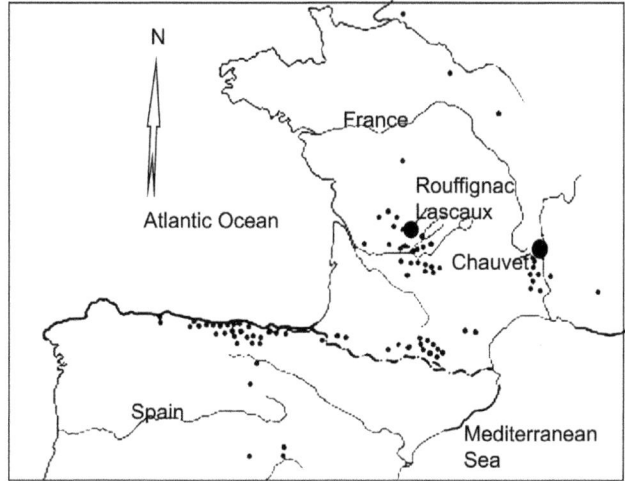

Figure 2.1 Distribution of major cave art and rock art sites in southern France and northern Spain. The Rouffignac and Lascaux caves are within thirteen miles of each other. Chauvet Cave is about 270 miles east of Lascaux.

placed many of the other caves at risk. It was never opened to the public and never will be: it will remain fully intact where all types of future research will still be possible. The French authorities have determined that only virtual visiting of the site, by way of a $70 million dollar replica, is the best option for transmitting and sharing knowledge fundamental to the cave, a unique artifact in the history of development of modern human thought.[15]

From the time that they have become available, the wall art in the caves has been the object of archaeological examination of the pigments other artifacts accompanying them. Although many hundreds of these caves and shelters have been discovered throughout the world, the greatest concentration is found in southern France and northern Spain (Figure 2.1).

2.2 THE PREHISTORIC CAVES OF FRANCE: CONTEXT

About 1 00 000 years ago, northern Europe was covered by a massive, mile-thick ice sheet that reached down to what are now the southern reaches of the Baltic Sea; its presence governed the frigid climate that early humans had to cope with.

The animals that abounded in the region, such as mammoths and wooly rhinoceros, had adapted over millennia to the harsh winters by growing thick, long fur coats and layers of insulating fat.

Modern humans arrived in Europe as early as 46 000 BP,[16,17] setting the stage for the cave art that flourished in the subsequent millennia. During this period, small bands of hunter-gatherers, possibly no more than 30 000 individuals in all of France, lived in rock shelters, called abris, pursuing a purely carnivorous diet. These early humans' major food source was reindeer.[18] Among the hundreds of sites in southern France and northern Spain, the virtual absence of reindeer among the animals, such as mammoths, mastodons, aurochs (wild cattle extinct since 1627), horses, and bears, in the rock art menagerie is an ongoing mystery.

In Europe, some 200 caves and rock shelters are known to contain art, the majority of which are in France and Spain, with other scattered sites in Italy, England, Romania, Portugal, Germany, Ukraine and Russia. Although prehistoric European cave and rock art are the most famous creative expressions prior to the advent of writing, recent archaeological work in Indonesia has revealed that rock art in Sulawesi predates the European cave paintings by about 11 000 years.[19] The Americas also have their fair share of Ice Age art.

The end of the Ice Age was marked by the gradual shrinking of the ice sheet beginning at about 15 000 BP, transforming the manner of human life over the following few thousands of years. As the tundra disappeared and the land became forested, the fauna also changed dramatically. Gone were the big game that sustained the Paleolithic hunters; in their place came, among other "small game," the lowly snail. The archaeological record shows that the major implements used in the hunt changed to those used to gather, trap and farm. Mesolithic garbage dumps turned up prodigious numbers of snail shells that can handily be radiocarbon dated (see Box 2.2) to give us an idea of when their owners became a menu item. Historian W. Scott Haine jokes that we have no record, however, of when garlic became essential to the dish.[20] With the advent of food that could be grown, gathered, and husbanded, Magdalenian art disappeared,

---

Box 2.2   Radiocarbon Dating

Carbon has three **isotopes**: C-12 and C-13, which are stable, and C-14, which is radioactive with a **half-life** of 5730 years. C-14 is produced by the action of cosmic rays on the nitrogen in the atmosphere and reacts with oxygen to produce carbon dioxide. All three isotopes are in the atmosphere in the form of carbon dioxide. Due to respiration and other means of exchange, all living creatures are in equilibrium with carbon during their lifetimes, but at the time of death, the exchange ceases, thus setting the radioactive clock to zero with a fixed amount, usually known or calibrated, of C-14 in place. Over time, the C-14 decays, and the measurement of the residual C-14 compared to the constant level in living material can be calculated back to the time of death. For example, if a wooden artifact is found to contain half the amount of C-14 it would normally have, then the decay process must have proceeded for the time of the half-life, 5730 years, marking the age of the wood. After about ten half-lives have passed, *i.e.*, about 57 000 years, the tiny amount of C-14 left undecayed would be very difficult to measure, so this marks the limit of this method of dating. For example, carbon from fossil fuels, obtained from plants that died millions of years ago, contains no C-14, so it is impossible to determine an age. This description applies only to creatures living on the earth's surface. Corrections have to be made for other situations such as fossil fuel consumption and nuclear testing, both of which alter the normal amount of C-14 in the atmosphere.

"killed by the snail," in the words of one French paleontologist. Presumably, there was no longer a need to invoke animal spirits for survival if, indeed, that was the purpose of impressing their images on cave walls.

2.3   PALEOLITHIC LIFESTYLE

The primary goal of human activity in the harsh Paleolithic environment was simply survival. That these little bands of humans succeeded against all odds could only be attributed to creativity, adaptability and inventiveness, especially in the use of the material resources at hand. It was essential that virtually every object and every activity contribute in some way to survival.

2.3.1   Fire

Ancient peoples perceived fire as a living being in which red was the color of life. Fire's "domestication by humans... undoubtedly constitutes the most important event in the history of humanity, an event that completely changed the conditions of existence and formed the basis of what can be called 'civilization.'"[21] It provided warmth and protection from the elements; the hearth was the place where people made tools and cooked their food. By extension, the hearth probably played some role in stone working, leather working, cosmetics, food preservation and medications. We are blessed that by heating minerals, color also found its way into spectacular art! But why?

2.3.2 Belief Structure

Color chemistry was not necessarily a "survival" technology, at least in terms of providing the necessities of life. Why was it so highly developed even among the ancient cave painters of tens of millennia ago? Perhaps the answer lies in a perceived mysterious world where survival depended not only on practical necessities but also on having the favor of certain spirits whose images, painted on cave walls, would allow a tribe to control the outcome of the hunt to assure survival for another winter. While we cannot second-guess the motives of ancient peoples, we can certainly infer a belief structure that was a powerful force in human activity, including belief in an afterlife – to say nothing of the danger involved in these activities.

2.3.3 Communal Cultural Expression

Human expression, no matter what the form, is intimately related to our cultural heritage. The artifacts – the wall paintings can be valuable markers of a culture that existed long before other modes of cultural transmission, such as writing, sculpture, or architecture developed. The subject matter of the paintings is also a window that opens onto a long-disappeared world where extinct species such as aurochs and mammoths roamed the plains of what we now call Europe. It is no wonder that their discovery mesmerized the world in 1900; a century later, scientists continue to be mesmerized by the unexpected discoveries found as they probe their composition.[4] Regardless of what future research uncovers, there will always be conflicting opinions on the motivations of Paleolithic artists that continue to stir our imaginations. However, there is no doubt that it was a major communal endeavor.[22]

2.4 LASCAUX CAVE. A SERENDIPITOUS DISCOVERY

Two of the most spectacular caves that have changed our whole perception of Ice Age art were Lascaux, discovered in 1940, and Chauvet, which came to light in 1994. Lascaux lay buried after its entrance collapsed 17 000 years ago. The apocryphal story tells us how an enterprising pooch named Robot chased a rabbit down a hole. Robot's master, 17-year-old Marcel Ravidat, intrigued, threw some rocks into the hole and back came hollow echoes from deep underground. The more prosaic version is that Ravidat noticed that an uprooted tree had revealed the hole. In any event, he returned with three friends, Simon Coencas, Jacques Marsal, and Georges Agniel – the four teenagers thought that they had found a legendary tunnel leading to some secret buried treasure. They forced their way down a narrow shaft that, 50 feet farther down, opened into a large cavern where they stared, open-mouthed, at some of the most astonishing cave art ever discovered. This event changed the apprentice auto mechanic's life forever.

Several days after the discovery, realizing that they couldn't keep this secret just among themselves, the boys led their teacher, Léon Laval (1885–1949), to the site. Laval first thought it was a hoax, but when he saw the paintings, his own experience in prehistory immediately signaled their authenticity. Very soon afterwards, the Abbé Breuil came and confirmed Laval's opinion (Figure 2.2).[23] The antiquities authority placed the site under protective surveillance

Figure 2.2 Abbé Breuil in Lascaux Caves at La Mouthe. Reproduced from ref. 23, https://wellcomecollection.org/works/ep8tjm4e, under the terms of the CC BY 4.0 license, https://creativecommons.org/licenses/by/4.0/.

at once but Lascaux did not open to the public until 1948. In fact, during World War II it was used as a munitions storage facility; one can only shudder if this cache had been discovered by the enemy or accidentally detonated by the Resistance.

2.4.1   An Extinct Menagerie Revealed

Lascaux is a relatively small cave with an entrance directly into a 60-ft long chamber called the "Hall of the Bulls," since its ceiling is covered with larger-than-life paintings of bulls, some of which are more than 15 feet long – a feature that marks Lascaux as "the Sistine Chapel of Prehistoric Art." In the next passageway, called the "Axial Gallery," we encounter more ceiling art: deer, aurochs, horses, ibex, and a bison. This cavalcade of animals is the best-known, best-loved, and most sophisticated grouping of all Ice Age art. Not only are the colors dazzling, but the execution is almost modern. There are animals shown in profile, but their foreparts are angled at a three-quarter perspective. A simple disconnect of a leg on a figure conveys the illusion of three-dimensionality. The sheer mastery of the drawings conveys an impression of a highly advanced civilization. If abstract art can be defined as the use of colors, forms, and patterns to achieve an effect, then these representations of animals are very close to the mark (Figure 2.3).[24]

Figure 2.3   The so-called Yellow Chinese Horse in the Lascaux Cave Axial Gallery (4.2 ft long) is a lovely example of pigment use. This horse, with its upright stiff brush-like mane, depicts a species native to the Mongolian steppes, Przewalski's horse, apparently prevalent in Europe during the Magdalenian era. Note the disconnected left legs, conveying an appearance of three-dimensionality. Reproduced from ref. 24.

2.4.2   Lascaux Sickness and Near Death

Lascaux continued its existence as both an underground museum and an archaeologist's treasure chest for only 15 years. Bright lights and troops of visitors, as many as 1200 per day, exuding sweat and body gases, forced its closure. Until that time, Ravidat acted as a guide. It was he who sounded the alert to the green algae (the "green disease") and calcium carbonate deposits (the "white disease") that was dooming this priceless art. Today, cave conservationists are battling a whole ecosystem of algae, black fungi and bacteria that threatens the integrity of Lascaux's paintings.[25]

2.4.3   Lascaux Reborn Again and Again – II, III, IV

Disappointed would-be visitors to Lascaux had to wait 20 years before Lascaux II, a worthy replica that captured about 80% of the original art, opened in 1983. Some purists decried the project claiming that no one would visit the Louvre to view a fake "Mona Lisa," but to many visitors, a fake Lascaux is better than no Lascaux. And, in fact, the visit, at least as far as this visitor is concerned, was quite satisfying – even the descending entrance into the replica mimicked the original. The underground chambers came as close as possible to reality, the visit was well-explained, there was enough time to see the preliminary exhibits and to ogle the artwork, which was splendidly done and to scale. In 2012, a traveling exhibit, Lascaux III, went on the road.

Today, Lascaux II has been superseded by Lascaux IV, a large, gray boxlike structure built on a hill outside the town of Montignac and a tad removed from the original site. It incorporates 90% of the art in a structure that models the original as closely as possible, modified only because of accessibility considerations.[26]

Naturally, people, and especially scientists and historians, were eager to find out as much as they could about these caves. Lascaux was a lucky find that supplied a rich collection of artifacts and a ready-made palette ripe for examination. This next section details the pigment finds there, and how modern science identified them, gleaned information about how they were processed, and how the Ice Age artists carried out their task.

2.5   ARCHAEOLOGICAL CHEMISTRY: THE CAVES UNVEILED

Over the past century, archaeologists examined the caves by virtually every means available. Newly analytical techniques have enabled the analysis of samples taken from the wall paintings and from the moveable artifacts found scattered throughout many of the caves.

The cavers' color palette was largely black (various forms of carbon black or of manganese oxides) and red (hematite or red ochre, Fe_2O_3), with occasional yellow pigments (goethite, or limonite FeOOH) as well. These colors defined the surface colors available to the Ice Age artist. However, this limitation does not mean that the colors were simple. The dozen or so pigments used at Lascaux ranged in shade from light yellow to deep black, all of them powdered minerals except for charcoal.[27]

Aside from determining the chemical nature of the pigments, archaeologists realized that in order to characterize them exactly, they had to also find out what they were mixed with (other pigments, colorless extenders), how they were prepared (selection, grinding, blending, heating), what binders were used to fix them to the wall (vegetable oil, animal fat, urine, blood, water) and, if possible, determine their geological sources and supply chain.

2.5.1   Lascaux's Color Palette

Archaeologists were very fortunate when they set up shop to examine the pigments at Lascaux. There was no need to seek out samples to analyze from the wall paintings themselves because the artists had kindly left behind over 150 mineral fragments found to be representative of the materials applied to the walls. The fragments were overwhelmingly black (105), followed by yellows (26), reds (24) and finally whites (3). But the really interesting thing was the color variation among them because almost all of them were complex.

Not only did the Lascaux artists mix their pigments: they also carried out some high-tech atypical processing. For example, they manufactured calcium phosphate by heating bone to over 400 °C, but then they further processed the product by mixing it with calcite, $CaCO_3$, and heating it up to 1000 °C (quite a feat with simple hearths) to produce tetracalcitic phosphate ($4CaO\cdot P_2O_5$). It is unclear what valuable property they sought in this white substance, but they certainly went to a lot of trouble to obtain it! Curiously, they used a lot of powdered quartz, but quartz contributes nothing to the color of a pigment, yet it is very difficult to grind – it is number 7 on the **Mohs hardness scale**. Grinding naturally occurring quartz produces amorphous silica which gives modern paints their "tooth" to priming coats because of its abrasive nature.[28] Quite possibly, the ancient artists discovered this property too.

2.5.2   The Blacks

The Lascaux black pigments turned out to consist mainly of manganese compounds. Regarding availability, southern France has many manganese mines and ore fields, especially in the Pyrenees. It may surprise even chemists to learn that mineralogists have distinguished about 20 different forms of naturally occurring manganese oxides based on their degree of **oxidation** (II, III, IV), chemical composition, and crystalline structure. Deposits of manganese dioxide, MnO_2, are very common; the two crystalline forms of MnO_2 are pyrolusite and polianite. Some other manganese minerals often found in deposits near Lascaux are manganite ($MnOOH$), bixbyite ($(Mn, Fe)_2O_3$), very rare hausmannite (Mn_3O_4), barium-containing romanechite ($Ba_2Mn_5O_{10} \cdot xH_2O$), cryptomelane ($K(Mn^{4+}, Mn^{2+})_8O_{16}$), and hollandite ($BaMn_8O_{16} \cdot xH_2O$). While some of these forms could be rendered useful as pigments by fine grinding, there is another amorphous form, called "wad,"[‡] or bog manganese, or "manganese ochre," that can be made into "crayons" and used as such with little advance preparation.

At Lascaux, the crayons contained a great variety of manganese oxides used singly or in mixtures like pyrolusite, romanechite and hollandite. Analysis of black samples from three different figures yielded three very different paint mixtures. A sample taken from the "chignon" (topknot) of the Great Bull in the Axial Gallery revealed a mixture of romanechite and iron oxide and a mixture of hollandite and cryptomelane for the bison, indicating pigments sourced from different sites.[29] But a second sample from the "chignon" yielded a great surprise that smacked of Neanderthal technology – see **Section 2.8** for this incredible news.

[‡]An early miner's term referring to various black mineral substances. They often called graphite deposits "wad," black lead, or plumbago, which has carried over into our own time – we refer to graphite-based pencils as lead pencils.

One burning question involved why these artists used manganese, likely gathered at a distance, rather than the obvious choice of dark and fine charcoal close to hand by charring any twig. Geological survey maps pinpointed the closest sources of manganese 40 km north, northwest of Lascaux; red and yellow ochre and white clay were available 20 km east of the site. But there were unmapped sources much closer to Lascaux: manganese only 5 km away, and ochres only a half km away. Additional analyses[30,31] showed that the yellow-orange-red pigments were hematite and its ochre precursors.

2.5.3   Pigment Processing

Heating yellow ochre (goethite, FeOOH) to form red hematite (Fe_2O_3) is a dehydration process that was well-known to our ancient artists/chemists. The ancient workers used heat not only to change the color, but often to prepare the substance for grinding; they knew from long experience that burned materials crumbled easily. An examination of the pigments from the cave of Troubat indicated that the heated pigments could be found in some archaeological layers of the cave but not in others, showing that the Ice Age artists varied their practice over time.[32]

The cave artists knew that they had to do long and laborious grinding of their minerals to control color and fluidity in the paint.[33] They also used systematic recipes, each one providing specific desired properties such as color, texture, and luster.

To apply the ground and mixed pigments to the wall, natural cave water was a good binder because of its high calcium content, though others of animal or vegetable origin did just as well: blood, urine, egg white, and even oil-based binders – upstaging the Renaissance artists by millennia.[34] The artists used macerated ends of twigs, fingers, the palm of a hand, and mossy tamping pads to apply the pigment. Some spectacular cave images show that people blew the pigment through tubes to make hand stencils and other forms.

Lighting a cave was a problem handily solved by using tallow candles equipped with juniper branch wicks. The lamps were simple but clever affairs: shallow cup-shaped stones capped with a lump of animal fat. Juniper branch wicks were an ingenious choice: the resin is self-contained in each seed and therefore burns slowly and cleanly, one seed at the time, emitting a minimum of smoke.

One thing that research has taught us about cave art is that its creation was not a simple matter. It was underpinned by sure knowledge of the available materials linked to precise geographic locations. Paleolithic artists explored selectively and chose deliberately to secure materials with the best physical and mechanical properties. Taken together, this thoughtful process could only be the result of long experience handed down over generations to members of the group, and certainly not an activity left to random discovery.[35]

Lascaux's discovery was spectacular. No one thought that any new discovery would be able to match it. But they were wrong. Grotte Chauvet's art set the Paleolithic clock back by 20 000 years and at the same time, corrected everything anyone had written to date about Paleolithic art. If Picasso had only superlatives for Lascaux, then how could he have expressed what was found in Chauvet? We'll never know.

2.6 DISCOVERY OF GROTTE CHAUVET

The bitter experience of Lascaux's fading images was fresh in the minds of the French authorities when it came time to deal with the sensational discovery of the Grotte Chauvet (Figure 2.1). On 18 December 1994, three very practiced spelunkers, Eliette Brunel-Deschamps, Jean-Marie Chauvet and Christian Hillaire, decided to follow their instincts and explore a heap of rubble under a rock shelter where they had recently sensed a draft. Moving the blockade pebble by pebble, they fashioned a crawl space so narrow that they could only squeeze through into open space by sacrificing some blood and skin to the rock walls. In Eliette's own words, "...chipping and removing rock and spending incredible amounts of energy, I was able to force the squeeze and come out in a flare where I was able to stand up. Ahead of me, the gallery stretched horizontally for a few more meters. I thus continued to advance, and then...a dream come true! I had arrived on a rocky outcrop, with a huge void under my feet. Holding onto the rock, looking down approximately 10 meters, I could distinguish the floor of a large chamber, barely lit up by the circle of light from my headlamp.

Guys, this is something big..."

2.6.1   Chauvet – Off Limits

Eliette's understated words continued to describe an almost mystical experience of awe and wonder at what she and her two companions saw that night – an overwhelmingly powerful moment that would remain etched in their hearts and memories forever.[36] By 29 December, experts from the Ministry of Culture for Prehistoric Art Affairs arrived to give judgment, later declaring that "no Paleolithic cave currently known anywhere in the world is in such good condition and shows such potential for research..." And the authorities made sure that this priceless patrimony would remain that way: on 15 January 1995, the first of several steel security doors was installed, and later in the month, video monitors came into play. The announcement of the discovery to the public came as nothing less than a bombshell. Never open to the public, on 22 June 2014, Grotte Chauvet was declared a UNESCO World Heritage Site.[37]

2.6.2   Faux Chauvet

Meanwhile, a structure similar in concept to Lascaux II came into being in 2016. Called the Caverne du Pont d'Arc, it is a true-to-scale replica of the original in all its details. The work of hundreds of engineers, artists, and architects, laboring for 8 years with 3-D maps, and high resolution scans and photographs, it is a modern marvel of high-tech replication. However, there was no attempt to reproduce the contours of the original cave in the way that the designers at Lascaux accomplished so effectively. The result makes one think of walking along a sinuous pathway through an airplane hangar separated by only a few feet from a group of 35 up ahead (with narration by a guide in French) and a group of 35 behind, clearly being guided in Italian. The general rule is "keep moving – this is a people-moving enterprise." There is no time to gawk, to take a second look, or even to look away for an instant without missing an important point. For this visitor, it was truly not an awe-inspiring experience. But again, Faux Chauvet is better than No Chauvet.

2.6.3   Chauvet's Images and Colors

If Lascaux is the "Sistine Chapel," then from its sheer quality and size, Chauvet is the "Louvre" of Ice Age art. It seems to have been sealed off by rock falls 28 000 years BP,[38] but artifact dating in the cave itself questions whether or not human access could have taken place at later dates.[39] More recent work presents a chronology acquired from the artworks and other data associated with the wall art and animal and human occupation. The results show that humans occupied the cave in two distinct interludes, from 37 000 to 33 500 BP and again from 31 000 to 28 000 BP, but there is no evidence that they occupied the cave beyond that date.[40]

Chauvet's images are the best-preserved and earliest executed in Europe, dating from the Aurignacian period. Over 1000 images of animals representing at least 14 different species, some never before seen in other instances of cave art, include a menagerie of the most dangerous and least hunted: 75 large felines, 76 mammoths, 65 wooly rhinoceros and 15 cave bears, a total of 63% of all the identifiable species in the cave. Among the herbivores are horses, bison, aurochs, ibex, reindeer, giant elks and musk oxen. One astounding almost life-sized figure in the cave is a finger engraving on a soft clay surface of a long-eared owl (*Asio otus*). The positioning of the wings shows that its head is turned around 180 degrees, the way that owls compensate for restraints in their eye movement. This detail did not escape the probing eye of the Paleolithic artist! The figures are exceptional for their great beauty. They are masterpieces that display the artists' skill in painting, form, engraving, and shading, as well as their portrayal of anatomical precision, movement, and three-dimensionality. Many border on the abstract where the artist uses a single deft line to capture the "essence" of mammoth or of bison.[41] Although Chauvet contains both black and polychrome drawings, the monumental character of the black compositions are responsible for the cave's great fame and its unique place in art history.[42] Googling either "Lascaux" or "Chauvet" in "Google Images" will open a unique world for you – it is well worth the journey.

Until Chauvet was discovered, a theory advanced by the Abbé Breuil and taken up by another great French prehistorian, André Leroi-Gourhan (1911–1986), proposed that figurative art was an evolutionary matter that developed from crude to more complex representations over the course of eons.[43] The artwork at Chauvet immediately demolished that idea; our concepts about human cognitive evolution was turned on its head virtually overnight as one can clearly see from the iconic horse panel from Chauvet, dated to about 32 000 years BP (Figure 2.4).[44,45] The dynamism

Figure 2.4   The famous horse panel of Chauvet. Reproduced from ref. 44
under the terms of the CC BY-SA 2.0 license, https://creativecom-
mons.org/licenses/by-sa/2.0/deed.en.

of these four juxtaposed images mark this as one of the greatest works of Paleolithic art. Each horse seems to be moving at
a different rate and with a different attitude: at a steady pace,
galloping with ears laid back, sleeping or snorting. It is unclear
if the panel represents four different horses or the same horse at
different periods.[46]

Chauvet's most spectacular images are the Horse Panel (Figure
2.4) and the Panel of Lions and Rhinoceroses (Figure 2.5).[47,48] Art
historians have studied these two panels intensively to specify
the techniques used in their execution, outline a catalog of the
forms used, and set them into a wider cultural context. Unlike
most of the other black drawings examined, the Chauvet artists'
choice of pigment has been charcoal, and not just any charcoal,
but only that derived from the wood of the pine tree, *Pinus syl-
vestris*.[49] All of the red drawings in Chauvet, as one might expect,
were done with red ochre.[50,51]

Visiting Rouffignac cave may be something of a letdown after
the magnificent displays at Lascaux and Chauvet. It is one of the
largest caves in southern France, about three miles deep, with
pitfalls every few feet. Curiously its images were not recognized
as anything but modern graffiti until 1956, when a few prehistorians decided to give it another look. And if you decide to look as

Figure 2.5   Grotte Chauvet: Lion and Bison Panel. Photograph: Claude Valette. This panel is unique in its realism and dynamism, as well as the fact that predators are almost never portrayed in Paleolithic cave art. Reproduced from ref. 48 under the terms of the CC BY-SA 4.0 license, https://creativecommons.org/licenses/by-sa/4.0/deed.en.

well, you will be visiting an authentic cave with images laid down by human hand 17 000 years ago. Get ready for a great ride!

2.7   ROUFFIGNAC

Only a few kilometers west of Lascaux lies the cave of Rouffignac, a site with several unique features. Like Altamira and Lascaux, it has a ceiling, called the Great Ceiling (or the Grand Plafond, in French) that is covered with the vast majority of the artworks. It is sometimes called the Mammoth Cave because more than 60% of the animals depicted are mammoths (Figure 2.6).[52] One noted paleontologist declared that the virtuosity of the artists was unquestioned. "Each animal outline comes," he exclaimed, "as one sure, swift swipe of the drawing instrument, at an approximately 5 : 1 scale, at least for the mammoths: a feat worthy of a Matisse!"[53] And, happily, Rouffignac is one of the few great French Paleolithic caves that are open to the public – it is even equipped with a little narrow-gauge railway that takes the visitor to the key sites.

When I visited Rouffignac in 2016, it was tempting to imagine swiping my finger on one of the images to see if the manganese

Figure 2.6   Detail from the "Great Ceiling," Cave of Rouffignac. This is the most complete drawing of any mammoth in the cave. Note also the ibex in the foreground. Reproduced from ref. 52.

oxide was still "wet." But such sacrilege committed against an image inscribed so many millennia ago was, to me, unthinkable. However, some 20th century prehistorians had no such scruples! And, indeed, the manganese is still as supple as the day the image was painted.

Rouffignac is a monochromatic cave, as far as pigments are concerned. All of the painted figures are drawn with black manganese oxide.[54,55] Unfortunately, the cave harbors no charcoal figures, and as you may recall from Box 2.2 on radiocarbon dating, you need some carbon to date an object. There is, in fact, nothing in the cave that would allow dating by a direct method. Archaeologists have "agreed" on an approximate date of about 13 000 BP, placing the cave squarely in the Magdalenian Era, but the search went on for a more exact technique. Eventually, U-series dating (Box 2.1) provided a provisional answer. What substance mentioned in the paragraphs below may have provided this answer? (Hint: It formed very slowly.)

Noting the lack of carbon in any of the figures analyzed, we can conclude that the manganese black material had never been subjected to burning, for if it had then particles of carbon would have been found mixed in with the manganese.[56] Anthropologists claim that the preferential selection of manganese, rather than carbon, as a black pigment directly associates its physical properties with social behavior.[57]

When Rouffignac's authenticity and age were challenged, the Abbé Breuil was a staunch defender. An impressive vindication of his judgment was the keen observational skills of the artists who knew the anatomy of extinct species so well that in one instance, we can view a mammoth with an anal plate, or operculum, residing on the Great Ceiling. (On the Rouffignac tour, you have to be prepared to look for this because the tour guide does not point it out.) No modern artist would have been aware of this evolutionary adaptation, a horny thickening of the inside base of the tail that effectively closed the anus, providing protection from the rigors of the glacial climate.

Several years later, another dramatic event confirmed Rouffignac's authenticity. The translucent covering that veils parts of some of the drawings, most obvious on what is known as the "great mammoth frieze," was characterized as calcite (calcium carbonate). When the Abbé observed it, though ignorant of its nature, he thought that it was due to a very slow process that would demonstrate definitively the great age of the paintings. He was absolutely right! This calcite efflorescence would have taken thousands of years of weathering action to develop, confirming the great age and authenticity of the figures that lay beneath it.

2.8   MANGANESE MINERALS IN OTHER CONTEXTS

Toying with the properties of manganese minerals seems to have been a favorite pastime of ancient "chemists." For example, between 3000 and 2000 BP, during the Etruscan civilization that flourished on the Italian peninsula, an unusual application of intense heat to a mix of pyrolusite (MnO_2) and hematite (Fe_2O_3) yielded an intense black pigment called jacobsite, $MnFe_2O_4$. To obtain this synthetic form of manganese black, the Etruscan "chemist" would have to heat pyrolusite to 450–500 °C to convert it to bixbyite (Mn_2O_3), then further heat a mixture of approximately 60–40% bixbyite/ hematite to over 950 °C to form the jacobsite. The technological advance of this process was the ability to enable the formation of black and red colors in a single oxidizing firing cycle.[58]

An astounding example of the non-decorative use of manganese oxides was found at the Neanderthal site of Pech-de-l'Azé, about 80 miles southeast of Rouffignac. There, the inhabitants collected lumps of manganese minerals for an apparently very different purpose. Worked about 60 000 years ago, these mineral pieces showed very clear abrasion marks as if turned on a grindstone. The archaeologists involved asked if these early inhabitants, popularly portrayed as brutish primates, had discovered a material that could facilitate combustion. To test this hypothesis, they tried to ignite wood turnings by heating and also sparking them with tinder, to no avail. However, when the wood was mixed with manganese dioxide powder, both when heated or placed near spark-lit tinder, ignition immediately happened. The combustion residue showed that manganese dioxide as pyrolusite was transformed into Mn_3O_4, hausmannite, releasing elemental oxygen in the process:

$$3MnO_2 \,(s) \rightarrow Mn_3O_4(s) + O_2 \,(g) \qquad (2.1)$$

It was this extra oxygen, shown as a product in eqn (2.1), that reduced the required temperature for ignition. Further experimentation showed that only materials predominantly composed of manganese dioxide were effective; other forms of manganese minerals showed no benefit. This experiment has most definitely changed our imagined perception of the individuals who inhabited these sites – from bumbling brutes to clever chemists![59]

However, the most amazing conclusion from this discovery is the puzzling identification of hausmannite in the chignon of the "Great Bull" of Lascaux (**Section 2.5.2**). The reporting team remarked that it has never been encountered before in prehistoric pigments.[60] Furthermore, its synthesis from more common forms of manganese oxides would require temperatures well beyond the reach of Paleolithic hearths. The authors and others surmised a possible biological origin or importation of the hausmannite from the high Pyrenees, 250 km away.[61] We propose that if the Neanderthals could achieve the high temperatures necessary to synthesize this unusual mineral, modern humans certainly could too. But in both cases, it would have been a byproduct in the act of kindling.

2.9   THE WIDER CONTEXT: PROJECT PIGMENTOTHÈQUE

In the 54 years between the discovery of Lascaux (1940) and Chauvet (1994), the French government realized that it needed a more sophisticated way to manage the research and conservation needs of the hundreds of Paleolithic sites in its care. Querying the coloring agents used during the Paleolithic era is not a simple matter. Analyzing the chemical composition is only the first step. How can these materials be recognized, defined, and described? Then many more questions arise. Where were they obtained? How were they used? Besides rock art, pigments were probably used for body painting, medicine and burials. What kinds of information do the pigments convey about the context? These are essential questions since colorants, present in every conceivable cultural context, have assumed a central position in current research on human behavior and the emergence of figurative expression. Consequently, no single discipline can address their breadth of meaning. Methods from archaeology, the geosciences, and the physicochemical sciences need to be harnessed to build a working hypothesis and procedural approach to understanding colorant processing and its place in the overall framework of Paleolithic activity. Built into this initiative is *Project Pigmentothèque* which has as its goal to build a reference database of natural manganese and iron-bearing minerals and a complementary collection from archaeological sites for comparative purposes.[62]

2.10   SOME CONCLUSIONS FROM THESE EXAMINATIONS

Groups working with the *Société préhistorique française* and publishing in their journal have carried out the major portion of investigations in several caves of the Ariège area during the 1980s to 1990s. Their work has given us a great deal of information not only about the nature and identification of the pigments used in the caves, but also of the sources of their raw materials and the technology underlying their use.[63,64] This approach led to the identification of workshops, *i.e.*, of socially organized groups in order to make a definite artistic product. These are some of their observations:

- Homogeneous mixtures needed long and laborious periods of grinding.
- The artists knew how to get good surface adhesion and hiding power for their pigments by using certain tried-and-true additives.

- The artists mixed their ingredients deliberately and selectively.
- The artists used three different recipes for the mineral fillers: F (feldspar, aluminosilicates with the general formula, $X(Al,Si)_4O_8$), B (biotite, a mica with the general formula $K(Mg,Fe)_3(AlSi_3O_{10})$ plus feldspar), and T (talc, hydrated magnesium silicate, $Mg_3Si_4O_{10}(OH)_2$). Each recipe gave a different desired effect in the finished work of art.
- The artists traveled far and wide in search of their raw materials.
- The artists scrupulously followed standard recipes to make up their paint pots.
- The introduction of biotite brings specific improvements to a liquid paint and may point to an evolution in the method of paint application.
- For the Magdalenians, the passage from recipe F to recipe B can be interpreted from this model of evolution in manufacturing techniques.

As for the identities of the pigments themselves, analysis yielded the following results for six of the caves examined (Table 2.2).[4]

Continuation of the work will be for the purpose of (a) better estimating how far a given workshop extended geographically

Table 2.2   Identities of red and black pigments from selected Paleolithic caves. Adapted from ref. 4 with permission from American Chemical Society, Copyright 2015.

Cave	Location	Red pigment	Black pigment
Rouffignac	Dordogne, France	None	Oxides of manganese
Lascaux	Dordogne, France	Hematite	Oxides of manganese
		Goethite	Carbon
Chauvet	Ardèche, France	Hematite	Carbon
Altamira	Cantabria, Spain	Oxides of iron	Oxides of manganese
			Charcoal or mineral
La Vache	Ariège, France	Oxides of iron + biotite	Oxides of manganese + biotite
Niaux	Ariège, France	None	Charcoal

and understanding the culture that gave rise to it; (b) tracing the evolution of techniques to create chronological benchmarks; (c) educating the public, and especially children, of the value of this cultural heritage. An RSC resource that enlarges on pigment synthesis to spark the interest of schoolchildren in this fascinating topic is available on the RSC website.[65]

2.11 FURTHER WORK

For the entire span of the 20th century – from 1902 until 1999 – examination of the materials in Paleolithic caves involved sample removal for analysis. But sensitive, reliable, and non-destructive techniques done in the field was the real need. The method that yields the greatest amount of information while disturbing an artifact hardly at all is micro-Raman spectroscopy, first developed by the Indian physicist, C. V. Raman (1888–1970).[66,67] Pioneered by University College London chemist, Robin J. H. Clark, this method was first attempted for use in a Paleolithic cave by a team from the Museum of Natural History (Paris) and the University of Toulouse.[68] It is now the method of choice for *in situ* analyses because it can deal with uneven surfaces, almost inaccessible art high up on a wall, and other difficult conditions. In addition, Raman instrument design is improving rapidly; portable models equipped with fiber optic probes are getting not only more powerful but also smaller.[69,70]

It is in this context that the work on the Chauvet cave is ongoing. The researchers are taking their time. They have produced partial reports[71,72] and we look forward to more complete information as it develops.

2.12 DO THESE CAVES CONTAIN ASTRONOMICAL MARKERS?

While chemists have been looking at the substances in the caves, an engineer and a theologian have been examining the animals in terms of their symbolic content – as possible date markers for the precession of the equinoxes and to record catastrophic events such as a meteor encounter. They present convincing evidence that the Shaft Scene at Lascaux records an encounter with a dated meteor stream 15 150 ± 150 years ago, which corresponds closely to the onset of a climate event recorded in a Greenland ice core. If this analysis can be borne out with further work, we are witnessing the intellectual capacities of Ice Age astronomers deserving of a doctoral degree – and we may even learn of other paleoclimatic events that could only be verified by this means.[73]

2.13   CONCLUSION

Our journey through the prehistoric caves has brought us into head-on contact with the first known cultural use of color in existence. We experienced first-hand that "Paleolithic art is a silent touch from distant ancestors, their marks a reminder of our own vitality and mortality... Despite our sometimes muddy and bloody way of arriving, we are sisters and brothers in time and space."[74] At the same time, their art "could have any number of meanings and uses, all of them invisible to us as we scrabble through the dirt for a unifying explanation."[75]

Furthermore, we are also aware that all is not perfect in the world of ancient rock art analysis. We have made some gross errors in the management of the caves. There are also issues of sample contamination; disconnects between the cave materials dated and the art produced are particularly problematic. Holistic integration of pigment recipe analysis, formal stylistic analysis, and direct timelines are three major areas where we'll see progress in the coming decades.[76] Included here is an online literature survey of rock art from 1864 to 2017; hopefully it will be updated occasionally.[77]

However, the argument regarding the replicas will not go away any time soon. The stringent restrictions on viewing the "real thing," especially among the so-called "rock stars" of rock art like Lascaux and Chauvet, have democratized the general public and even renowned art critics: no one except the experts associated with the research project can even hope to approach these precious monuments, and even they have very limited access. "Yes," as art critic Jonathan Jones has remarked, "this is all patronizing nonsense: no one wants to go halfway around the world to view a fake Rembrandt or a Seurat simulacrum." But there are plenty of genuine Seurats and Rembrandts around the world worth a visit – and what's more, the very act of viewing them will never even come close to destroying them. It is quite another matter with Paleolithic cave art. But there is another point to make:

Lascaux and Chauvet are not the only specimens of this magnificent art. There are plenty of other caves to visit in the area. They may not be as celebrated, but they are spectacular in their own right. Whether in person or online, with a bit of contemplation, we can come to realize that what those early artists invented was a style more elegant than Monet's, an acuity more precise than Audubon's, a technique more compelling than da Vinci's, an oeuvre more vital than Gaudí's, a language more eloquent than Shakespeare's, and all our yesterdays can trace their way back to that first syllable of recorded time.

In the next chapter, we will explore in more detail the use of a pigment surface that predates painting on cave walls by millennia. It is a fascinating journey that takes us to the depths of human motivation for applying colorants to our own bodies.

REFERENCES

1. T. Gray, *Elegy Written in a Country Churchyard*, 1751.
2. K. T. Lillios, *The Archaeology of the Iberian Peninsula: from the Paleolithic to the Bronze Age*, Cambridge University Press, Cambridge, 2020, p. 79.
3. M. Sanz de Sautuola, *Breves apuntes sobre algunos objetos prehistóricos de la Provincia de Santander*, Telesforo Martínez Blanca, Santander, 1880, http://simurg.bibliotecas.csic.es/viewer/image/CSIC000073342/1/, accessed June 2020.
4. E. Cartailhac, *L'Anthropologie*, 1902, **13**, 348–354.
5. M. V. Orna, Historic mineral pigments: colorful benchmarks of ancient civilizations, in *Chemical Technology in Antiquity*, ed. S. C. Rasmussen, American Chemical Society, Washington, DC, 2015, pp. 17–69.
6. As quoted in J. Thurman, *The New Yorker*, 2008, vol. 84((18), pp. 58–67.
7. Z. Stanska, *Daily Art News*, https://www.dailyartmagazine.com/pablo-picassos-bulls-road-simplicity/, accessed August 2020.
8. H. Moissan, *C. R. Chim.*, 1903, **136**, 144–146.
9. É. Cartailhac and H. Breuil, Les peintures et gravures murales des cavernes pyrénéennes. III. Niaux, in *L'Anthropologie XIX*, ed. G. Masson et Cie, Librairie de l'Académie de Médicine, Paris, 1908, pp. 14–46.

10. J. Marti, *Informe sobre los estudios realizados en las cuevas de Altamira*, Consejo Superior de Investigaciones Científicas, Madrid, 1977. Cited in M. Muzquiz Pérez-Seoane, *Análisis artístico de las pinturas rupestres del gran techo de la cueva de altamira: materiales y técnicas: comparación con otras muestras de arte rupestre*, Doctoral Dissertation, Faculty of Fine Arts, Universidad Complutense de Madrid, Madrid, Spain, 1988, p. 208.

11. J. M. Cabrera-Garrido, *Les matériaux des peintures de la grotte d'Altamira*, Actes de la 5ème Réunion Triennale de l'ICOM, Zagreb, 1978, pp. 1–9.

12. A. W. G. Pike, D. L. Hoffmann, M. García-Diez, P. B. Pettitt, J. Alcolea and R. De Balbín, *et al.*, *Science*, 2012, **336**, 1409–1413.

13. J. A. Lasheras, C. de las Heras and A. Prada, Altamira and its future, in *The Conservation of Subterranean Cultural Heritage*, ed. C. Saiz-Jimenez, CRC Press, New York, 2014, pp. 145–164.

14. J.-M. Geneste and M. Mauriac, The conservation of Lascaux Cave, France, in *The Conservation of Subterranean Cultural Heritage*, ed. C. Saiz-Jimenez, CRC Press, New York, 2014, pp. 165–172.

15. J.-M. Geneste and M. Bardisa, The conservation of Chauvet Cave, France, in *The Conservation of Subterranean Cultural Heritage*, ed. C. Saiz-Jimenez, CRC Press, New York, 2014, pp. 173–183.

16. J.-J. Hublin, N. Sirakov, V. Aldeias, S. Bailey, E. Bard and E. Delvigne, *et al.*, *Nature*, 2020, **581**, 299–302.

17. H. Fewlass, S. Talamo, L. Wacker, B. Kromer, T. Tuna and Y. Fagault, *et al.*, *Nat. Ecol. Evol.*, 2020, **4**, 794–801.

18. W. L. Voegtlin, *The Stone Age Diet*, Vantage Press, New York, 1975, pp. 61–62.

19. A. Brumm, A. A. Oktaviana, B. Burhan, B. Hakim, R. Lebe and J.-X. Zhao, *et al.*, *Sci. Adv.*, 2021, 7(3), eabd4648.

20. W. S. Haine, *The History of France*, ABC-CLIO, Santa Barbara, CA, 2nd edn, 2019, pp. 27–28.

21. M. Pastoureau, *Red – the History of a Color*, Princeton University Press, Princeton, 2017, p. 23.

22. L. Haking, Tracing upper palaeolithic people in caves: methodological developments of cave space analysis, applied to the decorated caves of Marsoulas, Chauvet and Rouffignac, southern France, *M.A. Dissertation, Archaeology*, Stockholm University, 2014.

23. Abbé Breuil in Lascaux Caves at La Mouthe, https://wellcomecollection.org/works/ep8tjm4e?wellcomeImagesUrl=/indexplus/image/M0010911.html, accessed June 2021.
24. A horse from Lascaux. Replica in the Brno museum Anthropos, 2009, https://commons.wikimedia.org/wiki/File:Lascaux,_horse.JPG, accessed June 2021.
25. F. Bastian and C. Alabouvette, *Int. J. Speleol.*, 2009, **38**(1), 55–60.
26. Lascaux IV by Snøhetta, *Architectural Record*, Montignac, France, https://www.architecturalrecord.com/articles/12474-lascaux-iv-by-sn%C3%B8hetta, accessed June 2020.
27. A. Leroi-Gourhan, *Sci. Am.*, 1982, **246**(June), 104–113.
28. NIIR Board, *Modern Technology of Paints, Varnishes and Lacquers*, Asia Pacific Business Press, Delhi, 2nd edn, 2007, pp. 164–165.
29. E. Chalmin, M. Menu and C. Vignaud, *Meas. Sci. Technol.*, 2003, **14**, 1590–1597.
30. C. Couraud and A. Laming-Emperaire, in *Lascaux Inconnu*, ed. A. Leroi-Gourhan and J. Allain, Centre Nationale de la Recherche Scientifique, Paris, 1979, pp. 153–170.
31. O. Ballet, A. Bocquet, R. Boucher, J. M. D. Coey and A. Cornu, Etude technique des poudres colorées de Lascaux, in *Lascaux Inconnu. XIIème Supplement à Gallia Préhistorique*, 1979, pp. 171–174.
32. M.-P. Pomiès, M. Barbaza, M. Menu and C. Vignaud, *L'Anthropologie*, 1999, **103**(4), 503–518.
33. P. Vandiver, Paleolithic pigments and processing, *M.S. Thesis, Materials Science and Engineering*, MIT, Cambridge, 1983.
34. C. Pepe, J. Clottes, M. Menu and P. Walter, *C. R. Acad. Sci., Ser. II: Mec., Phys., Chim., Sci. Terre Univers*, 1991, **312**, 929–934.
35. J. Vouvé, J. Brunet and F. Vouvé, *Stud. Conserv.*, 1992, **37**(3), 185–192.
36. E. Brunel, J.-M. Chauvet and C. Hillaire, *The Discovery of Chauvet-Pont D'Arc Cave, UNESCO World Heritage Site*, Éditions Equinoxe, Saint-Remy-de-Provence, 2014, p. 19.
37. D. Baffier, *Bull. Soc. Préhist. Fr.*, 2005, **102**(1), 11–16.
38. B. Sadier, J.-J. Delannoy, L. Benedetti, L. Didier, D. L. Bourlès and S. Jaillet, *et al.*, *Proc. Natl. Acad. Sci. U. S. A.*, 2012, **109**(21), 8002–8006.
39. J. J. Alcolea González and R. de Balbín Behrmann, *L'Anthropologie*, 2007, **111**, 435–466.

40. A. Quiles, H. Valladas, H. Bocherens, E. Delqué-Količ, E. Kaltnecker and J. van der Plicht, *et al.*, *Proc. Natl. Acad. Sci. U. S. A.*, 2016, **113**(17), 4670–4675.
41. J. Clottes, *The Chauvet-Pont d'Arc Cave*, Éditions le Dauphiné Libéré, Veury, 2015, p. 34.
42. G. Tosello and C. Fritz, *Bull. Soc. Préhist. Fr.*, 2005, **102**(1), 159–171.
43. A. Leroi-Gourhan, *Sci. Am.*, 1968, **218**(2), 58–73.
44. Etologic horse study, Chauvet cave, https://commons.wikimedia.org/wiki/File:Etologic_horse_study,_Chauvet_cave.jpg, accessed June 2021.
45. M. Gay, Développement de nouvelles procédures quantitatives pour une meilleure compréhension des pigments et des parois des grottes ornées préhistoriques, *PhD Thesis, Chemistry*, Université Pierre et Marie Curie, Paris, 2015, p. 17.
46. R. White, *L'art préhistorique dans le monde*, La Martinière, Paris, 2003, p. 79.
47. The Chauvet-Pont d'Arc Cave, Ardèche, https://archeologie.culture.fr/chauvet/en, accessed June 2020.
48. Panneau des Lions, Grotte Chauvet, https://commons.wikimedia.org/wiki/File:18_PanneauDesLions(PartieDroite)BisonsPoursuivisParDesLions.jpg, accessed June 2021.
49. I. Thery-Parisot and S. Thiébault, *Bull. Soc. Préhist. Fr.*, 2005, **102**(1), 69–75.
50. C. Ferrier, E. Debard, B. Kervazo, N. Aujoulat, D. Baffier and A. Denis, *et al.*, Approche taphonomique des parois des grottes ornées, in *Pleistocene Art of the World, Symposium 6 "Datation et taphonomie de l'art pléistocène"*, dir. J. Clottes, Actes du Congrès IFRAO, Tarascon-sur-Ariège, September 2010, pp. 1071–1094.
51. C. Fritz and G. Tosello, *Palethnologie*, vol. 7, 2015, http://journals.openedition.org/palethnologie/876, accessed October 2020, DOI: 10.4000/palethnologie.876.
52. Grotte de Rouff mammut, https://commons.wikimedia.org/wiki/File:Grotte_de_Rouff_mammut.jpg, accessed June 2021.
53. L.-R. Nougier and R. Robert, *The Cave of Rouffignac*, trans. David Scott, George Newne, Ltd, London, 1958. p. 55.
54. J. de Sanoit, D. Chambellan and F. Plassard, *ArcheoSciences*, 2005, **29**, 61–68, https://www.cairn.info/revue-archeosciences-2005-1-page-6.htm, accessed June 2020.

55. P. Graziosi, Analyses chimiques des peintures de la grotte de Rouffignac, *La Nature*, Dunod, revue mensuelle no. 3258, 1956.
56. E. Chalmin, C. Vignaud and M. Menu, *Appl. Phys. A*, 2004, **79**, 187–191.
57. T. Moutsiou, Colour in the Palaeolithic, in *The Value of Colour. Material and Economic Aspects in the Ancient World*, ed. I. S. Thavapalan and D. A. Warburton, Edition Topoi, Berlin, Berlin Studies of the Ancient World 70, 2020, pp. 55–68.
58. F. Schweizer and A. Rinuy, *Stud. Conserv.*, 1982, **27**(3), 118–123.
59. P. J. Heyes, K. Anastasakis, K. W. de Jong, A. van Hoesel, W. Roebroeks and M. Soressi, *Sci. Rep.*, 2016, **6**, 22159.
60. E. Chalmin, F. Farges, C. Vignaud, J. Susini, M. Menu and G. E. Brown Jr, Discovery of unusual minerals in paleolithic black pigments from Lascaux (France) and Ekain (Spain), *Conference Proceedings 882: X-ray absorption fine structure*, ed. B. Hedman and P. Pianetta, American Institute of Physics, College Park, MD, 2007, pp. 220–222.
61. V. Finlay, *The Brilliant History of Color in Art*, Getty Publications, Los Angeles, 2014, pp. 9–10.
62. *Project pigmentothèque*, https://www.researchgate.net/project/Pigmentotheque, accessed June 2020.
63. D. Buisson, M. Menu, G. Pinçon and P. Walter, *Bull. Soc. Préhist. Fr.*, 1989, **86**, 183–192.
64. J. Clottes, M. Menu and P. Walter, *Bull. Soc. Préhist. Fr.*, 1990, **87**, 170–192.
65. *Prehistoric Pigments*, https://edu.rsc.org/resources/CMP00004139/pdf/1540.article, accessed October 2020.
66. R. J. H. Clark, *Chem. Soc. Rev.*, 1995, **24**, 187–196.
67. R. J. H. Clark, *Chem. N. Z.*, 2011, (January), 13–20.
68. D. C. Smith, M. Bouchard and M. Lorblanchet, *J. Raman Spectrosc.*, 1999, **30**, 347–354.
69. C. Capel Ferrón, S. E. Jorge Villar, F. J. Medianero Soto, J. T. Lopez Navarrete and V. Hernández, Applications of Raman and infrared spectroscopies to the research and conservation of subterranean cultural heritage, in *The Conservation of Subterranean Cultural Heritage*, ed. C. Saiz-Jimenez, CRC Press, New York, 2014, pp. 281–292.
70. P. Vandenabeele, H. G. Edwards and J. Jehlička, *Chem. Soc. Rev.*, 2014, **43**, 2628–2649.

71. V. Feruglio and D. Baffier, Le rouge à Chauvet-Pont-d'Arc, in *Les chemins de l'art aurignacien en Europe*, 16–18 September 2005. *Das Aurignacien und die Anfange der Kunst in Europa*, Colloque International—Internationale Fachtagung Aurignac 2005, 379–392, Éditions Musée – Forum Aurignac, 4, Aurignac, France, 2007.

72. N. Aujoulat, D. Baffier, V. Feruglio, C. Fritz and G. Tosello, Les techniques de l'art parietal, in *La Grotte Chauvet. L'art des origines*, ed. J. Clottes, Éditions du Seuil, Paris, 2010, pp. 152–160.

73. M. B. Sweatman and A. Coombs, *Athens J. Hist.*, 2019, 5(1), 1–30.

74. R. D. Guthrie, *The Nature of Paleolithic Art*, University of Chicago Press, Chicago, 2005. p. 460.

75. S. Kriss, Pocket Worthy, https://getpocket.com/explore/item/what-the-caves-are-trying-to-tell-us?utm_source=pocket-newtab, accessed February 2021.

76. H. Salomon, *Bulletin de l'Association Scientifique Liégeoise pour la Recherche Archéologique*, 2013–2015, vol. 28, pp. 75–94.

77. L. Marymor, *Paleolithic Rock Art: A Worldwide Literature Survey Extracted from the Rock Art Studies Bibliographic Database for the Years 1864–2017*, https://www.mdpi.com/2076-0752/7/2/14/htm, accessed July 2020.

CHAPTER 3

Body Art in All Its Parts: Cosmetics Gone Wild[†]

Tattoos are souvenirs. They're road maps of where your body's been.[1]

John Irving

3.1 INTRODUCTION

The walls and ceilings of Paleolithic caves were not the first painted surfaces. As self-conscious hominins emerged from the savannahs and forests, their search for identity and beauty found a place on their own epidermis.

3.2 SEND IN THE CLOWNS!

Some months ago, while I sat in the predictable rush hour traffic on the George Washington Bridge, a fender-bender in an adjacent lane brought us all to a sudden halt. Out of the car immediately ahead leaped a costumed clown in full grotesque greasepaint, fists ready to take on the world. My gut reaction

---

[†]Glossary entries and chapter cross-references are in boldface.

March of the Pigments: Color History, Science and Impact
By Mary Virginia Orna
© Mary Virginia Orna 2022
Published by the Royal Society of Chemistry, www.rsc.org

was dread horror! I was the victim of a sensation summed up in a little known English word – coulrophobia – that is fraught with meaning in psychology as well as in show business. The fact that this word even exists witnesses to its usefulness – a morbid fear of clowns. How many films have we seen where the scary clown is the major theme? One of the most horrific in this genre is "Joker." The dozens of preceding films on the same theme suggest that not only is this phobia deeply embedded in Western culture, but even fills a psychic need. There must be something in the hideously painted face that causes many people to shrink instinctively in horror, and yet go back for more.

But there's another side to clowns. Stephen Sondheim's famous song is not even about clowns – it uses the clown idea as a metaphor for all that distracts and turns people away from the real issue when things get sticky.‡ And, as if clowns did not already have a bad rap, recent research shows that clown visits to sick children have more of a deleterious effect than a beneficial one.[2]

Clown makeup is just one facet of a human phenomenon that has many purposes, expressions, and meanings. Decorating one's body can lead, as with clowns, to enjoyment, distraction, horror, or dread. Body makeup may be used to intimidate the enemy, attract lovers, mask aging, hide defects, protect the skin, celebrate life, placate the gods, signify belonging, cure an ailment, disguise your identity or simply "beautify." The term "war paint" has been applied to indigenous tribes on every continent as well as to heavily made-up women with a sexual purpose. It particularly refers to two entrepreneurial women, Elizabeth Arden and Helena Rubinstein, who built cosmetics empires. What they purveyed gave rise to a book, *War Paint*[3] and a subsequent similarly-titled Broadway musical. "Painted Man" can mean either a fantasy novel or a band; we all know what "painted woman" means. And the human skin is the recipient of it all.

This chapter is a history of the human skin, that wonderful organ we all inhabit. Its care and adornment have occupied all peoples through all ages. Pigments were the first cosmetics, also doubling many times as medicines. We'll look at these pigments and their properties, and then how various civilizations used

---

‡Stephen Sondheim wrote this song for "A Little Night Music." Sending in the clowns is theatre jargon for doing something that will distract the audience from realizing that things are going awry and that a few jokes are needed.

them for skin protection, eye protection, hair coloring, war paint, group identity and stage and screen.

3.3 THE SKIN WE ARE IN

Our fondness for cosmetics predates written history and, in fact, any other form of art.[4] Long before prehistoric artists decorated karstic caves, they had found a surface far more plastic and convenient to harbor their dramatic instincts: their own bodies. The human body is a unique expressive medium. Any physical change to the body's surface or its features carries a special significance. It seems that decorating it is written right into our very own DNA.[5] Anthropologists have found that for many indigenous peoples, a painted face is a sign of an individual's human dignity. Body makeup, even in the 21st century, remains a fundamental part of group identification and culture.[6]

Our skin is our body's largest organ. It is our protective boundary that sets each of us apart from the rest of the world. It has been likened to a "complex, diverse ecosystem... a dynamic interface" that connects us to the world around us.[7,8] The average 200 lb male hefts about 30 lb (13.6 kg) of skin measuring 22 square feet (about 2 square meters) in area from head to toe. It varies in thickness and permeability from our very thin eyelids to the tough soles of our feet. Its three distinct layers, the epidermis, dermis and subcutaneous layer, each have their specific functions: protection, vapor and liquid transport, cell rejuvenation, sensation, hair growth, sweat and oil production, energy storage, body temperature control and insulation, to name a few.

Skin is also semipermeable[9] – it has pores that allow for communication with the outside world. On average, each square inch (6.5 cm²) of skin contains about 400 pores, each with a diameter of 50 micrometers (μm) or 0.002 in. While such small pore size serves as a barrier for most types of particles, they are wide open gates for molecular and ionic substances such as acids, bases and salts. The average skin pore size is about 2500 times larger than the largest known atom or ion. This fact alone signals the fact that no substance applied to the skin will remain on the surface for long. Pain medication administered in patches and toxic lead salts smeared on the cheeks will both invariably find their way into the body – for weal or for woe.

3.4   COSMETICS: ORNAMENTAL PIGMENTS

The ancient Greeks gifted us with the word "cosmetic," derived from their word for "ornamental" (κοσμητικός), appropriately enough. The concept links both tangible, external values (ornaments) with intangible values related to health, beauty, and magic that later branched into the disciplines of medicine, cosmetology, and religion. We can define a cosmetic product as any substance or preparation in contact with the human body for the purpose of changing its appearance.[10] Among the first cosmetic products were the red, yellow, black and white surface mineral pigments available to prehistoric humans.

Hematite, or red ochre (Fe_2O_3) was found in association with goethite, or yellow ochre (FeO(OH)), from which it was sometimes made by heat treatment. Black pigments were either manganese oxides (*e.g.*, MnO_2) or various forms of carbon such as charcoal or boneblack. The whites were calcium carbonate ($CaCO_3$) in several crystalline forms, talc and quartz. In addition to decorating cave walls, there is also a growing body of evidence that these pigments, especially red ochre, were present in other hominid-inhabited sites stretching back 10 00 000 years. Figure 3.1 shows the variety of colors obtainable from a single rock of surface-deposited ochre (Box 3.1).

3.4.1   Red Ochre

Red ochre, Fe_2O_3, is the most common red mineral in every land mass around the globe.[12] It is no wonder that early modern humans, and Neanderthals before them, collected it and put it to

Figure 3.1   Ochre-bearing rock (background); yellow ochres (left), charcoal and carbon black (right), red ochres and mixes with black (center).

Box 3.1 A Note on Cosmetics in General

Pampering the skin has been a priority from ancient times, and those who could went to any length to do so. Nero's wife, the Empress Poppaea, took a herd of she-asses everywhere she went so that she could bathe daily in their milk, reputed to keep her skin baby-like forever. However, most ancient cosmetics contained a variety of **inorganic** and **organic** chemical compounds with varying degrees of toxicity. Inorganic salts containing **heavy metals** are now banned, at least in countries that adhere to strict labeling requirements. Organic colorants are vetted for their adverse effects, usually for allergic reactions and carcinogenicity, and very few are permitted in foodstuffs; there is a bit more latitude for cosmetics, but recall those wide open pores in the human skin! There are entire disciplines devoted to using cosmetic technology to improve the qualities and marketability of cosmetics. These include rheology, surface chemistry, and the science of emulsions, oils, fats, waxes, thickeners and gums. Most substances used as cosmetic binders are derived from plants, an inexhaustible source of raw materials that are generally safe and non-toxic. "Saving face" is currently a multibillion-dollar business considered "essential" by about two-thirds of the human race regardless of gender or economic status. Cosmetics are not only here to stay: their presence will keep growing in our lives.[11]

---

use for a variety of purposes we can only guess about. What we do know is that they deliberately collected it because of its color.[13] The first indications of red ochre use date back nearly one million years where it is present at early Oldowan sites and South African caves.[14] *Homo erectus* carried lumps of red, brown, and yellow ochre to rock shelters in France 3 00 000 years ago. Fifty thousand years later, the entire surface of a similar shelter in the Czech Republic was found completely covered with red ochre powder, with the grinding tools lying nearby.[15] A team examining the finds from the Maastricht-Belvédère archaeological site in the Netherlands reported a similar timeframe for transport and manipulation of red ochre by early Neanderthals.[16] Evidence for a 1 00 000 year-old ochre-processing industry at Blombos Cave, South Africa was confirmed by dating the quartz sediments in which the ochre containers were buried. Archaeologists were able to infer that finely powdered ochre was mixed with crushed bone, stone and quartz in a liquid matrix and then stored in giant abalone shells. This investigation documents deliberate planning, production and curation of a pigmented compound and the use of containers, clear evidence of an elementary knowledge of chemistry and the ability for long-term planning. However, how the mixture was applied is unknown. The authors speculate on possibilities including decoration and skin protection.[17] There are many other possibilities: tanning animal hides; symbolic or ritual use; medicinal use; mixing it with adhesives; or for tattooing or painting the body.[18] While some of these possibilities are speculative and may never be resolved, there are numerous examples that ochre was used on dead bodies since it has been found in burial sites from at least 3 00 000 BP.[19] Among the most famous was the so-called "Red Lady" of Paviland who was inhumed with red ochre in southwest Wales about 32 000 BP.[20] Since the color red evokes the realities of life, death and blood, red ochre could have been used in ritual body decoration to mark status, rank, age, or gender.[21] In fact, there are many practices, even today, among indigenous peoples that concern family taboos, initiation rites, and gender distinction that involve red ochre gathering and usage.[22]

In 19th century rural England, red ochre, locally called reddle, supplied by the reddleman, was an important commodity used by farmers to mark their sheep in preparation for a fair. No description of red ochre and its social burden has ever surpassed Thomas Hardy's portrait of Diggory Venn, the reddleman, in *The Return of the Native*:[23]

> *Reddle spreads its lively hues over everything it lights on, and stamps unmistakably, as with the mark of Cain, any person who has handled it half an hour. A child's first sight of a reddleman was an epoch in his life. That blood-coloured figure was a sublimation of all the horrid dreams which had afflicted the juvenile spirit since imagination began. 'The reddleman is coming for you!' had been the formulated threat of Wessex mothers for many generations.*

3.4.2 Carbon Black

Black derived from charred plants is universally available, makes for a durable pigment and can be readily produced as a very fine powder. Any black pigment that predominantly contains the element carbon, regardless of its source, can be called carbon black. The common forms are boneblack, lampblack, vine black and charcoal black: the latter is well-known as the black pigment of choice for some of the Ice Age caves of Europe. Since it is found in every hearth that has ever been excavated, its presence is unremarkable, and so it is never mentioned in terms of burial sites and workshop areas except when referencing radiocarbon dating. However, we know that it was used as a body decoration from at least the Neolithic era (5000 BP), but its prior use for that purpose is shrouded in time past and needs artifact verification.[24]

3.4.3 Expanding the Palette

Our cultural heritage is intimately related to how our distant ancestors have expressed themselves in the activities of their daily lives. Knowing the identities of the materials used in these activities and how they were prepared can be valuable indications of the sociological context in which they were created.

As civilizations advanced and technology developed, in addition to the ochres, more and more pigments became available through trade, the development of mining, and in some cases, synthesis. Ancient peoples exploited pigments for a variety of purposes, including body pigmentation, architectural and funerary decoration, ritual activities, magic and medicine. They were mostly inorganic compounds containing extremely toxic **heavy metals**. In addition to carbon black, Table 3.1 [24] presents a list of the major pigments known to have been used as cosmetics in the ancient world.

With the rise of complex societies, the full-time specialist appeared for the first time. It is no surprise that a practical consequence of intentional manipulation and study of newly-available materials, as well as a greater knowledge of those that were close to hand, should give rise to the first artificial pigments. The poster-child example of this development is the invention of the oldest known synthetic pigment, Egyptian blue (**Chapter 4.6.8**).

3.4.4 Three Iconic Red Pigments: Cinnabar, Minium, Realgar

With the rise of societal complexity, funerary practices changed drastically in order to meet new communal needs. Society had become hierarchical, and people's social standing was reflected in their burial and the posthumous preparation of their bodies for the afterlife. Red – for its important association with life,

Table 3.1 Partial list of pigments used for cosmetic purposes in the ancient world.

Pigment	Formula/Name	Color	Source/Properties/Usage
Cerussite (white lead)	$PbCO_3$ Lead carbonate	White	Face cream in Bronze Age Greece through post-Elizabethan times.
Red Ochre	Fe_2O_3 Iron oxide	Red	Universally available and utilized; (Figure 3.1).
Cinnabar (mineral) Vermilion (synthetic)	HgS Mercury sulfide	Red	Chief source: Almadén mine, Spain; vermilion artificially prepared; (Figure 3.2).
Galena	PbS Lead sulfide	Black	Documented use in Egypt from 5000 BP.
Henna	Lawsone, $C_{10}H_8O_3$	Brown red	From leaves of *Lawsonia inermis*; Figure 3.4 contains lawsone molecular structure.
Magnesia alba	$MgCO_3$ Magnesium carbonate	White	Pigment or extender.
Malachite	$CuCO_3 \cdot Cu(OH)_2$ Basic copper carbonate	Green	Oldest known bright green pigment; curative cosmetic in pre-dynastic Egypt.
Minium	Pb_3O_4 Lead tetroxide	Red	Red lead (Figure 3.2).
Realgar	$\alpha\text{-}As_4S_4$ Tetraarsenic tetrasulfide	Red-orange	Sometimes called "ruby sulfur" (Figure 3.2).
Woad	$C_{16}H_{10}N_2O_2$ **Indigo**	Blue	Blue pigment extracted from the woad plant, *Isatis tinctoria*.

Figure 3.2 Left to Right: Cinnabar (Almadén Mine, Ciudad Real, Spain); Minium (Old Yuma Mine, Saguaro National Monument, Arizona, USA); Realgar (Shimen Mine, Hunan Province, China). Reproduced from ref. 25–27 under the terms of the CC BY-SA 3.0 license, https://creativecommons.org/licenses/by-sa/3.0/deed.en.

blood and death – remained the pigment of choice. But the discovery of new minerals produced a broader range of reds. The ever-desirable properties of more subtle tonality, greater intensity and more polished luster could be fulfilled by the advent of mercury-containing cinnabar,[25] lead-containing minium[26] and arsenic-containing realgar (Figure 3.2).[27,28] Restricted access to these new materials soon became a mark of distinction for the social upper classes who could afford them.[29]

For pigment use, the three red crystalline substances in Figure 3.2 had to be finely ground, mixed with an extender like talc or magnesia alba and then with a binding material, usually a fatty substance like animal tallow or a tree resin. A smaller particle size usually yields a lighter-colored pigment, so the intensity of the color changes on treatment.[§] It is easy to see how each of these three pigments could exhibit different tonalities and lusters, but each of them would also proclaim the high status of the wearer, either living or dead. Only the highest echelons of society could afford cinnabar, a commodity that cost 70 sesterces per pound, an ancient Roman soldier's monthly salary.

3.4.4.1 Cinnabar and Vermilion. Readily available in Europe, the Middle East, and beyond, this highly-prized, rich, red compound of mercury, HgS, mercury(II) sulfide, is the principal ore of the element mercury. Called κιννάβαρι by the Greeks, *zinjifrah* by the Arabs, it was later Romanized to cinnabar, a word purportedly of Indian origin used to denote dragon's blood, a red resin.[30] The main source of cinnabar in the ancient world was the famous mine of Almadén, in Spain, the largest mercury mine in the world. The Romans routinely sentenced fractious slaves to work in the mine, which was really a death sentence given mercury's well-known toxicity. When crushed and ground, cinnabar releases elemental mercury, which is volatile and far more toxic than the cinnabar itself. Cinnabar's widespread use in the ancient world, from at least the 10th millennium BCE, has prompted anthropologists to posit motives associated with rituals and other types of symbolic activities. They have found that at various times it has been used to signify blood, victory, and success, exemplified

---

[§]As the particle size of the pigment approaches the wavelength value of the impinging radiation, scattering of the radiation increases so that the viewer increasingly sees the total spectrum of white light, which washes out the intrinsic color of the pigment.

by its use in Roman triumphal processions. Association with the reality of life-death duality and the concept of immortality most likely led to its important role in burial rituals in Neolithic cultures. Some believe the color was prized because of its vibrant permanence, which is so unlike the blood it visually resembles. Pliny says that it is held in such esteem in Ethiopia that their nobles were in the habit of staining their entire bodies with it.[31]

Cinnabar's synthetic counterpart, vermilion, takes its common name from a misnomer. It is derived from the Latin *vermes*, meaning worm, originally designating the kermes insect found in the ilex or evergreen oak and used for the preparation of a red dye. From kermes, in its turn, the words crimson and carmine are derived. Vermilion can be handily made by heating a mixture of its elemental components, mercury and sulfur, for a very long time in a sealed container. This so-called "dry method" is attributed to 4th century BCE Chinese technicians. In late antiquity, alchemists interested in turning base metals into gold were also interested in substances like liquid mercury and cinnabar because of their unusual physical properties.[32] Numerous recipes on how to synthesize vermilion from its elements appear in medieval treatises such as the *Mappae Clavicula*,[33] but the empirical "stoichiometry" hardly ever called for enough mercury to produce a satisfactory yield. Although the correct stoichiometric (molar) ratio is 1:1, the ratio by weight is approximately 6:1. However, the ancient recipes erroneously called for mercury:sulfur weight ratios from 1:6 to 1:1.[34,35] Apparently the impractical proportions of the ingredients have a basis in Aristotelian four-element philosophy.[36] An exception is "The Book on How to Make Colours" which gives a recipe in virtually correct stoichiometric proportions, and a consequent satisfactory yield when carried out in a modern chemistry laboratory[37] However, correct stoichiometry is not enough. Vermilion synthesis is a very complex process that also depends for its quality and yield on external factors such as pot design, temperature control and the grinding process of the reactants.[38] One important point that some of these recipes make is that there are two forms of vermilion, one black and the desired one, red. HgS exhibits at least two crystalline modifications, called polymorphs: cinnabar (or vermilion, α-HgS) and metacinnabar (or *aethiops* mineral, β-HgS). Red α-HgS is stable at low temperatures and black β-HgS at higher temperatures.[35] Although it is a moot point when vermilion is used as body paint, its tendency to darken into the β-form can be devastating to an artwork unless it is kept indoors. When α-HgS is exposed to the ultraviolet (high energy) rays of the sun, darkening will quickly take place.[39] A mitigating factor regarding cinnabar's toxicity when used as a body paint is that it is one of the least soluble compounds of mercury known, so it is not likely to penetrate much further into the skin than the epidermis.

3.4.4.2   Minium.   Red lead, or minium, takes its ancient name from the River Minius, or Miño, which marks the border between Portugal and Galicia, Spain, where it was first mined in ancient times. It is also called lead tetroxide because of its formula, Pb_3O_4 ($2PbO\cdot PbO_2$); it can also be obtained just by heating litharge, PbO. The pigment is bright scarlet and has good hiding power, a great asset both for easel painting and for body painting.[40] Its major use in the Middle Ages was for painting in illuminated manuscripts; it gave its name to the style of painting (minium → miniature) because of its widespread presence in this type of painting. The name "miniature" has nothing to do with measurement. The pigment is not very stable when exposed to light and air. It was once confused with cinnabar, which Pliny called *"minium;"* his term for red lead was *"minium secundarium."* Once the confusion was cleared up and the pigment was recognized as a substance distinct from HgS, minium henceforth was the name applied. Because of its rarity, it came in as a poor second in one's choice of red body paint. However, it was particularly prized as a face blush.[41]

3.4.4.3   Realgar.   From ancient times, realgar, As_4S_4, was a common pigment that was traded over long distances to make paints, inks, and dyes. It is a member of the group of compounds called arsenosulfides occurring in hydrothermal and volcanic ore deposits. Because of its brilliant red color and beautiful well-formed crystals, it looks very much like a gemstone, occasioning its "street" name of "ruby sulfur" or "ruby arsenic." However – a gemstone it is not. It stands a number two on the **Mohs hardness scale**, making it just a little harder than talc and softer than almost all other mineral substances. For this reason, it was prized as a colorant because it could be ground so easily into a finely divided powder. Since arsenic exists in many compounds exhibiting a valence of three, such as As_2O_3 and As_2S_3, realgar's formula may seem curious. Since arsenic is a metalloid, it has the capacity of forming covalent bonds with itself, and in realgar, four arsenic atoms are covalently bonded to each other with an overall charge of 8+. So, in combination with four S^{2-} ions, it forms electrically neutral As_4S_4. Historically, it has been used as a pigment, depilatory, poison, ingredient in explosives and fireworks, a ritualistic "medicine," and a cosmetic. Its name derives from the Arabic, Rahj al ghār, meaning "powder of the mines."

3.5 TOXICITY

Ironically, habitual use of these pigments on the body, lips, cheeks, and around the eyes would launch the wearer on a steep slope toward the mortuary. It is well documented that people were not stingy in their cosmetic use – they literally slathered it on, mainly to demonstrate their affluence. But all three heavy metals, mercury, lead and arsenic, are among the most toxic substances known. They do the most damage by interfering with the basic metabolic processes of the human body by producing reactive oxygen species that cause oxidative damage to proteins and DNA.[42] From there, it is a cascade effect: they inhibit enzyme reactions, impair heme production, interfere with DNA repair and cause membrane damage, protein dysfunction and inflammation in the arteries and veins.[43,44] External symptoms were, among others, nausea, muscle cramps, skin lesions and darkening of the skin. The poor relations who were constrained to showing off with hematite may have fared no better. While we may think of iron as a benign metal, absolutely necessary for synthesizing hemoglobin, toxicity from iron overload was also possible. Recent research has nailed the culprit in producing oxidative stress and altered antioxidant capacity mechanisms that induce neurotoxicity: the Fe(II) ion introduced into the brain through concentrated ambient ultrafine particles in the air causing the oxidation and modification of lipids, proteins, carbohydrates, and DNA.[45,46]

3.6 EGYPTIANS: FAMOUS FOR USE OF MALACHITE, GALENA, KOHL AND HENNA

Ancient Egyptians have a very colorful past well documented by their own writings on papyri and on their monuments and tombs. Much of the literature on papyri reveals a society intent on maintaining social balance by observing the societal moral contract rigidly controlled by the Pharoah.[47] Since 1788, only about 50 000 papyri – out of an estimated total of 4 00 000 preserved in collections – have been translated, and a majority of them are quite fragmentary.[48] Chemists are very fortunate that two key Greek papyri from Thebes, the *Leyden Papyrus X* and the *Stockholm Papyrus*, came to the attention of chemist-scholar Earle R. Caley. Caley translated them from Greek and published both of them in the *Journal of Chemical Education* in 1926 and 1927 respectively.[49,50] Together they constitute the earliest and most direct transmission of the nature and extent of chemical knowledge as of the third century CE, including their extensive use of materials used for fakery.

It seems that the ancient Egyptians were equally adept at another form of mild deception: the wholesale use of cosmetics, and particularly galena, kohl, malachite and henna, as part of their daily practice. There are numerous papyri filled with cosmetics recipes because the Nile Valley was filled with all kinds of natural resources for their manufacture. Egypt has a long history of mining dating back to predynastic times (about 6000 BP). Major sources of zinc, lead, tin and copper can be found in Upper Egypt; the copper deposits in the Timna Valley are legendary.[51]

From these mines came some of the most beautiful minerals to meet the eye, both literally and figuratively, as we will see in this next section.

3.6.1 The Eye Has it

Eye paint is inarguably the most distinguishing feature of Egyptian facial imagery as evidenced from tomb paintings, mummy decoration, and even modern popular films. The eye's mystical and religious properties were associated with the sky god, Horus, the sun god, Ra, and the moon god, Thoth. Egyptians believed that the use of kohl conferred protection from the gods; given the multiple eye problems that they suffered, this protection was probably one of their major health concerns, endemic to this very day.[52] It is no wonder that from cradle to grave, Egyptians of both genders and every level of society habitually dabbed their eyes with green malachite and black kohl. Figure 3.3[53] gives an example of the eye's appearance after kohl application. Let us take a look at these pigments in some detail.

Figure 3.3   Woman holding a sistrum. Deir El-Medina, Egypt. Middle Kingdom (*ca.* 1250–1200 BCE). Reproduced from ref. 53 under the terms of the CC BY-SA 3.0 license, https://creativecommons.org/licenses/by-sa/3.0/deed.en.

3.6.1.1 Bright Green Malachite.   Malachite, the green copper ore and oldest known eye paint in history, occurs both in the Eastern Desert of Egypt and in the Sinai copper deposits. It takes its name from the Greek μαλαχίτιδα, mallow plant, since its green color resembles the leaves of the plant.[54] Malachite is not to be confused with malachite green, a synthetic organic dye first prepared in 1877. If you did, your timing would be off by about 7000 years. Malachite the mineral, a green copper carbonate hydroxide, is a brilliant crystalline semiprecious gemstone that lies midway on the **Mohs hardness scale**. As such, it takes a great deal of effort to grind it into the fine powder suitable for eye decoration that Egyptians found so enchanting.[55]

In the Early Dynastic Period (3150–2600 BCE), malachite was very popular with both sexes, usually being liberally spread all the way from the eyebrow to the base of the nose. In the Middle Kingdom (2050–1652 BCE), green cosmetics continued to be used for the brows and the corners of the eyes. Egyptian under-takers made sure that all the equipment was there to prepare the cosmetic: it has been found as lumps of ore set beside palettes and grinding pebbles, ready for the afterlife occupant to complete his or her eye *toilette*.[56] The noted British Egyptologist, Flinders Petrie notes that malachite was the commonest mineral found in late predynastic (*ca.* 3100 BCE) burials. Eighteen examples were found in one such site, whereas in earlier sites, only two examples were found.[57]

3.6.1.2 Galena and Kohl. In popular parlance, galena and kohl are often understood to be synonymous. As we shall see, they can also be quite different depending on location, method of manufacture, and supplier. Masking the nature of these two substances is another form of deception, but in this case, not so mild.

Galena, a mineral consisting of lead sulfide, PbS, was mined locally in the Eastern Desert region and near the Red Sea. However, instrumental analysis of some predynastic tomb specimens suggests that some lead ores were imported from Syria and Anatolia.[58] Often found mixed with antimony compounds in natural deposits, its main cosmetic feature and why it is so prized is its gray-black metallic hue.[59] Galena is a relative latecomer to the cosmetic world, supplanting malachite only very gradually in late predynastic times and lasting up into the Coptic era (2500 BCE – 700 CE). By the time of the New Kingdom (1500–1300 BCE), it had superseded malachite almost entirely. It is quite possible that long experience and observation of galena's therapeutic effects as an eye cosmetic gave rise to this gradual change.

Gravesites often held the crude form of galena in small linen or leather bags, whereas the prepared form, kohl, could be found in shells or small vases. When kohl is found in a compact mass rather than in powder form, its original material was probably a paste that had dried up over time. Analyses of 58 of these tomb samples show that 64% of them were galena (37), 6 were manganese oxide (MnO_2), 5 were brown ochre (probably a mix of $FeO(OH)$ and Fe_2O_3), 4 were malachite ($CuCO_3 \cdot Cu(OH)_2$), 2 were lead carbonate ($PbCO_3$), and one each were black oxide of copper (CuO), magnetic iron oxide (Fe_3O_4), antimony sulfide (Sb_2S_3), and chrysocolla ($Cu_2H_2Si_2O_5(OH)_4 \cdot xH_2O$).[60] This finding puts to rest the misconception that kohl is primarily an antimony preparation, an error possibly propagated by Pliny who wrote of a Roman eye cosmetic called *stibi* (from *stibium*, the Latin name for antimony).[61]

What is very clear from the mix of substances all thought to be kohl, is that kohl could be anything at all. The traditional purposes were thought to be for eye beautification, but also for prevention or treatment of eye diseases, and mitigation of sun glare.

The modern practice of kohl usage is widespread in Northern Africa, the Middle East and the Far East. Besides its cosmetic appeal, it is a popular folk medicine (known historically as "Al-Kuhl," "Surma," or "Ithmid") that is applied to the conjunctival surfaces of the eyelids rather than to the outside. It is customary to even apply kohl to infants and children. An early-1990s study found that daily use of kohl was three times more likely than just occasional use.[62] It was found, not surprisingly, that lead blood levels were, on average, 2.5 times higher in infants on whom kohl was applied than not.[63] In the absence of label requirements in the regions where kohl use is prevalent, no one is certain about the amount of lead burden being imposed on vulnerable children, nor have there been any comprehensive studies done to characterize the chemical contents of kohl from various sources. A team of researchers from Walter Reed Army Medical Center (Washington, DC, USA) and King Faisal Specialist Hospital (Riyadh, Kingdom of Saudi Arabia) collected 21 different samples of kohl as homemade powders (6), commercial preparations (9) and natural stone kohl (6) and determined their content by **energy dispersive X-ray spectroscopy**. They found that two-thirds of the samples contained lead ranging from 2.9% to 97%. Three of the homemade samples were predominantly composed of 90–97% lead, two were mainly iron and carbon, and one contained antimony. Seven of the commercial specimens contained lead ranging from 24–97%; while five specimens contained carbon as the main colorant. As for the stone kohl, three of the samples were almost pure galena, PbS; in two others, iron was the dominant element.[64] A later study found that of 12 commercially available black cosmetic specimens, six consisted of galena and six of amorphous carbon.[65] Since kohl preparations continue to be made at home or as unlabeled products by small companies in India and the Middle East, its continued unregulated use may constitute a major health hazard, especially among infants and young children.

A lengthy review article on kohl came to three important conclusions: (1) Details published on the chemistry of ancient eye makeup and results of the other studies on the chemical composition of kohls provided reasonable support to conclude that the major constituent of kohl (surma) from the very beginning was galena (lead sulfide) and that "Al-Kuhl" (Kohl), "Surma" and "Ithmid", all indicate only one substance, galena. (2) Lead sulfide from kohl as an eye cosmetic has not been demonstrated to cause toxic injury. (3) Lead, after application of galena based kohl, is not absorbed through the transcorneal route and that a number of studies, both in animals and in humans are available which indicate that kohl is not responsible for increased blood lead concentration.[66]

Confirmation of the possible reasons for conclusions 2 and 3 came from a French study in 2010. Some Egyptian makeup jars at the Louvre Museum contained two unexpected lead-bearing compounds that are not naturally occurring: phosgenite $(Pb_2Cl_2CO_3)$ and laurionite $(Pb(OH)Cl)$. Both Pliny the Elder[67] and Dioscorides[68] describe their deliberate industrial-scale synthesis based on the common starting materials of litharge (PbO), salt (NaCl) and a mix of two commonly available carbonates $(Na_2CO_3$ and $NaHCO_3)$. Both phosgenite and laurionite are thermodynamically favored as products as long as the system's pH can be maintained acidic enough to avoid the formation of competing species like lead hydroxides.

This difficult process must have been driven by the overwhelming benefit offered by the hard-won products. Theorizing on the idea that since micro concentrations of common divalent cations, such as Ca^{2+}, can activate certain essential biological functions, the Louvre investigators looked at the effect that sub-micro Pb^{2+} concentrations might have on oxidative stress responses. They were astounded by the result: Pb^{2+} led to a steady 240% increase of nitrogen monoxide, NO, an important immune system messenger that signals infection to immune cells and increases blood flow in capillaries. These data suggest that the presence of lead-based cosmetics, like the deliberately manufactured phosgenite and laurionite, in constant contact with the eye's lachrymal fluid was a potent defense against bacterial infections through the spontaneous reaction of the wearer's own immune cells. In the Nile Valley environment, where eye disease was a constant threat, this protection could far outweigh any negative effects of lead toxicity.[69] It would be inexcusably anachronistic to posit that the ancient Egyptians were aware of the beneficial effects of nitric oxide, but they most certainly knew of the therapeutic effects of kohl usage. Capitalizing on this idea, some pharmaceutical companies even today manufacture ophthalmic products based on nitric oxide activity.[70]

The French team noted that hieroglyphic inscriptions on the Louvre cosmetic containers distinguished among the one containing black makeup, galena (labeled "medesmet") and two others containing the lead chlorides and carbonates (labeled "eye lotion to be dispersed; good for eyesight"). Eye problems were apparently a chronic problem for the Egyptians. Medical papyri list over 100 recipes for treating problems of the eyelids, iris and cornea, as well as trachoma and conjunctivitis.[71]

Kohl's reputation as a substance that confers both beauty and health upon one of the body's most important assets spread far beyond ancient Egypt as the centuries rolled on. It found its way eastward to ancient Mesopotamia, India and South Asia; northward to Greece, Rome and all of Europe; southward to Africa, and eventually to the New World. It is now a multibillion-dollar business with specialty products such as eye liner, eye shadow and mascara in a variety of formulations and colors. By law in many countries, the cosmetic contains non-toxic ingredients based on carbon powder mixed with oils and waxes. A combination of good health, enhanced beauty, and shrewd marketing has made eye cosmetics a winning mix, all due to the first known large-scale chemical process. It is so appropriate that the word "kemē," which refers both to the land of Egypt and the black earth of the Nile Valley, should come down to us as our own "chemistry."[69,72]

By its very name, we know that another Egyptian chemical wonder, henna, clothes itself in red and orange. Henna, "hinna," in Arabic-speaking Persia, means "to dye red." However, some of its other properties can accompany you from the cradle to the grave. Henna is equally at home in your hair or on your hide.

3.6.2 Henna

What miraculous material can not only add luster and shine to your hair but also condition it, promote hair growth, prevent dandruff, deter head lice and cover up your grizzle? Only a 5000 year-old natural product that has found new life not only at teen tattoo parties, but also is an antidiarrheal agent. Henna is also thought to cure jaundice, headache, eczema and acid reflux. This magic marvel was well-recognized at least 3500 years ago by its prominent place in the famous *Ebers Papyrus*,[73] a medical compendium of Egyptian Thebes.

The henna plant, *Lawsonia inermis*, a small thorny flowering tree sometimes called mignonette or Egyptian privet, is the sole species of the genus *Lawsonia*.[74] It projects a different face in its many different growth phases and environments due to the different levels of its active ingredient, lawsone (structural formula shown in Figure 3.4), present in each one. The cultivation conditions, the plant part used, and the method of preparation all play a role in the amount of lawsone available as color.

While near neighbors in Lybia, Tunis, Syria and Canaan were using henna for body art, the more serious Egyptians were more attentive to its therapeutic effects as an anti-inflammatory and for skin diseases.[75] However, to modern Western imaginations,

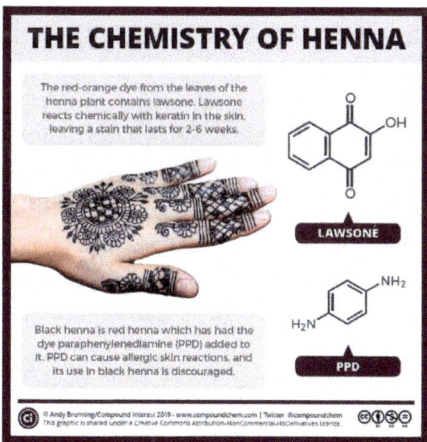

Figure 3.4   Left: Ground henna leaves (Punjab, India). Photo: Mary Virginia Orna.; Right: Henna, red and black, in a nutshell. Graphic courtesy of Andy Brunning.

the word "henna" conjures up only one vision: that of the red-orange color of henna-dyed hair. And indeed, that seems to have been its major use in the mummification process.

Henna's chemistry is tricky. Lawsone, the orange-red dye-active compound known to chemists as 2-hydroxy-1,4-naphthoquinone, is captive in the leaves of the henna plant in the form of three precursor molecules called hennosides. They all consist of the basic lawsone structure bound to glucose molecules, which give them the stability needed to function within the plant. Round henna leaves have a natural greenish color as shown in Figure 3.4, left. To produce the dye, typically called "red henna," powdered and dried leaves are mixed into a paste with water and a weak acid such as lemon juice or vinegar and let stand for 6 to 24 hours. The hennosides break down by this action into the unstable glucose-free 1,2,4-trihydroxynaphthoquinone which then air-oxidizes to form lawsone. Mixed with some essential oils such as eucalyptus or lavender oils, it is now ready in the form of a paste to be applied to the hair or skin.

Next comes some more delicate chemistry. Hair dyes are classified according to their permanence into temporary, semi-permanent non-oxidative, demipermanent and permanent. Temporary hair dyes, usually consisting of large molecules that cannot penetrate beyond the outer hair cuticle, wash out with the first shampoo. Semi-permanent non-oxidative dye consists of small dye molecules that can penetrate beyond the outer cuticle into the hair cortex, but they do not chemically bond with the hair, so they easily wash out after three to six shampoos. The two latter types of hair dyes require the action of hydrogen peroxide (H_2O_2) on small molecules that penetrate into the cortex and then couple to form larger dye molecules that are permanently locked into the body of the hair fiber.[76] Henna, a natural dye, does not fit into any of these categories: it is a permanent dye that does not require the use of H_2O_2. Here is the mechanism: lawsone is a small molecule containing ketone groups that we can schematically represent as LAW–C=C–C=O. Hair is a proteinaceous material called keratin containing amino groups, –NH_2. We can represent hair as H_2N–KER. Lawsone can gradually migrate into the outer layer of the skin or hair and bind to the protein to form a compound known as a Schiff base:[77]

$$\text{LAW–C=C–C=O} + \text{H}_2\text{N–KER} \rightarrow \text{LAW–C(NH–KER)–CH–C=O} \quad (3.1)$$

The paste should be left on the hair for a few hours. The longer the application, the darker the stain. Since the dye is now chemically bonded to the hair, the dye is permanent, or "fast." This was accomplished without the use of harsh chemicals that can break and denature the hair strands.[78] Since application to the skin involves the same chemical reaction with the skin protein, henna is a permanent stain on the body as well. However, both the hair color and skin color fade with time, in about two to six weeks.

Red henna is a completely natural product as long as the label on the package claims that the material is 100% henna. In recent years, itinerant tattoo artists have popularized another form of henna called "black henna." This works to their advantage because the black color is quite popular and also its use reduces the time of the process from 4–6 hours to 1–2 hours. It also gives a longer lasting effect. But "black henna" may not contain any true henna at all. In place of true henna, some tattoo artists substitute **indigo** (from the *Indigofera tinctoria* plant) because it is dark blue and with the addition of the dye intermediate *para*-phenylenediamine, or PPD for short, a very nice black can be produced (Figure 3.4 right).[79] PPD is a white to purplish crystalline powder that air-oxidizes, turning first red, then brown, and finally black, conferring its own color on the henna-PPD or indigo-PPD mix. There is evidence that PPD, in the process of air oxidation, can **polymerize**, gradually forming a brown color, later turning to black.[80,81] The luscious black color completely disguises the henna, and yet it is the henna molecule, if present, that binds to the skin protein to produce a permanent dye.[82] The concern over the use of PPD in "black henna" stems from the fact that it is a known potent contact allergen and its use is banned in Europe and North America. The customer, usually a tourist, is often unaware of its presence. By the time a possible reaction develops, the artist is long gone or the sufferer is already back home.[83] Potential tattoo recipients are advised to ask to see the ingredient label for the ink or paste; if there is no label, it is wise to say "no go." There are other ways to find out if the product contains PPD: if the paste is an intense black, if it sets very quickly, if it does not fade within about three weeks, and if the paste has little to no scent (Box 3.2).[84]

While the Egyptians were famous for their cosmetic use of lead, we'll see in the next section that the Romans were not far behind.

Box 3.2   Hair's the Problem

More than half the women worldwide and up to 80% in developed countries do not think that Mother Nature did a good job on their hair. They constitute the multibillion-dollar market in hair-coloring products which, despite recent warnings about health and safety, will not go away any time soon.[85] The hair dye revolution started in 1907 when, at the request of a Parisian barber, French chemist Eugène Schueller (1881–1957) began to investigate ways of permanently correcting Mother Nature's "error." Schueller eventually founded the cosmetics giant L'Oréal. For two years, he tried out trial-and-error products on his own hair, and then he discovered the color-creating properties of PPD.[86,87] PPD contains two amine groups, $-NH_2$ (Figure 3.4, right). Oxidizing these groups (usually with hydrogen peroxide, H_2O_2) to imines, $=NH$, makes the PPD reactive enough to couple with another dye intermediate to produce a colored molecule. By judiciously selecting the right proportions of these intermediates, he could fine-tune the hair dye color to suit anyone's whim. He added an alkaline ingredient, often ammonia, NH_3, to the mix to get the hair's outer layer, the cuticle, to open up so that the dye molecules would lodge in the hair's inner layer, the cortex. H_2O_2 finished the job by oxidizing the hair's natural colored dye, melanin, to a colorless product. Since the new colored dyes were deeply entrenched in the hair's body, no ordinary shampoo could wash them out – they were there to stay until the hair grew out. Hair-coloring chemistry has not changed much since Schueller's day, but it is constantly being monitored for carcinogenicity and other ills, and modifications are slowly taking place.[88]

3.7   GRECIAN (OR ROMAN?) FORMULA

To the ancient Greeks, the use of cosmetics by men was the height of effeminacy because they were deemed to be a powerful symbol of the feminine. In fact, if a man were to exhort his wife to renounce cosmetics, this would signal her elevation to his side as a man-like partner, thus reassigning her place in the gender hierarchy.[89] Judging from the popularity of the men's cosmetic, Grecian Formula (© Combe, Inc.), we might say that its use signals a perpetuation of the Greek attitude into the modern era. Why would a man want to use a product that disguised his use of a cosmetic? In this case, a gradual darkening of graying hair assumes that by its very gradation, no one would notice what was going on. Why not just dye your hair and get it over with? Oh, no, that would be acting like a woman!

When I first learned how Grecian Formula worked, I was appalled. It was essentially a preparation of lead acetate $((CH_3COO)_2Pb)$ that, when applied to the hair, reacts with the sulfur-containing keratin in hair strands, depositing a film of black lead sulfide (PbS) that, with continued use, accumulates to a desired darkness. Shades of galena! It has found a brand-new use! Or has it? There was actually a "Roman formula" commonly used by aging Roman Senators: just dip a lead comb into vinegar and brush it often through your hair. A lot less expensive – and it worked![43] It seems that the Egyptians may have imitated the Romans in this regard: a team from the *Centre de Recherche et de Restauration des Musées de France* reproduced an ancient hair coloring recipe from Greco-Roman times (*ca.* 332 BCE – 395 CE) that used lead salts. They found it remarkable that the keratin structure could control PbS **nanoparticle** growth inside a hair strand.[90] Lead-based hair dye had a long lifetime over the centuries¶ and lead acetate was still permitted in the US formula until it was finally banned in 2017. (Canada and Europe got the jump on the USA by a whole decade.) But Grecian Formula © lives on – the new pomade contains bismuth citrate, $C_6H_5O_7Bi$ – a water soluble salt that on reaction with the sulfur in the keratin forms Bi_2S_3, pleasingly black and hopefully benign.

Reflecting on the use of a gradually activated hair coloring, we can ask how this was discovered. By accident? By experiment? By technology transfer? The Romans' partiality for lead vessels would have put them into contact with lead acetate sooner or later – just pour a bit of vinegar into a lead bowl and wait a while. We also have to speculate on lead acetate ingestion. It is no accident that its common name is sugar of lead – it was a sweetening agent in an era when the only alternative was honey.

Our next foray into natural product cosmetic chemistry takes us farther north to visit the "barbaric" Britons who favored a dark blue pigment that could be put to a variety of uses.

3.8 A TALE OF WOAD

A single line in Julius Caesar's *Gallic Wars* has bound woad inextricably with the blue dye that garbed Briton warriors in Roman times, and possibly long before that. "All the Britons stain themselves with woad, thus presenting a horrifying aspect in battle."

¶See, for example, N. Belcher, The barbers' and hairdressers' private recipe book, embracing recipes for hair producers, hair renewers, tonic dressings, hair oils, hair dyes, pomades, toilet soaps, shaving soaps, shaving creams, perfumery and colognes, instantaneous grease extractor, washes, and lotions, *etc.*: with a treatise upon the diseases of the human hair, their causes, mode of treatment, and their remedies, Rockwell and Rollins, Boston, 1868.

Despite the fact that Arthur Goldynge's 1565 translation of Caesar's *vitrum* to woad was quite a stretch, other writers have seconded this statement. Pomponius Mela, at the time of the Emperor Claudius, makes a more general statement: "They stain their bodies with woad, either because they like the look of it, or for other reasons."[91] Neither of these comments is meant to be complimentary. In fact, blue was a color despised by the Romans. Pliny associated the hue with savages, midnight orgies, and dishonesty. No one wore blue clothing unless they were eccentric or in mourning. Even persons born with blue eyes were suspect. So, culturally, woad and all those who had anything to do with it were barbaric in the eyes of the ancient Romans.[92]

Yet woad has a long and glorious history both in the British Isles and on the European continent. Indigenous to Turkey and the Middle East, the woad plant, *Isatis tinctoria*, was widely cultivated for its dye, possibly from Neolithic times; it has even been tentatively identified on mummy wrappings from the Fifth Dynasty (2400 BCE),[93] but definitely in Europe from the La Tène period (5th – 1st centuries BCE) on.[94] It falls into the same chemical category as its near relatives, chemically identical **indigo** and Tyrian purple (dibromoindigo) – a vat dye that depends on a **reducing agent** for its solubility as a near-colorless dyebath; when it re-oxidizes, it forms its characteristic insoluble blue pigment (**Chapter 12.5**). In addition to tattooing, one author suggests that the Britons may have used this property of vat dyes to do whole-body immersions and exit from it dyed blue – an experience that could have been associated with the magical and mysterious.[95]

It is doubtful if Caesar ever got far enough north to eyeball one of his "Picts," or painted men (from the Latin "picti"). It is equally doubtful that Briton warriors were still dyeing themselves blue at the time of the film "Braveheart," throwing Caesar's remark 1000 years off-kilter. What we *do* know from the literary record is that numerous classical authors attest to the body painting and tattooing. Ovid speaks of "woad-blue Britain,"[96] and Pliny[97] speaks of a vegetable dye that Britons used to stain the whole body, making them look like "Ethiopians." (Woad can turn black over time and woad-gatherers' hands were often stained black after harvesting the plant.)[98] The archaeological record is quite thin on woad cultivation in Britain: the earliest example was found at Dragonby (Lincolnshire) as seeds and seed pods in an Iron Age pit.[99] Woad cultivation gradually spread throughout Europe from

Neolithic times. Amiens, in northern France, was a great center of the woad trade, memorialized by a statue of woad traders on its 13th century cathedral façade (Figure 3.5).[100]

In Britain, woad was grown commercially until about 1930. Its existence, either as plant or dyeworks, was unwelcome in many quarters. Gardeners call the woad plant a "gross feeder," that is, it sucks up all the mineral nutrients in a soil, rendering it barren. Woad growers would traditionally rent their space, strip it bare, and then move on. The one bright light in the woad plant's cultivation was in its destruction: its ashes were valued for their high potash content, but at what price! Woad exacted a further high price from society in terms of its works. Woad fermentation, often accomplished by composting with manure and urine, released such noxious odors that it was considered a public nuisance,[101] polluting the atmosphere and the water for miles around.[102] In Elizabeth's reign, competition from imported indigo prompted

Figure 3.5 Woad Merchants. Amiens Cathedral, 13th Century. Reproduced from ref. 100 with permission from Wellcome Collection under the terms of the CC BY 4.0 license, https://creativecommons.org/licenses/by/4.0/deed.en.

legal protection[103] even though a 1587 proclamation forbade woad manufacture within three miles of any royal residence.[104] The woad dyers specialized in this one dyestuff throughout the medieval and early modern eras.[105] Imported indigo from the *Indigo tinctoria* plant gradually supplanted widespread woad production, but it remains a specialty colorant to this day.

Our cosmetics journey has taken us through the lands of red, black, green and blue, each with its own cultural significance and symbolism. Now we arrive at white, actually a non-color that reflects all of the wavelengths of the solar spectrum back to the viewer. Traditionally, white in the Bible and in literature stands for innocence, purity and goodness. But, if in white all the colors are present, does that mean that all the other meanings, including evil and passion, are also hiding there? Achieving a perfect white for cosmetic purposes, especially when applied to the face, symbolized perfect beauty, eternal youth, moral power and noble status – fit for a queen like Elizabeth I of England. However, the hidden threat was anything but symbolic as we shall see in the tale to follow.

3.9   CERUSSITE. PALE BEYOND PALE

The element lead, Pb, is a bluish-gray shiny metal that is both very soft and very dense. While that description does not sound very exciting, when lead decides to hitch itself up to other elements or groups of elements, it instantly changes personality. As we have already seen in Figure 3.2, minium, Pb_3O_4, is a beautiful bright red. Other lead compounds can be bright orange, yellow, black and white.[106] Cerussite, or white lead (Table 3.1), has been the darling of all lead-based cosmetics, at least in Europe. Taking its name from the Latin, "cerussa" (meaning white powder), its chemical name is lead carbonate, $PbCO_3$. It has been valued for centuries for its pure whiteness and creamy texture, making it a favorite face cream since the Bronze Age (1350–1100 BCE). As a face cream, cerussite's uses included blemish removal, skin toning, and wrinkle removal. It is quite possible that the wholesale use of cerussa as an habitual face mask contributed to the lead burden already suffered by the Roman population through the use of leaden liquid containers and lead pipes in their plumbing and the custom of adding sugar of lead to wine.[107] However, there is no conclusive proof that lead poisoning directly contributed to the fall of Rome.[108]

As Europe entered the so-called Dark Ages, cosmetic use fell off drastically. This trend was due not only to the general chaotic nature of mass migration, barbarian invasions and crushing poverty, but also to the growing power of the church, which expressly forbade their use. As time went on, however, exceptions were made, especially to women of higher rank, so that they could protect their marriages by maintaining an attractive appearance. Gradually, women began to acquire cosmetics again, and the big winner in the comeback was cerussite. The growing recognition among the rising upper classes that a dark complexion denoted work in the fields drove many women to adopt the white lead face mask once again.[109,110] As the use of cerussite picked up, so did treatment for gastrointestinal problems, but it took a long time to make the connection.[111]

As the Middle Ages were giving way to the Renaissance, the celebrated author of the painters' handbook, *Il Libro dell'Arte*, Cennino Cennini (*ca.* 1370 – *ca.* 1440), issues this warning to women inclined to embellish their complexions: "But I advise you, if you desire to preserve your complexion for a long period, to be accustomed to wash yourself with water from fountains, rivers, or wells, and I warn you that if you use any artificial preparation your face will soon become withered, and your teeth black, and in the end women get old before the natural course of time and become the ugliest old hags possible. This is quite sufficient to say on this subject."[112] But no, Cennino, it is not sufficient – as we shall see, faded beauty is one threat, but a much more dire one was in store for those who habitually used cerussite.

White face mask application reached its peak during the reign of Elizabeth I in England (1558–1603). Sensitive not only to the inherent theatricality of sovereignty, but also to the unmistakable signs of her own aging process, Elizabeth did her best to camouflage the latter and enhance the former. In fact, her face paint was so thick that at one point, some wag remarked that he could cut off a curd of cheesecake from each cheek. Despite this declaration of unambiguous power, Elizabeth had her critics, not least of whom was the Bishop of London who inveighed against her practice from the altar. Apparently he desisted when the queen's subtly-veiled threat to "fitte him for heaven" reached his ears.[113]

Elizabeth's open use of facial cosmetics signaled legitimacy to other women, and soon, women of noble rank and courtly circles followed suit. However, herein lies the paradox of cosmetics: the rub is not *that* a person uses cosmetics, but *who* that person is. Put the same face on an ordinary housewife and she is immediately labeled deceptive and unfaithful.[114] As a 16th century commentator put it, "A woman that paints puts up a bill that she is to be let."[115]

Cerussite users continued to turn a blind eye to the fatigue, joint pain and nausea that accompany lead poisoning, attributing most of these symptoms to exhaustion due to an active social life. In the American colonies, increasing social sophistication gave rise to the first face powders, all containing lead, arsenic, or bismuth. Imperial Royal Cream Wash Balls Soap, one of the most popular, contained a hefty dose of white lead. By 1735, it was touted as "the most beneficial of all cosmetics."[116] The wake-up call was a long time in coming: 157 years after Elizabeth's death, in 1760, Maria, the Countess of Coventry, succumbed to lead poisoning at the ripe old age of 27. Seven years later, the famous actress Kitty Fisher perished *via* the same killer.[117]

In the Victorian era, white lead use gradually fell out of favor, not so much because of toxicity, but simply because it was no longer fashionable. Its use picked up for a time toward the end of the 1800s, but horror stories about its toxicity in the form of skin pustules, lead palsy, and gruesome death drove some medical crusaders to have it banned. By 1884, they had largely succeeded, but other toxic substances, like mercury, often took their place. Physicians began to lobby for full disclosure of contents for both foods and cosmetics, but it was not until 1938 that the Food, Drug and Cosmetic Act (FD&C) became law in the United States.[117]

Moving from day-to-day life to stage and screen gave rise to very different cosmetic needs. Ludwig Leichner and Max Factor were certainly up to the task, as we shall see next.

3.10 GREASEPAINT FOR STAGE AND FILM

Technological advances in stage lighting (See Box 3.3: Theatre Lighting) set every actor's face awash with a brilliant radiance that obliterated all color and left a great white stare. The usual

---

Box 3.3   Theatre Lighting

Given the high-tech trappings of the modern stage, including sophisticated lighting techniques that can transform any stage set instantaneously, it is hard to imagine what the stage of a century ago looked like. In 1800, the only lighting available was candlelight, and in a vast space, we know how far the light from a single candle will go. In England, the classic land of gaslight technology, some improvement came with theatre gaslighting. An important development was Goldsworthy Gurney's (1793–1875) discovery that a block of calcium oxide (lime) emitted a brilliant white light when heated. This phenomenon was quickly exploited to construct optical systems, using focusing lenses, for use in lighthouses and in the theatre as spotlights. This invention, having entered our vocabulary as "limelight," was the first practical incandescent system, although it required constant supervision and by its very nature was extremely hazardous. The fact that the system was widely used in theatre spotlights gave rise to the metaphorical use of its name to mean public attention or fame.[122] Limelight gave way, eventually, to electric lighting. In each of these innovations, the actor on stage was thrown into a view that revealed every nook, cranny, and flaw of costuming or complexion. Since each of these lighting systems exhibited a different spectral irradiance curve (**Chapter 1**.3), each was richer in intensity for different hues of the visible spectrum, requiring makeup that complemented what was being reflected to the viewer's eye. Actors required heavier makeup to look their best on stage, and thus the science of theatrical makeup, underpinned by "greasepaint," was born.

---

theatre makeup, some powder and rouge, simply did not work in the garish glow of the arc light. Actors began to mix highly colored mineral pigments with a base of animal tallow or wax to counteract this effect. Stage makeup became a largely do-it-yourself affair until Ludwig Leichner (1836–1912) divined the opportunity to formulate and commercialize the first standard greasepaint in 1873 in Berlin. Leichner was both an opera singer and a student of chemistry, a perfect mix of talents to do the job. He knew what was needed on stage; he knew how to get it in the laboratory. One of the problems was that the hot lights caused perspiration which in turn caused makeup to run – so he carefully formulated his oil bases to hold steady when blasted with onstage lighting temperatures and the natural temperature of the human body. His original 2 : 1 mix of lard (m.p. 105 °F, 40.6 °C) and mutton suet (m.p. 135 °F, 57.2 °C) held up very well under those conditions; eventually he experimented with other additives like beeswax and white wax. His colorants ranged from yellow ochre to burnt sienna, depending on the type of complexion needed, and eventually his formulas became a paint-by-number affair that every actor could count on.

What Leichner was to the stage, Max Factor (1877–1938), an enterprising Polish fugitive, was to film. What worked for the stage, Factor immediately recognized, would never work in the movies. Stage makeup, a word coined by Factor, was a coarse affair that counted on the great distance between the audience and the player. For example, using fake beards and eyebrows might work on the stage, but the audience would immediately spot them in a film. Factor knew this and in a bold move approached film director Cecil B. DeMille, who was shooting "The Ten Commandments" at the time, with wigs and beards that he had made with real hair and were genuine-looking. DeMille countered that he could not afford to buy them, but could he rent them? Factor thought for a moment and agreed if DeMille would hire his three sons as extras (at $3.00 per day); their real job would be to collect the hair-pieces at the end of each day and make sure they were in good order. The deal was struck, and Max Factor soon became a Hollywood fixture.

Factor realized that in a film, your face was up close and personal to the viewer at all times. In addition, colors have to be made in a graduated, color-corrected scale over the entire spectrum, taking into account needs for black and white shadowing, to meet the demands of the TV and motion picture cameras. Normal "street colors" are not dense enough in color, *i.e.*, they lack the required intensity and opacity to accomplish the coverage needed.[118] For example, movie actors had to wear dark brown lipstick because it photographed red on film, but it was horrible-looking in reality.[119] Factor was a genius at hyping such brand names as Pan-Cake, Pan-Stik (Figure 3.6)[120] and Erace; he was equally adept at maintaining trade secrets regarding their ingredients. Since the FD&C Act did not come into play until 1938, the year of Factor's death, he personally did not have to deal with it, but his successors had to comply. However, we can be quite certain that the use of heavy metal mineral pigments was a thing of the distant past – there was no need once the vibrant synthetic organic pigments invented by clever chemists flooded the color market. And any standard textbook on cosmetics can fill in the blanks.[121]

Figure 3.6 1952 Max Factor Advertisement for Pan-Stik. Reproduced from ref. 120.

No modern city would be complete without its handy set of tattoo parlors in every neighborhood. Lest you think that this need for permanent and obvious identity markers is recent, read on. You will find that it actually predates every other cosmetic practice we have explored.

3.11 TATTOOING

Application of woad to the skin's surface is an example of temporary makeup. Any type of makeup that can be applied in the morning and removed later in the day falls into this category. All of the pigments listed in Table 3.1, with the exception of henna, are forms of temporary makeup. Tattooing is quite another matter.

In 1769, Captain James Cook (1728–1779) was the first to note the word "tatu" in his journal as the Marquesan word for the sound of tapping, which he heard as people marked their skin with pigments. The word "tattoo" later entered the English language as a verb meaning "to inject pigments into the skin by puncturing a design," and a noun for the design itself. However, the idea has broadened somewhat with the dermatological categories of traumatic, amateur, professional, medical and cosmetic tattoos. Traumatic tattoos are unwanted permanent marks made under the skin through road accidents or puncture injuries. Amateur tattoos are those administered by oneself, one's friends, or by street vendors or other unlicensed individuals. Professional tattoos may be either cultural or modern. Certain cultures have time-honored methods and designs that are part of an ethnic group's heritage. Modern tattoos are done by a licensed artist under controlled and sanitary conditions with regulated pigments and often very artistic designs. Medical tattoos are permanent guides for radiation therapy and are placed by a physician. Cosmetic tattoos, often called permanent makeup, are used for lining eyes, hair, lips and eyebrows, or sometimes for covering over an unwanted previous tattoo.[123]

3.11.1 Tattoo History

Ötzi, the Iceman, is probably the world's most celebrated find of human remains in the world. Examining this 5300-year-old "mummy" has yielded priceless information on what life may have been like in the early Bronze Age. Intact skin has allowed scientists to discern 61 tattoos in 19 different clusters. These markings mainly line up along the lower back in locations that modern acupuncturists would target for joint and spinal therapy. While it is hazardous to posit the purpose of these tattoos, they are at least solid evidence of the practice over five millennia.[124] Evidence, though not of human skin, for tattooing practice goes back to the Upper Paleolithic era (*ca.* 15 000 BP) when people would decorate themselves with ochre and threaded shell beads: what appears to be tattoo kits have been found among these artifacts.[125] Tattooed Egyptian mummies from the 11th Dynasty (4000 BP) vie for age with some found in the western deserts of China.[126] Tattooed Eskimo bodies frozen in snow have been recovered with "embroidered" faces accomplished by impregnating a thread with soot and drawing it through the skin with a bone needle.[127] Prehistoric Europeans like the Celts, Picts, and Germans continued the practice of tattooing as marks of honor, or belonging or for religious reasons. However, in classical Greece, we see the first break with this idea in the effort to dissociate civilization from barbarism: tattoo markings gradually came to be associated with criminal status, slavery or serfdom. Among the Romans, branding and tattooing also came to be associated with bondage or lawbreaking.[128]

The first written documentation on tattooing is a proscription in the Bible: "You shall not make any cuttings in your flesh for the dead, nor tattoo any marks on you. I am the Lord." (Leviticus 19:28). Apparently, this prohibition did not deter the Jews, nor the Christians that followed. Tattooing became so widespread that in 787 CE, Pope Adrian I specifically banned the practice. That still did not make any difference. A couple of hundred years later, in the aftermath of the Battle of Hastings in 1066, King Harold's dead body had to be identified by his lover, Eadgyth, who was the only person familiar with the markings on his private parts. And the ever-informative Marco Polo documents its practice wherever he traveled in the Far East.[129,130] Japan has a long history of full-body tattooing. During the Edo period (1600–1800 CE), it became quite commonplace and the practice spilled over into the following century (Figure 3.7).[131] History and the archaeological record point to a universal human penchant to somehow embellish nature's gift with culturally motivated meaning.

Figure 3.7   Tattooed Japanese Men, *ca.* 1870. Hand-colored photo by Beato Felice (1834–1907); Reproduced from ref. 131.

3.11.2   Modern Tattoo Practice

Fast forward to the 21st century. While tattooing may seem to have dropped out of sight among the Western general populace over the past few centuries, it has always been close to the surface of our consciousness. It has enjoyed continued great popularity in modern India and the Far East. Reports of anthropologists working among indigenous peoples around the world have documented an ongoing culture of tattooing. Returning servicemen, particularly sailors, have almost always sported visible and invisible marks of their worldwide travels, often acquired after a weekend of camaraderie to signify "belonging" or just for something to do. Gang members have often been walking billboards for which brotherhood or warrior clan they belonged to. Gang tattoos not only identify membership; they also message commitment and allegiance to one's own gang, intimidation and threat to others.

Up until the 1960s, tattoos were largely unusual among ordinary private citizens, at least in Europe and North America. But the situation has shifted drastically in the past decades both in terms of ease of access and cultural acceptance. Tattoos are increasingly seen to be works of art, legitimate body adornment on an equal footing with any other kind of makeup. No longer back street operations, tattoo shops advertise their wares in the best of neighborhoods to an upwardly mobile clientele. A 2015 Harris poll documented that in the United States, 31% of women and 27% of men over all age groups paraded at least one tattoo. In the U.K., about 20% of the population is tattooed. Illustrative of tattooing's emergence from sub-culture to popular culture in Great Britain is the 173% increase in the number of tattoo parlors between 2004 and 2014.[132]

Nevertheless, parents may not be terribly enthusiastic about their teens' version of self-expression, especially when they arrive home already pierced or advertising a bright red rose (or something less benign) on the left arm. Advice from health science personnel involved with adolescents interested in body art starts with suggesting that parents initiate early conversations with their children, stay away from scare tactics, which are largely ineffective, and recognize that their social values may not coincide with their children's. After all, even Barbie has a butterfly tattoo.[133]

The tattoo process, to be indelible, involves the injection of a colored material intradermally with a needle, or set of needles, moistened with the tattoo colorant. The procedure is necessarily painful and there is often bleeding as well. Since skin penetration is involved, the process should also be sterile in order to prevent infection. Other risks include skin injury and contact dermatitis at the tattoo site and possible pigment transport to other locations in the body such as the liver and lymph nodes. However, the greatest risk is the long-term residence of highly toxic and potentially carcinogenic substances between the dermis and the epidermis of one's skin. Many, if not most, tattoos seem to be tolerated quite well, but the identities of the colorants, their pharmacology, their purity and their eventual biodistribution in the tattooed individual is largely unknown.

3.11.3   Tattoo Colorants

The colored substances that are suitable for tattooing are not dyes, although they are often called that. These are pigments, that is, insoluble chemical compounds that must be suspended in a liquid, called the carrier, or vehicle, in order to be injected. They are variously called inks, dyes, pigments, chromophores or colorants. They may be mineral pigments or plant derivatives or synthetic colorants or a mixture of any of these. Commercially prepared tattoo colorant formulas are proprietary and do not carry a list of ingredients on their label. Their use is largely unregulated. Table 3.2 contains a partial list of possible colorants, although it is not exhaustive.[134] Virtually any colored substance can be used. The carriers are unregulated as well. They may be something as simple and harmless as purified water, or they may contain other chemicals such as alcohols glycerine or detergents.[135]

Use of these and a growing list of other pigments (about 113 documented in late 2018) will continue to be problematic as tattoo practitioners increase along with tattoo popularity. Tattoo specialists tend to purchase pre-prepared colorant mixes, but then many of them add in their own concoctions as well, making for a very complex mix difficult to analyze. Although a number of these colorants are approved for use in cosmetics, none are approved for injection into the skin. In fact, the trend at present is turning toward the vibrant colors of the long-lived **azo pigments** (containing one or more azo group, $-N=N-$) which are primarily manufactured for use as printing inks and automobile paints. This fact is particularly vexing because azo colorants have not been approved as cosmetics, that is, for surface application to the skin, whereas in a typical azo colorant tattoo, as much as 250 mg of the dye can be injected subcutaneously to cover a skin area of 100 cm^2. Azo dyes are known to decompose into carcinogens. The medical literature contains multiple case reports of malignant melanomas forming after tattooing,[136,137] yet they now comprise about 80% of all tattoo colorants. The color green presents an interesting chemical problem. Green tattoo pigments often contain chromium(III) oxide, Cr_2O_3, sometimes called Guignet's Green, considered harmless in itself. However, it is manufactured by reduction of chromate (CrO_4^{2-}) and dichromate $(Cr_2O_7^{2-})$ compounds, both of which are powerful oxidizing agents and skin irritants as well. Trace amounts of these starting materials in Guignet's Green have been found to produce severe

Table 3.2　Partial list of chemical substances commonly used in tattooing.

Common name	Chemistry	Remarks/Toxicity
RED		
Cinnabar	Mercury sulfide, HgS	Highly toxic; FD&C Hg limit: 3 ppm (for Hg vapor)
Cadmium red	Cadmium selenide, CdSe	Highly toxic, Known carcinogen
Common rust	Iron oxide, Fe_2O_3	None known
Naphthol-AS	Produced from coal tar derivatives	Risk of allergic reaction; skin irritant
ORANGE		
Cadmium seleno-sulfide	Mix of red CdSe with yellow CdS	Known carcinogens; highly toxic
Azo dyes	Azo dye (−N=N−)	Decompose to compounds adverse to health
YELLOW		
Cadmium yellow	CdS	Known carcinogen; highly toxic
Yellow ochre	FeO(OH)	None known
Chrome yellow	Lead chromate, $PbCrO_4$	Known carcinogen; highly toxic. FD&C Pb limit: 10 ppm; allergenic; irritant
Curcuma yellow	Curcuma (sometimes called turmeric or curcumin) from a plant of the ginger family	None known
GREEN		
Guignet's Green	Chromium Oxide (Cr_2O_3)	Skin irritant; mutagenic; may be allergenic due to chromate impurities
Malachite green $[Cu_2(CO_3)(OH)_2]$	Basic copper(II) carbonate, $CuCO_3 \cdot Cu(OH)_2$	None known
Ferrocyanides and Ferricyanides	Mixture of potassium ferrocyanide (yellow) and ferric ferrocyanide (Prussian Blue)	None known
BLUE		
Azurite	Basic copper carbonate, $CuCO_3 \cdot Cu(OH)_2$	Skin irritant
Lapis lazuli; natural ultramarine	Complex aluminosilicate clathrate, $(Na,Ca)_8(AlSiO_4)_6(SO_4,S,Cl)_2$	None known

(continued)

Table 3.2   *(continued)*

Common name	Chemistry	Remarks/Toxicity
Egyptian blue	Calcium copper tetrasilicate, $CaCuSi_4O_{10}$	May be harmful if swallowed
Cobalt blue	Cobalt aluminate, $CoO \cdot Al_2O_3$	None known
Thalo blue	Copper phthalocyanine	None known
VIOLET		
Manganese Violet	Manganese ammonium pyrophosphate	None known
Quinacridone	Fusion of acridone and quinoline	Mild skin irritant
WHITE		
Titanium white	Titanium dioxide, TiO_2	Suspected carcinogen
Lead white	Basic lead carbonate, $2PbCO_3 \cdot Pb(OH)_2$	FD&C Pb limit: 10 ppm; highly toxic, known carcinogen
Barium white	Barium sulfate, $BaSO_4$	None known
Zinc white	Zinc oxide, ZnO	None known
BLACK		
Soot, boneblack	Carbon, C	None known
Magnetic iron oxide	Fe_3O_4	None known
Iron oxide	Iron(II) oxide, FeO	None known
Logwood	Heartwood of *Haematoxylon campechianum*	None known

allergic reactions.[138] Ignorance of principal, as well as ultra-small amounts of, ingredients can do untold harm. All the more reason to *think before you ink!*[139]

As of 31 October 2019, the US Food and Drug Administration (FDA) has provided the following safety and regulatory background on the use of tattoo colorants:[140]

> *FDA considers the inks used in intradermal tattoos, including permanent makeup, to be cosmetics. When we identify a safety problem associated with a cosmetic, including a tattoo ink, we investigate and take action, as appropriate, to prevent consumer illness or injury. The pigments used in the inks are color additives, which are subject to premarket approval under the Federal Food, Drug, and Cosmetic Act. However, because of other competing public health priorities and a previous lack of evidence of safety problems specifically associated with these pigments, FDA traditionally has not exercised regulatory authority for color additives on the pigments used in tattoo inks. The actual practice of tattooing is regulated by local jurisdictions.*

This last part of the quotation is very telling. The FDA has not, and more than likely, will not regulate tattoo colorants. However, according to Helmenstine,[134] the FDA is examining tattoo inks to determine the chemical composition of the inks, learn how they react and break down in the body, and whether there are short- and long-term health hazards involved. As more is learned and publicized, it might be possible to build a database of this information, but we will never overcome the proprietary nature of the business.[141]

Regarding other tattoo issues, as of 2011, only 14 states satisfactorily met the criteria for sanitation, training, and infection control in the tattoo industry.[142] To keep up with the evolving "state of the body art," almost every state has had to enact laws addressing its practice. As of 2018, 45 states have passed laws prohibiting tattooing of minors.[143] Permanent makeup, a type of tattoo, is included in all of the previous discussion. Since the procedure involves injection of a colorant *via* a needle into your skin to look like eyeliner, lip liner, eyebrows, or other makeup, it is considered tattooing.

3.11.4   Toodle-oo Tattoo!

While interest in tattooing has been on the rise among young people, not only does that interest fall off abruptly as life situations
change, but there is genuine regret among many for having gotten
one at all. This is not a new development either – tattoo removal
recipes have abounded for thousands of years. They were often
drastic, painful, gruesome, and mainly ineffective. Take, for example, one of the first reports by a certain Scribonius Largus in 54 CE:
"make an ointment with heads of white garlic, ground up cantha-
rides (blister beetles), sulfur, bronze coins, beeswax and oil." We
can surmise that the active ingredient was the cantharidin in the
beetle brew that has a great track record for wart removal; as for tattoos, it could probably remove superficial markings, but not much
else. However, it was probably less gruesome than treatment with
strong acids like nitric acid and sulfuric acid that would inevitably
cause scarring, or worse. Historically, other substances, more or less
unpleasant, like milk, acids, salt, pepper, lye, turpentine, decomposed urine and pigeon excrement numbered among the desper-
ate measures tried.[144] By the early Middle Ages, tattoo removal was
virtually a medical specialty. Tattooing fell into disuse by the 11th
or 12th century, likely because of ecclesiastical disapproval – it took
three centuries for Adrian I's prohibition to finally hit home.

Renewed interest in tattoo removal took many forms. For example, between 1963 and 1966, as many as 70% of tattooed sailors
in the Royal Navy had deep regrets, especially when they were
preparing for marriage. At the time, virtually the only option was
the services of a skilled surgeon and a two-week hospital stay. In
previous decades, application of solid potassium permanganate
powder ($KMnO_4$) was sometimes effective due to its powerful oxidizing capacity, but not every dye was susceptible. Another possibility was an intradermal injection of a tannic acid suspension,
followed by some harsh rubbing over the area with a silver nitrate
($AgNO_3$) stick. This treatment produced a black, leathery crust of
silver tannate which promotes dermal chemical irritation which,
in turn, brings about the skin's rejection of the pigment.[145]

Although the first report of tattoo removal by laser treatment
was in 1965,[146] numerous other methods such as dermal abrasion
and cryosurgery continued to be used into the 1970s.[147] As late
as 1991, treatment with trichloroacetic (CCl_3COOH) acid accompanied by surgical excision continued to be offered.[148] Today, the
Quality-switched (QS) laser is considered the gold standard in

tattoo removal treatment. It is a safe, rapid, effective and virtually painless procedure compared with continuous wave lasers and the non-laser procedures that have been used in the past and documented above.[149]

The difference between Q-switching and other forms of tattoo removal, including earlier forms of laser treatment, is twofold: (1) The laser machine is capable of delivering very short pulses of energy (durations of about 10 to 100 nanoseconds) at very high power, typically in the Gigawatt range, that pulverizes the pigment particles with minimal heat generated in the surrounding tissues. (2) The laser system is ideally equipped with settings for four laser wavelengths that operate on the principle of selective photothermolysis, that is, the settings are chosen for maximum energy absorption by the target pigments.[150] Table 3.3 summarizes the action of these different laser types (Box 3.4).

Black and generally dark pigments are relatively easy to remove with laser light. Green and orange are more difficult to remove, and yellow is the most difficult of all. For laser light to be effective,

Table 3.3   Principal Q-switched laser systems used in tattoo removal.

Type of laser	Operating wavelength (nm)	Color of target pigment	Spectral region where energy is absorbed
Neodymium: yttrium-aluminum-garnet (QS Nd:YAG)	532	Red, black	Green-blue-violet
Ruby (QSRL)	694	Blue-green, black	Orange-yellow
Alexandrite (QS Alex)	755	Blue, black	Infrared-red-orange
Neodymium: yttrium-aluminum-garnet (QS Nd:YAG)	1064	Black	Infrared

---

Box 3.4   LASER

LASER is an acronym for Light Amplification by Stimulated Emission of Radiation, which is exactly what a laser is and how it works. A typical laser consists of a gain medium, that is, a material like a gas or a crystal that can be excited from a ground state into higher energy states by an external source. When the system returns to the ground state, light is emitted and then amplified by multiple reflections between two mirrors. The light emitted has a single wavelength, that is, it is monochromatic; it is also coherent and unidirectional, that is, the light waves are in phase and highly focused. The pulsed output necessary to deliver rapid-fire impulses to tattoo dye particles in the skin is accomplished by a technique called Q-switching. For example, a Q-switched ruby laser uses a synthetic ruby crystal as its gain medium and produces pulses with a deep red color at a wavelength of 694.3 nm.

it must be absorbed by the pigment. The light energy is transmitted to the pigment particles which are physically broken up into microgranules which can then migrate from the dermis into the body and eventually be removed *via* the lymphatic system and normal elimination processes. Yellow pigments reflect almost the entire red, orange, yellow and green portions of the visible spectrum; they absorb only blue and violet light, presenting a very narrow energy window to the laser specialist. On the other hand, the darker pigments absorb almost the entire visible spectrum and are therefore receiving the full blast of pulverizing energy.

Other colorants, particularly iron oxide, the anatase crystalline form of titanium dioxide, and red pigments in general undergo what is called "paradoxical darkening" when exposed to laser light, *i.e.*, while they have generally been used as "brighteners" in tattoo inks, they show their "dark side" when it comes to attempted removal.

Depending upon the type of pigment and the amount to be removed, full removal needs between six and twenty visits spaced about a month apart. In other words, tattoo removal is not so easy – it is nothing like what is depicted in Figure 3.8.[151]

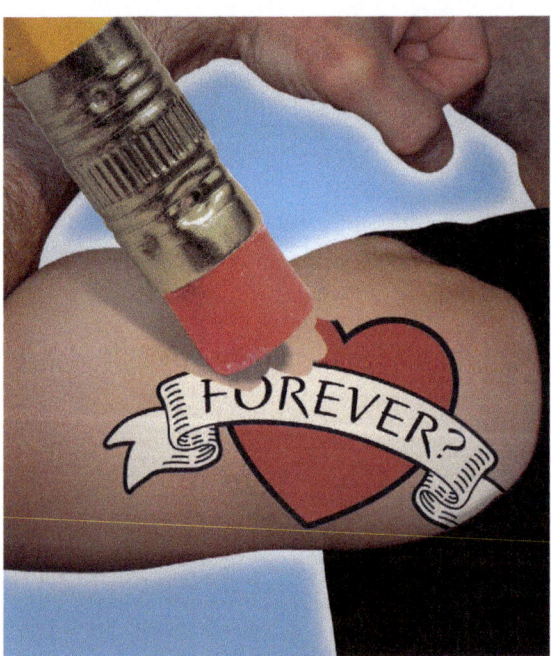

Figure 3.8   Erasing a tattoo is very hard work. Reproduced from ref. 151 with permission from US FDA.

3.12   CONCLUSION

Yes, we've come a long way, baby, from the mists of the Aurignacian era into the Klieg lights of the 21st century. What was an "industry" of found objects and colorful nuggets has become a $140-billion enterprise worldwide, and yet even in a time of joblessness and want, it is still considered "essential." This chapter has barely touched the many other areas of cosmetic and colorant use, particularly some of the more common ones like lipstick and rouge.[152,153] Another consideration as we move further on into the 21st century is the issue of "green chemistry," and how this concept can be applied to the creation of sustainable cosmetics. The cosmetics industry is leaning more and more toward incorporating naturally-derived ingredients in their products; it would be incumbent upon the consuming public to better understand their activity and composition.

We have seen how the Egyptians took the lead in creative chemistry applied to unique cosmetics. In the next chapter, we will see how they took advantage, once again, of the natural resources in the Nile Valley to build one of the greatest civilizations of all time and to fashion one of the oldest synthetic pigments known.

REFERENCES

1. John Irving as told to Mel Gussow, *New York Times*, 1 August 2001, Section E, p. 1.
2. C. Foyle, *Foyle's Further Philavery*, Chambers Harrap Publishers, Edinburgh, 2008, p. 41.
3. L. Woodhead, *War Paint: Madame Helena Rubinstein and Miss Elizabeth Arden: Their Lives, Their Times, Their Rivalry*, Turner Publishing, Nashville, 2004.
4. I. Watts, Red ochre, body painting, and language: Interpreting the Blombos ochre, in *The Cradle of Language*, ed. R. Botha and C. Knight, Oxford University Press, New York, 2009, pp. 62–92.
5. L. C. Parish and J. T. Chrissey, *Clin. Dermatol.*, 1988, **6**(3), 1–4.
6. A. Hope and M. Walch, *The Color Compendium*, Van Nostrand-Reinhold, New York, 1990, p. 45.
7. J. Hamblin, *Clean: The New Science of Skin*, Riverhead, New York, 2020.

8. B. Jarvis, *The New Yorker*, 3 & 10 August 2020, 66–70.
9. M. Hoffman, Picture of the skin, https://www.webmd.com/skin-problems-and-treatments/picture-of-the-skin, accessed April 2021.
10. M. R. Belgiorno, *Cosmetics in Archaeometry and Aphrodite. Proceedings of the Seminar 13th June 2013, CNR, Rome*, ed. M. R. Belgiorno, DeStrobel, Rome, 2017, pp. 45–93.
11. For a comprehensive history of cosmetics, fragrances, and skin-care products, please see F. J. González-Minero and L. Bravo-Díaz, *Cosmetics*, 2018, **5**, 50–59.
12. G. Rapp, *Archaeomineralogy*, Springer, Heidelberg, 2009, pp. 207–212.
13. E. Hovers, S. Ilani, O. BarYosef and B. Vandermeersch, *Curr. Anthropol.*, 2003, **44**(4), 491–522.
14. M. D. Leakey, *Olduvai Gorge Volume III: Excavations in Beds I and II, 1960–1963*, Cambridge University Press, Cambridge, 1971.
15. J. B. Hutchings, Color in anthropology and folklore, in *Color for Science, Art and Technology*, ed. K. Nassau, Elsevier, New York, 1998, p. 198.
16. W. Roebrooks, M. J. Sier, T. Kellber Nielsen, D. De Loecker and J. M. Pares, *et al.*, *Proc. Natl. Acad. Sci. U. S. A.*, 2012, **109**(6), 1889–1894.
17. C. S. Henshilwood, F. d'Errico, K. L. van Niekirk, Y. Coquinot, Z. Jacobs and S.-E. Lauritzen, *et al.*, *Science*, 2011, **334**, 219–222.
18. M.-P. Pomiès, M. Barbaza, M. Menu and C. Vignaud, *L'Anthropologie*, 1999, **103**(4), 503–518.
19. E. E. Wreschner, R. Bolton, K. W. Butzer, H. Delporte, A. Häusler and A. Heinrich, *et al.*, *Curr. Anthropol.*, 1980, **21**(5), 631–644.
20. M. Sommer, *Bones and Ochre: The Curious Afterlife of the Red Lady of Paviland*, Harvard University Press, Cambridge, 2007.
21. A. Marschak, *Curr. Anthropol.*, 1981, **22**(2), 188–191.
22. V. Finlay, *Color. A Natural History of the Palette*, Random House, New York, 2004, pp. 39–60.
23. T. Hardy, *The Return of the Native*, Macmillan Co., London, 1919, p. 92.

24. V. Sherrow, *For Appearance' Sake: the Historical Encyclopedia of Good Looks, Beauty and Grooming*, Oryx Press, Westport, CT, 2001, pp. 49–51.

25. Rob Lavinsky, iRocks.com, http://www.irocks.com/, https://commons.wikimedia.org/wiki/File:Cinnabar-180777.jpg, accessed June 2021.

26. Rob Lavinsky, iRocks.com, http://www.irocks.com/, https://www.mindat.org/photo-232909.html, accessed June 2021.

27. Rob Lavinsky, iRocks.com, http://www.irocks.com/, https://commons.wikimedia.org/wiki/File:Realgar-Picropharmacolite-114514.jpg, accessed June 2021.

28. M. L. Vásquez de Ágredos Pascual, Painting the skin in ancient Mesoamerica, in *Painting the Skin: Pigments on Bodies and Codices in Pre-columbian Mesoamerica*, ed. E. Dupey García and M. L. Vásquez de Ágredos Pascual, University of Arizona Press, Tucson, 2018, pp. 11–23.

29. M. V. Orna, Historic mineral pigments: Colorful benchmarks of ancient civilizations, in *Chemical Technology in Antiquity. American Chemical Society Symposium Series 1211*, ed. S. C. Rasmussen, American Chemical Society, Washintgon, DC, 2015, pp. 17–69.

30. A. H. Church, *The Chemistry of Paints and Painting*, Seeley, Service, Co., London, 3rd edn, 1901. p. 165.

31. E. R. Caley, *J. Chem. Educ.*, 1928, **5**(4), 419–424.

32. E. Spindler, *The Story of Cinnabar and Vermilion (HgS) at the Met*, https://www.metmuseum.org/blogs/collection-insights/2018/cinnabar-vermilion, accessed June 2020.

33. C. S. Smith and J. G. Hawthorne, *Trans. Am. Philos. Soc.*, 1974, **64**(part 4), 1–128.

34. F. Yu, *Chinese Painting Colors: Studies of Their Preparation and Applications in Traditional and Modern Times*, trans. J. Silbergeld and A. McNair, University of Washington Press, Seattle, 1988, p. 5.

35. S. Kroustallis and R. Bruquetas, Paint it red: vermilion manufacture in the middle ages, in *Making and Transforming Art: Technology and Interpretation*, ed. H. Dubois, J. H. Townsend, J. Nadolny, S. Eyb-Green, S. Kroustallis and S. Neven, Archetype Publications, London, 2014, pp. 23–30.

36. S. Bucklow, Paradigms and pigment recipes: vermilion, synthetic yellows and the nature of egg, *Z. Kunsttechnol. Konserv.*, 1999, **13**, 140–149.

37. M. J. Melo and C. Miguel, The making of vermilion in medieval Europe: historically accurate reconstructions from The Book on How to Make Colours, in *Fatto D'Archimia: History and Identification of Artificial Pigments*, ed. S. Kroustallis and M. Del Egido, Ministerio de Educación, Cultura y Deporte, Madrid, 2012, pp. 183–196.

38. C. Miguel, J. V. Pinto, M. Clarke and M. J. Melo, The alchemy of red mercury sulfide: the production of vermilion for medieval art, *Dyes Pigm.*, 2014, **102**, 210–217.

39. R. L. Feller, *Report and Studies in the History of Art*, 1967, vol. 1, pp. 99–111.

40. R. J. Gettens and G. L. Stout, *Painting Materials: a Short Encyclopaedia*, Dover Publications, New York, 1966, pp. 152–154.

41. O. Wilner, *Class. J.*, 1931, **27**(1), 26–38.

42. N. Ercal, H. Gurer-Orhan and N. Aykin-Burns, *Curr. Top. Med. Chem.*, 2001, **1**, 529–539.

43. L. Patrick, *Altern. Med. Rev.*, 2006, **11**(2), 114–127.

44. Z. Fu and S. Xi, *Toxicol. Mech. Methods*, 2020, **30**(3), 167–176.

45. D. A. Cory-Slechta, M. E. Sobolewski, E. Marvin, K. Conrad, A. Merrill and T. Anderson, *Toxicol. Pathol.*, 2019, **47**(6), 019262331987840.

46. R. J. Ward, F. A. Zucca, J. H. Duyn, R. R. Crichton and L. Zecca, *Lancet Neurol.*, 2014, **13**(10), 1045–1060.

47. T. A. H. Wilkinson, *Writings from Ancient Egypt*, Penguin Books, London, 2016.

48. Duke University, Durham, NC, Duke Papyrus Archive, https://library.duke.edu/rubenstein/scriptorium/papyrus/texts/world.html, accessed June 2020.

49. E. R. Caley, *J. Chem. Educ.*, 1926, **3**(10), 1149–1166.

50. E. R. Caley, *J. Chem. Educ.*, 1927, **4**(8), 979–1002.

51. Interactive Map: Resources of Ancient Egypt, http://www.ancientegypt.co.uk/geography/explore/fea.html, accessed October 2020.

52. T. Wilkinson, *A World Beneath the Sands: the Golden Age of Egyptology*, W. W. Norton, New York, 2020, p. 118.

53. Walters Art Museum, http://thewalters.org/, https://commons.wikimedia.org/wiki/File:Egyptian_-_Wall_Painting_-_Woman_Holding_a_Sistrum_-_Walters_329.jpg, accessed June 2021.

54. Online Etymology Dictionary, https://www.etymonline.com/word/malachite, accessed June 2020.

55. G. Brunton and G. Caton-Thompson, *The Badarian Civilization and the Predynastic Remains Near Badari. British School of Archaeology in Egypt*, Bernard Quaritch, London, 1928, pp. 85–87.

56. B. G. Aston, J. A. Harrell and I. Shaw, Stone, in *Ancient Egyptian Materials and Technology*, ed. P. T. Nicholson and I. Shaw, Cambridge University Press, Cambridge, 2000, pp. 5–77.

57. W. M. Flinders Petrie, *Prehistoric Egypt*, British school of archaeology in Egypt, Bernard Quaritch, London, 1920, p. 43.

58. Z. A. Stos-Gale and N. H. Gale, *Revue d'Archéométrie*, 1981, **5**, 285–295.

59. C. Regner, *Schminkpaletten*, Harrassowitz Verlag, Wiesbaden, 1996, p. 26.

60. A. Lucas, *J. Egypt. Archaeol.*, 1930, **16**(1/2), 41–53.

61. Pliny the Elder, *Natural History*, trans. J. Bostock and J. T. Riley, Henry George Bohn, London, 1855–1857, XXXIII, vol. 33, p. 34.

62. A. M. Al-Kawajah, *Trop. Geogr. Med.*, 1992, **44**(4), 373–377.

63. A. Nir, A. Tamir, A. Zelmik and T. C. Iancu, *Isr. J. Med. Sci.*, 1992, **28**(7), 417–421.

64. A. F. Al-Hazzaa and P. F. Krahn, *Int. Ophthalmol.*, 1995, **19**, 83–88.

65. A. D. Hardy, R. I. Walton, R. Vaishnay, K. A. Myers, M. R. Power and D. Pirrie, Egyptian eye cosmetics ("kohls"): past and present, in *Physical Techniques in the Study of Art, Archaeology and Cultural Heritage*, ed. D. Creagh and D. Bradley, Elsevier, New York, 2006, vol. 1, pp. 173–203.

66. Z. A. Mahmood, S. M. S. Zoha, K. Usmanghani, M. M. Hasan, O. Ali and S. Jahan, *et al.*, *Pak. J. Pharm. Sci.*, 2009, **22**(1), 107–122.

67. H. Rackham and Pliny the Elder, *Natural History*, Livre XXXIII, 1952, pp. 106–110, http://www.attalus.org/info/pliny_hn.html, accessed June 2020.

68. T. A. Osbaldeson, Dioscorides Pedanii, *De Materia Medica*, Book 5, no. 102, IBIDIS Press, Johannesburg, 2000, http://www.cancerlynx.com/dioscorides.html, accessed June 2020.

69. I. Tapsoba, S. Arbault, P. Walter and C. Amatore, *Anal. Chem.*, 2010, **82**(2), 457–460.

70. Z. A. Mahmood, I. Azhar and S. W. Ahmed, Kohl use in antiquity: effects on the eye, in *Toxicology in Antiquity*, ed. P. Wexler, Elsevier, New York, 2nd edn, 2019, p. 93–104; 99.

71. P. Walter, P. Martinetto, G. Tsoucaris, R. Bréniaux, M. A. Lefebvre and G. Richard, *et al.*, *Nature*, 1999, **397**, 483–484.

72. R. J. Buckley, *Ophthalmic Physiol. Opt.*, 2012, **32**, 443–445.

73. A. Hallmann-Mikołajczak, *Arch. Hist. Filoz. Med.*, 2004, **67**(1), 5–14.

74. M. A. Miczak, *Henna's Secret History: the History, Mystery and Folklore of Henna*, Writers Club Press, New York, 2001.

75. C. Cartwright-Jones, *Henna in the Ancient Egyptian Pharmacopoeia: the Ebers Papyrus*, The henna page, http://www.hennapage.com/henna/encyclopedia/medical/ebers.html, accessed June 2020.

76. S. Aparecida da France, M. Ferrera Dario, V. Brigatto Esteves, A. Rolim Baby and M. V. Robles Velasco, *Cosmetics*, 2015, **2**, 110–126.

77. V. K. Tandon and H. K. Maurya, *Tetrahedron Lett.*, 2009, **50**, 5896–5902.

78. F. R. Gallo, G. Multari, G. Palazzino, G. Pagliuca, S. M. M. Zadeh and P. Cabral Nya Biapa, *et al.*, *Rev. Bras. Farmacogn.*, 2014, **24**, 133–140.

79. A. Brunning, Compound Interest 2019, The chemistry of henna, https://www.compoundchem.com/2019/08/12/henna/, accessed June 2020.

80. B. Venkateswarlu, K. Ravishankar, B. Chandrakanth, K. Mounika, B. Saritha and G. Shravani, *et al.*, *Int. J. Chem. Pharm. Sci.*, 2011, **2**, 1–8.

81. A. J. Blake, P. Hubberstey and D. J. Quinlan, *Acta Crystallogr.*, 1996, **C52**, 1774–1776.

82. A. Al-Suwaidi and H. Ahmed, *Int. J. Environ. Res. Public Health*, 2010, **7**(4), 1681–1693.

83. P. J. Almeida, L. Borrego, E. Pulido-Melián and O. González-Díaz, *Contact Dermatitis*, 2011, **66**, 33–37.

84. Health Canada, PPD in "Black Henna" temporary tattoos is not safe, https://www.canada.ca/en/health-canada/services/cosmetics/black-henna-temporary-tattoos.html, accessed June 2020.

85. C. E. Eberle, D. P. Sandler, K. W. Taylor and A. J. White, *Int. J. Cancer*, 2020, **147**(2), 383–391.

86. L'Oréal: our history, https://www.loreal.com/en/group/culture-and-heritage/l-oreal-history/, accessed June 2020.

87. M. Bar-Zohar, *Bitter Scent: the Case of L'Oréal, the Nazis and the Arab Boycott*, Dutton Books, London, 1996.

88. M. M. Bomgardner, *Chem. Eng. News*, 2020, **98**(17), 28–33.

89. V. Yates, Men's cosmetics in Plato and Xenophon, in *Essays in Global Color History: Interpreting the Spectrum*, ed. R. B. Goldman, Gorgias Press, Piscataway, 2016, pp. 139–161.

90. P. Walter, E. Wellcome, P. Hallégot, N. J. Zaluzec, C. Deeb and J. Castaing, *et al.*, *Nano Lett.*, 2006, **6**(10), 2215–2219.

91. Tacitus, *Germania and Agricola*, Notes by P. Frost, Whittaker & Co., London, 1861, pp. 91–92.

92. M. Pastoureau, *Blue. The History of a Color*, Princeton University Press, Princeton, 2001, pp. 26–27.

93. J. Balfour-Paul, *Indigo – Egyptian Mummies to Blue Jeans*, Firefly Books, London, 2011, p. 13.

94. V. Zech-Materne and L. Leconte, *Veg. Hist. Archaeobot.*, 2010, **19**(2), 137–142.

95. G. Carr, *Oxf. J. Archaeol.*, 2005, **24**(3), 273–292.

96. Publius Ovidius Naso, *Amores II*, vol. 16, p. 39.

97. Pliny the Elder, *Natural History*, **vol. XXII**, p. ii.

98. C. D. Plowright, *J. R. Hortic. Soc.*, 1901–1902, **26**, 33–40.

99. M. van der Veen, *Oxf. J. Archaeol.*, 1993, **12**(3), 367–371.

100. Wellcome Collection gallery, 3 April 2018, https://commons.wikimedia.org/w/index.php?title=File:Statues_at_Amiens_Cathedral;_woad_merchants_Wellcome_M0001818.jpg&oldid=326474331, accessed June 2021.

101. J. B. Hurry, *The Woad Plant and its Dye*, Oxford University Press, London, 1930, p. 22 ff.

102. D. V. Thompson, *The Materials and Techniques of Medieval Painting*, Dover, New York, 1956, pp. 135–139.

103. E. F. Armstrong, *J. R. Soc. Arts*, 1939, **87**(4527), 295–298.

104. S. nicDhuinnshleibhe, *A Brief History of Dyestuffs and Dyeing. Presented at Runestone Collegium*, 19 February 2000, http://kws.atlantia.sca.org/dyeing.html, accessed June 2020.

105. J. Edmonds, *The History of Woad and the Medieval Woad Vat*, 3rd reprint, Published by John Edmonds, 1st edn, 2006, p. 12. Available from: Lulu.com.
106. Shutterstock Table of Lead Compounds' Colors, https://www.shutterstock.com/image-vector/table-lead-compounds-colors-characteristic-inorganic-724710460, accessed October 2020.
107. P. Grandjean, Lead in Danes: historical and toxicological studies, *Environ. Qual. Saf., Suppl.*, 1975, **2**, 6–75.
108. L. Cilliers and F. Retief, Lead poisoning and the downfall of Rome: reality or myth? in *Toxicology in Antiquity*, ed. P. Wexler, Elsevier, New York, 2nd edn, 2019, pp. 221–229.
109. L. Eldridge, *Face Paint: The Story of Make-up*, Abrams, New York, 2015, Section 1.
110. K. Krawczynski, *Daily Life in the Colonial City*, Greenwood Press, Denver, 2013, p. 401.
111. M. A. Angeloglou, *A History of Make-up*, Macmillan, New York, 1971, pp. 49–125.
112. Cennino Cennini as quoted in A. P. Laurie, *The Materials of the Painter's Craft*, T. N. Foulis, London, 1910, p. 183.
113. S. Houston, *Skin deep in Painting the Skin: Pigments on Bodies and Codices in Pre-Columbian Mesoamerica*, ed. E. Dupey García and M. L. Vásquez de Ágredos Pascual, University of Arizona Press, Tucson, 2018, pp. vii–viii.
114. F. Karim-Cooper, *Cosmetics in Shakespearean and Renaissance Drama*, Edinburgh University Press, Edinburgh, 2012, pp. 43–44.
115. S. Gosson, *Pleasant Quippes for Newfangled upstart Gentlewomen*, ed. T. Richards, St. Martin's Lane, London, 1596, Tilley no. W663.
116. G. Vail, *A History of Cosmetics in America*, Toilet Goods Association, New York, 1947, p. 74.
117. J. A. Witkowski and L. C. Parish, *Clin. Dermatol.*, 2001, **19**, 367–370.
118. V. J.-R. Kehoe, *The Technique of Film and Television Make-up*, Hastings House Publishers, New York, 1958, p. 16.
119. F. E. Basten, *Max Factor: The Man Who Changed the Faces of the World*, Arcade Publishing, New York, 2008, p. 62.

120. Max Factor, *Lori Nelson Uses Pan-Stik*, 1952, https://commons.wikimedia.org/w/index.php?title=File:Lori_Nelson_uses_Pan-Stik_by_Max_Factor,_1952.jpg&oldid=351985092, accessed June 2021.

121. G. K. Sharma, J. Gadiya and M. Dhanawat, *A Textbook of Cosmetic Formulations*, Pothi.com e-book, ISBN: 9781365355912, https://store.pothi.com/book/ebook-gaurav-kumar-sharma-textbook-cosmetic-formulations, accessed June 2020.

122. R. Adunka and M. V. Orna, *Carl Auer von Welsbach: Chemist, Inventor, Entrepreneur*, Springer, Cham, 2018, pp. 57–60.

123. N. Goldstein, *Clin. Dermatol.*, 2007, **25**, 417–420.

124. M. Samadelli, M. Melis, M. Miccoli, E. E. Vigl and A. R. Zink, *J. Cult. Herit.*, 2015, **16**(5), 753–758.

125. M. Corballis, *The Recursive Mind: The Origins of Human Language, Thought and Civilization*, Princeton University Press, Princeton, 2011, p. 214.

126. A. Deter-Wolf, B. Robitaille, L. Krutak and S. Galliot, *J. Archaeol. Sci. Rep.*, 2016, **5**, 19–24.

127. K. Sperry, *Am. J. Forensic Med.*, 1991, **12**(4), 313–319.

128. J. Caplan, *Hist. Workshop J.*, 1997, **44**, 106–142.

129. V. Lautman, *The New Tattoo*, Abbeville Press, New York, 1994, p. 10.

130. *The Travels of Marco Polo, the Venetian*, ed. T. Wright, George Bell & Sons, London, 1886, p. 268.

131. Felice, Beato (1834–1907); Tattooed Japanese men, *ca.* 1870. Hand-colored photograph, https://commons.wikimedia.org/w/index.php?title=File:Beato,_Felice_(1834_–_1907)_-_Tattooed_japanese_men_-_ca._1870.jpg&oldid=254099428, accessed June 2021.

132. Skins and needles, Royal Society for Public Health, Vision, Voice and Practice, https://www.rsph.org.uk/uploads/assets/uploaded/97c182fb-3d70-472c-90ef36ded8da1b63.pdf, accessed June 2020.

133. M. Armstrong, *Am. J. Nurs.*, 1999, **99**(6), 80.

134. M. A. Helmenstine, Tattoo ink chemistry, ThoughtCo, 11 February 2020, https://www.thoughtco.com/tattoo-ink-chemistry-606170, accessed July 2020.

135. M. A. Helmenstine, Tattoo ink carrier chemistry, ThoughtCo, 11 February 2020, https://www.thoughtco.com/tattoo-ink-carrier-chemistry-608403, accessed July 2020.

136. V. Kluger and N. Koljonen, *Lancet Oncol.*, 2012, 13((4), E161–E168. https://www.thelancet.com/journals/lanonc/article/PIIS1470-2045(11)70340-0/fulltext, accessed April 2021.

137. K. Navrazhina, B. Goldman and M. C. Leger, *Cureus*, 2018, **10**(7), e2975.

138. L. J. A. Lowenthal, *Arch. Dermatol.*, 1960, **82**, 237–243.

139. E. Engel, F. Santarelli, R. Vasold, T. Maisch, H. Ulrich and L. Prantl, *et al.*, *Contact Dermatitis*, 2008, **58**, 228–233.

140. US Food and Drug Administration, *Tattoos and Permanent Makeup Fact Sheet*, https://www.fda.gov/cosmetics/cosmetic-products/tattoos-permanent-makeup-fact-sheet, accessed July 2020.

141. C. De Cuyper and D. D'hollander, Materials used in body art, in *Dermatological Complications with Body Art: Tattoos, Piercings and Permanent Make-up*, ed. C. De Cuyper and M. L. Pérez-Cotapos S, 2nd edn, Springer, Cham, 2018, pp. 21–48.

142. V. P. Carlson, E. J. Lehman and M. Armstrong, *J. Environ. Health*, 2011, **75**(3), 30–37.

143. National Conference of State Legislatures, *Tattooing and Body Piercing: State Laws, Statutes and Regulations*, https://www.ncsl.org/research/health/tattooing-and-body-piercing.aspx. accessed July 2020.

144. C. Arellano, D. A. Leopold and B. B. Shafiroff, *Plast. Reconstr. Surg.*, 1982, **70**(6), 699–703.

145. R. W. B. Scutt, *Br. J. Plast. Surg.*, 1972, **25**, 189–194.

146. I. Goldman, R. G. Wilson, P. Hornby and R. G. Meyer, *J. Invest. Dermatol.*, 1965, **44**, 69–73.

147. B. M. Prinz, S. R. Vavricka, P. Graf, G. Burg and R. Dummer, *Br. J. Dermatol.*, 2004, **150**, 245–251.

148. P. Hall-Smith and J. Bennett, *Br. Med. J.*, 1991, **303**, 397–398.

149. W. Kirby, A. Desai and T. Desai, *Skin & Aging*, 2010, **8**, 38–40.

150. J. C. Mao and L. M. DeJoseph, *Facial Plast. Surg. Clin.*, 2012, **20**, 125–134.

151. Tattoo removal is not so easy. FDA photo illustration by Michael J. Ermarth, https://commons.wikimedia.org/w/index. php?title=File:Tattoo_Removal_Not_So_Easy_(8413853175). jpg&oldid=283849324, accessed June 2021.

152. T. Riordan, *A History of the Innovations that Have Made us Beautiful*, Broadway Books, New York, 2004.

153. J. Cunningham, Color cosmetics, in *Chemistry and Technology of the Cosmetics and Toiletries Industry*, ed. D. F. Williams and W. H. Schmitt, Blackie Academic and Professional, London, 2nd edn, 1996, pp. 149–182.

CHAPTER 4

The Tombs of the Pharaohs: Egypt's Legacy to Civilization[†]

And so sepúlchred in such pomp dost lie, That kings for such a tomb would wish to die.[1]

John Milton

4.1 INTRODUCTION

Our cultural heritage from this wellspring of ancient civilization is immeasurable. Ancient Egyptian pigment usage is replete with "firsts" that testify to the innovative spirit and ingenuity that inspires us today.

4.2 MUMMY POSTMORTEM: THE LIFE AND DEATH OF MUMMY BROWN

Ancient Egyptian civilization has left a profound impact on our own. Language, agriculture, craftsmanship, art, chemistry, metallurgy and construction are all areas that have enriched our world. However, most of our knowledge of these accomplishments

---

[†]Glossary entries and chapter cross-references are in boldface.

March of the Pigments: Color History, Science and Impact
By Mary Virginia Orna
© Mary Virginia Orna 2022
Published by the Royal Society of Chemistry, www.rsc.org

comes to us through the Egyptians' preoccupation with death and burial. This chapter necessarily begins with examining that reality.

Mummies have a reputation for surviving the centuries intact. So, why are there so few mummies left from Egypt's Dynastic Period (3000–2649 BCE) and Old Kingdom (*ca.* 2649–2130 BCE)? We can point to grave-robbers, perhaps, but the simple fact is that early attempts at mummification during these periods were largely unsuccessful. Despite elaborate efforts to preserve the bodily form and its tissues with resin-soaked linen, the corpse still decomposed and the result was a mere skeleton enclosed in an intricately decorated cover. This problem made the journey into the afterlife uncertain since possessing a bodily form to accomplish it was a prerequisite of the ancient Egyptian religion. The first definitive success was the burial of Queen Hetepheres (*ca.* 2600 BCE) at Giza: her body had been immersed in a solution of natron (see Box 4.1).

As the dynasties and centuries rolled on, the Egyptians became much more proficient at the mummification process. By the time Alexander of Macedon imposed Greek rule in the fourth century BCE, the Egyptians had perfected the process, which proceeded seamlessly into the Roman era beginning in 30 BCE. Engaging in a little experimental archaeology, the great Egyptologist, Alfred Lucas (1867–1945), determined that immersing a corpse in solid natron for about 40 days was enough

---

Box 4.1 Natron

Possibly chemistry had its beginnings when the Egyptians began to exploit one of its natural resources, the salt deposits at the Wadi Natrun. When the salt-containing lakes in this dried-up valley, located between Cairo and Alexandria, begin to shrink in the summer months, a substance called natron becomes abundantly available. Natron is a mixture of salts of sodium (*natrium*, in Latin, from which we derive sodium's chemical symbol, Na). Called *natrun*, its Arabic name, it contains mainly sodium carbonate decahydrate, $Na_2CO_3 \cdot 10H_2O$, and sodium bicarbonate, $NaHCO_3$, with some sodium chloride, $NaCl$ and sodium sulfate, Na_2SO_4 present as minor ingredients. Natron is an excellent drying agent and because it is the salt of a weak acid and a strong base, it provides an alkaline (basic) environment which can greatly limit bacterial activity. The ancient Egyptians knew how to put this perfect "mummifying" medium to good use – to desiccate and sterilize the body all in one fell chemical swoop. They also used it for making glasses and glazing, for preserving food, for gold refining and for medicine.

for near "perpetual" preservation – about 300 years, as it turned out. Despite the ban on mummification by Theodosius I (347–395) in 392 CE, the process continued well into the Christian era and only came to an end with the arrival of Islam in Egypt in 641 CE.

4.2.1   Medicinal Mummy

Due to such an effective, long-lived and universal practice for both humans and animals, Egypt was a land literally crammed with mummies. This great abundance of archaeological material afforded opportunities not only for the scientists bent on analyzing them for ancient diets and diseases, but for less scrupulous folk as well. But the etymology of the word "mummy" is both enlightening and surprising. Using the term exclusively to denote the chemical preservation of bodies is incorrect. The word, probably derived from either Persian or Arabic, is "mumia," meaning "pitch" or "bitumen," which oozed down from mountains in black, oily streams and was prized, as "mummy," for its medicinal properties. Because mummified bodies often look quite black, the word "mummy" became the very name for this blackness.[2] As early as the twelfth century, and possibly before, Egyptian merchants found a handy supply of what they believed to be bitumen, in mummy carcasses. Realizing trade in mummies was a lucrative practice, they encouraged the idea that mummy's beneficial properties were due precisely to its origins in human bodies, and voilà! Medicinal Mummy was born.[3]

Security was always a big problem at Egypt's tomb cities. From the 19th dynasty on in the New Kingdom (*ca.* 1200 BCE) in the Valley of the Kings, all tomb workers had to live in the walled city of Deir El-Medina. Guards could easily frisk them as they moved in and out of the single gate. There were all kinds of built-in deterrents to theft within the tombs themselves, from hidden pits to large slabs of stone designed to trap intruders. Nevertheless, tomb robbery was a way of life throughout Egypt's long history, especially since so much gold and precious jewelry was entombed with the sojourners to the afterlife. In the Middle Ages, millions of mummies literally lay around for the taking: another form of "black gold."

4.3 DECIPHERING THE ROSETTA STONE

A lengthy history of Egypt is beyond the scope of a book on pigments, but some background is necessary to identify the players in this centuries-long drama. Islamic rule lasted about 1300 years through several caliphates, sultanates, and the Ottoman Empire. A very significant, but brief, intervention by the French from 1798 to 1801 resulted in the uncovering of the Rosetta Stone (found near the port city of Rosetta, now Rashid).[4] The stone, actually a stele, one of many similar official decrees of the 2nd century BCE, contained three forms of the Pharaoh's message in hieroglyphics, in a relative of ancient Egyptian Coptic text called Demotic, and in ancient Greek. When the British, after much haggling, took possession of the stone (as well as all of Egypt for almost a century), they presented it to the reigning monarch, George III, who decreed it should go on display in the British Museum – and there it has remained since 1802, an icon of modern history and the most-visited attraction in the museum. Many scholars set to work to decipher the Demotic text and translate the Greek with the intention of unlocking the door to the hieroglyphics. Thomas Young (1773–1829), famous for establishing the wave theory of electromagnetic radiation, provided the necessary conceptual framework for this work by demonstrating that the Demotic text was both phonetic and ideographic. This discovery paved the way for Jean-François Champollion's (1790–1832) successful translation that launched the discipline of Egyptology[5] and Egyptomania soon became all the rage.[6] Were it not for this breakthrough, our knowledge of Egypt's long history under the Pharaohs might still be lost in the mists of time. Figure 4.1[7] shows a giant replica of the Rosetta Stone as the floor of the courtyard in Champollion's birthplace in Figeac, in the Occitaine region of southern France.

Thanks to Champollion, experts have been able to read the hieroglyphic inscriptions by the thousands in the tombs of the Pharaohs and on papyrus rolls found under the sands of every conceivable venue in Egypt. Cracking the code, for example, revealed that the hieroglyph for "color" could mean existential reality; it translates to "being," "nature," and "character," among other ideas.[8] Thus, these transcriptions and translations enabled a comprehensive understanding of the culture and practice of this ancient civilization – but not without a bit of contention. Even though more than 200 years have passed since the French

Figure 4.1   Place des Écritures, Figeac, France. A replica of the Rosetta Stone
by Joseph Kosuth. Champollion's birthplace now houses the
Musée Champollion. Reproduced from ref. 7.

lost the Rosetta Stone to the British, there have been factions on
both sides who want to restage the battle of Waterloo. As an example, this festering wound revealed itself in 1972 when the stone
was exhibited in Paris between the pictures of the two men who
made its secrets known to the world: Thomas Young on one side
and Jean-François Champollion on the other. Some of the French
who attended the exhibit complained that Young's portrait was
larger than Champollion's; on the other hand, those from the
British faction were outraged that Champollion's portrait was
patently larger than that of Young. So much depends on the lens
one uses to view reality: the organizers of the exhibit had worked
carefully to make sure both pictures were of exactly equal size.[5]

Among the important opportunities afforded from the study
of Egyptology, and pertinent to the remainder of this chapter,
was the examination of the nature of the pigments covering the
painted walls of Egyptian tombs and observation of the details of
mummification practices.

4.4   MUMMY REVISITED

Our 21st century encounters with mummies are restricted to
museum exhibits or to horror movies (*e.g.*, "The Mummy's Curse"),
but one does not need fiction to appreciate the true horror of

mummy exploitation. Western Christians have been enamored of mummies ever since returning Crusaders gushed about their curative properties when taken either as a *digestif* or to heal external bruises. Relying on the "mysterious life source" passed on by the everlasting dead,[9] it soon became commonplace to engage in "medicinal cannibalism," *i.e.*, to take a refreshing draft of mummy juice or a chomp from a mummified hand – far preferable to the torments of medieval physicians. In the sixteenth and seventeenth centuries, mummy was the most popular and sought-after drug in Europe.[10] Consequently, mummy trafficking was quite lucrative; it became widespread from about 1500 until the end of the nineteenth century. British merchants, through bribery, managed to smuggle tons of mummified flesh into England, peaking in the late 1590s.[11] About 1590, the French physician, Ambroise Paré (1510–1590), declared that mummy was "the very first and last medicine of almost all our practitioners against bruising."[12] Seventeenth century aficionados remarked about the "black, hard, shine like pitch, having much such a smell, but more pleasant."[13] As the century wore on, "true" mummy became much more difficult to obtain. The shortage, as might be expected, generated a thriving trade in counterfeits: many bodies fresh from the gallows found their way into apothecaries' shops well into the eighteenth century.[14] It must be said that not everyone approved of this aberrant craze. The philosopher Sir Thomas Browne (1605–1682) decried this "dismal vampirism" that "exceeds in horror the black banquet of Domitian,‡ not to be paralleled except in those Arabian feasts, wherein Ghoules feed horribly."[15] Apparently to no avail: the practice continued for at least two more centuries. There were even some clumsy attempts to cover up past history to deal with this unease. For example, in 1968, historian Erwin Ackerknecht (1906–1988) claimed that mummy as a drug had been wholly discredited by the 1580s, but many Europeans were still using it as such well into the 1780s.[12]

4.4.1 Mummy Brown

Since Europeans blissfully wined and dined on mummy, it's not too much of a stretch to also paint with it. If mummy ingestion can turn off even the most hardened cynic, then grinding

---

‡In 89 CE, the Roman Emperor Domitian staged a banquet in a room completely painted in black and adorned with tombstones inscribed with his guests' names; it was a dark practical joke resolved with gifts of silver skeletons. In Arabian popular legend, ghouls often stalked the desert, preying upon unwary travelers.

up cured and dried body parts to color your visual *opus magnum* should not give you pause. Mummy Brown grew into a profitable item of commerce, first among apothecaries as there was always a fine line between pigments and drugs; they often doubled for one another. So it was almost natural that "colourmen" like Roberson & Co. would transform a "medicine" into a luxury pigment destined for elite paintboxes (Figure 4.2).[16] Astonishing as it may seem, a 1903 issue of London's *Illustrated Mail* carried an article entitled "Pictures Painted with Mummies," and as late as 1904, people were still advertising in London's *Daily Mail* for "a mummy to make into pigment."[17]

Mummy seems to have made its debut into works of art from as early as the close of the sixteenth century.[18] Enthusiastic artists, eager to try out fancy new pigments, went to bizarre and lavish lengths to experiment with this bitumen-like powder. They became addicted to its lovely transparent brown color when added to oil, especially when they found it was almost impossible to duplicate without a skillful blending of many other pigments. They even put in special orders for muscle, deemed to make the finest mummy with a wonderful "silken feel" that went onto the canvas with "a kind of sensual ease." Mummy was also quite versatile: artists could use it as a glaze or a simple daub to "capture the buttery tones of shadows or the dark swaths on water."[14]

Although treatises on significant occurrences of certain pigments are loath to finger actual paintings containing mummy,

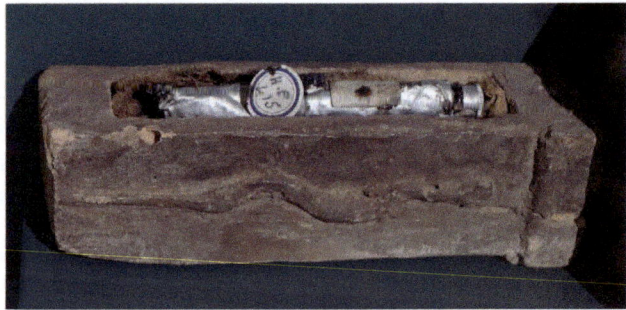

Figure 4.2   A tube of Mummy Brown in a coffin that was probably originally made for an eel or snake. Photo by Geni, May, 2019. Reproduced from ref. 16 under the terms of the CC BY-SA 4.0 license, https://creativecommons.org/licenses/by-sa/4.0/deed.en.

one can surmise from the documented palettes of certain well-known artists that mummy figured amidst the browns in their works. Among these purported mummy users are Eugène Delacroix (1798–1863), Martin Drölling (1752–1817), Sir William Beechey (1753–1839), Lawrence Alma-Tadema (1836–1912) and Edward Burne-Jones (1833–1898).

But by the late-1800s, Mummy Brown's reputation was about to crack along with the many actual paintings containing it. Color theorist George Field (1777?–1854) remarked rather trenchantly:[19]

[L]ittle reliance should be placed on this brown...[N]othing is to be gained by smearing one's canvass with a part, perhaps, of the wife of Potiphar. With a preference for materials less frail and of a more sober character, we likewise hold... that it is not particularly prudent to employ without necessity these crumbled remains of dead bodies, which must contain ammonia and particles of fat in a concrete state and so be more or less apt to injure the colours with which they may be united.

Symptomatic of a growing distaste for Mummy Brown, once its contents were recognized for what they really were, is the tale recalled by a teenaged Rudyard Kipling (1865–1936). In the late 1870s, he was visiting Edward Burne-Jones, his uncle by marriage, when a fellow artist, Alma-Tadema, noted he had recently been invited to a colourman's workshop to view a mummy prior to its transformation into minced Mummy Brown. Startled to realize that Mummy Brown was the genuine article, Burne-Jones insisted on giving his own tube a decent burial in the back garden.[20]

About Mummy Brown, art historian Philip McCouat remarked:[21]

[A]wareness of its grisly origins (which appears to have somehow been forgotten by some), and the increasing respect for mummies' scientific, archaeological, anthropological and cultural importance, added to the significant reduction in the number of mummies available, usage of the pigment fell away dramatically in the early twentieth century.

4.4.2 Demise of Mummy Brown

According to the Smithsonian Institution, Roberson & Co., the major supplier of Mummy Brown in the UK, sold its last tube in 1964. Its managing director remarked: "We might have a few odd limbs lying around somewhere, but not enough to make any more paint. We sold our last complete mummy some years ago... Perhaps we shouldn't have. We certainly can't get any more."[22]

In checking for the identity of mummy in lists of pigments, we find it mentioned as a synonym for asphalt based on the erroneous idea that this substance had been used for embalming Egyptian mummies, as pointed out earlier by David.[2] Mumia is a term used in a sixteenth century painters' manual, but it is related neither to asphalt nor to flesh (*carne*),[23] whereas it is referred to as *carne momia* (mummy flesh) as a synonym for asphalt for the first time in the early eighteenth century.[24] It seems none of the earlier treatises on mummy mention that it contains asphalt; this claim came up only in the eighteenth century when, due to its popularity, artists were insisting on finding out what it contained.[25] Oil paints now sold under the name of mummy consist of bituminous earths (degradation products of petroleum) like Van Dyke brown.[26] Other synonyms are tar, pitch, Bergharz and Antwerp Brown. Siddall states that despite their obvious talents as opaque, velvety black pigments, they are poorly recorded.[27] They are assumed to be natural organic pigments difficult to describe or characterize because their properties depend upon where they were found and how they were produced.[§] In 2017, Mummy Brown joined the modern world as September's "Color of the Month" in color consultant Donald Kaufman's high-end paint line.[28] And the wheel has come full circle regarding the non-destructive handling of archaeological artifacts. Whereas mummies were destroyed with reckless abandon for at least the past eight centuries, new non-invasive sensing techniques have been developed to harvest protein-based materials that may provide clues to cause of death, ancient burial practices, and details that will help conservators to stop, or at least slow down, degradation.[29]

[§]Sometimes, Mummy Brown is referred to as "*Caput mortuum*," or "death's head," but in many other instances the latter also refers to a purple color called Cardinal Purple. "Caput mortuum" will be discussed in **Chapter 7.4.1.6.**

If you try to find the term Mummy Brown among the list of pigments developed by the ancient Egyptians, you will never find it because Mummy Brown *consists* of ancient Egyptians. It was not a gift to the modern world, but plunder wrested from the ancient world. Let us hope mummies are allowed to rest in peace (at least, the few that remain intact). Perhaps the mummy's curse will be lifted for good.

4.5 PIGMENTS' ROLE IN ANCIENT EGYPTIAN CULTURE

Ancient Egyptian culture, a 4000-year-old continuous chronicle of monumental achievements in virtually every area of human endeavor, is considered to be a major wellspring of western civilization. The peoples responsible for this extraordinary success story were once humble farmers and shepherds who migrated, possibly in Neolithic times, from a region presently encompassing Israel, Jordan and Turkey. This information, long speculated upon, was finally confirmed by DNA analysis of 90 ancient mummies from the pre-Ptolemaic, Ptolemaic and Roman periods.[30,31] Herodotus (*ca.* 484–425 BCE), who visited Egypt 2400 years ago, remarked: "The Egyptians have certainly discovered more things that are wonderful than all the rest of mankind."[32] He estimated the Egyptian priests had a written history dating back to at least 11 000 years before his time, although that timeframe seems fairly exaggerated.[33] The narrow strip of land referred to now as ancient Egypt, about 750 miles long and no more than 20 miles wide, has an area that falls roughly midway between the U.S. states of Maryland and West Virginia. Called *Khema*, "The Black Land," with reference to its dark, alluvial soil, it identified with that discipline, chemistry, for much of its history through its ingenious transformation of material substances *via* exploitation of natural resources, metallurgy, stone working, viticulture and brewing, beekeeping and honey collecting, pottery and faience, papyrus invention and manufacture, and cosmetics and pharmaceuticals.[34]

Color literally defined the Egyptian cultural landscape. Without color, there was not art. Without color, the natural and divine worlds simply did not exist. Within this culture, ancient Egyptian painters played a godlike role: through their work, they recreated the natural and divine worlds according to a set form.[35]

The flat, linear, block-like Egyptian paintings depicting every-day activities in the context of eternity are familiar to even the casual visitor to a museum's Egyptian wing; Figure 4.3 is an example. Here we see the individuals characteristically in profile set on a baseline among other registers of the painting.[36] The color of the pigment was what conveyed a symbolic meaning to the work, however. The yellow shades in Figure 4.3[37] express life and growth since this color comes from the sun.[38] Depending on the context, red can have both positive and negative connotations. According to the *Ebers Papyrus*,[39,40] the goddess Isis appeals to be protected from all things red, *i.e.*, everything evil or dangerous. However, in Egyptian symbolism, red is also associated interchangeably with golden yellow, the color of the sun disk, so appears appropriately enough in this harvest scene, especially in depicting the luscious fruit in the bottom register.[41]

Red and yellow share the same word in Egyptian color terminology, *dšr*. The other three basic color terms are *km* (black),

Figure 4.3   Peasant Couple Harvesting Papyrus. In the cemetery of Deir el-Medina, this wall painting is in the vaulted tomb chamber of Sennutem, a necropolis officer of the early Ramesside period (*ca.* 1290 BCE). Reproduced from ref. 37.

symbolizing both death and resurrection, *ḥd* (white), symbolic of omnipotence and purity and *wȝḏ* (green), denoting fertility and vegetation.[42,43] The paucity of color terms is curious because the Egyptian palette was far richer in the available number of colors and pigments of each type. Certainly, at least in this case, their language and their practice were unmoored from one another.

4.5.1 Work of the Tomb Painters

Tomb painters worked together in a type of labor union that regulated their materials and their pay. Art, rather than being a matter of creativity, was standardized and tightly controlled.[44] These conclusions seem to be borne out by Earle R. Caley's comments on the two famous third century CE papyri, the Leyden Papyrus and the Stockholm Papyrus[45,46] on the ancient chemical arts. On the one hand, the recipes contained in these documents are often abbreviated and incomplete, indicating that users were already familiar with the nature of the processes. On the other, the recipes' magical formulas tended to show that the chemical arts in ancient Egypt were in the hands of the priestly caste.[47] However, remarkably, some artists broke out of the mold by deliberately using pigment mixtures to achieve various visual effects during the reigns of Thuthmosis IV and Amenhotep III (*ca.* 1419–1370 BCE).[48]

The work of the tomb painters is our most important source of information about the daily life of the ancient Egyptian. Coupled with our present knowledge of how to decipher the hieroglyphics accompanying the paintings, thanks to Champollion's work, Egyptologists have written volumes regarding this remarkable civilization. The identities of the pigments themselves constitute another precious source. Thanks to modern scientific instrumentation such as **X-ray fluorescence (XRF)** and **X-ray diffraction (XRD)**, chemists are now writing their own volumes on pigments, documenting their sources, syntheses, usage, and in some cases, degradation. The application of pigments on the tomb walls was quite simple: the mineral pigments, usually in the form of cakes or bricks, were mixed with water and vegetable gums or resins. They were then brushed on quickly before the mixture hardened into an unworkable mass.[49]

4.5.2   Egyptian Pigments Unveiled

Over a ten-year period, from 1991 to 2000, the British Museum carried out a pigment analysis project aimed at conservation and treatment of painted papyri in its collection. The project selected representative samples of each color range for each historical phase identified in the history of color illustration on papyrus (1550 to 30 BCE). Each sample was analyzed using the methods named in **Section 4.5.1** and added to the database of almost 1400 pigment samples from wall paintings and tombs analyzed by the *Max-Planck Institute für Kernphysik* at Heidelberg.[50]

Table 4.1, culled from these results,[51,52] consists of representative pigments listed by color group. The approximate dates of usage are based on analysis of actual objects from tombs, papyri, and other sources. With the availability of new pigments beginning in the Romano-Egyptian (after 30 BCE) period, the palette expanded to encompass organic as well as inorganic mineral pigments.

4.6   ANCIENT EGYPTIAN CHEMISTS: SHORT ON DOCUMENTATION, LONG ON PRACTICE

Several of the listed pigments with special and interesting histories have been selected for discussion; they are marked with an asterisk in Table 4.1. Obviously, this is such a massive topic that it would take more than one volume to deal with all of them. However, it must be remembered the ancient Egyptian artists were doers, not recorders – they did not write down their recipes. We can only determine the identities of these pigments and possibly how they were manufactured by actual examination of the artifact.

4.6.1   Huntite

Huntite, a magnesium calcium carbonate, $3MgCO_3 \cdot CaCO_3$, is a remarkable pigment. Identified as a mineral only as recently as 1953 CE,[53] Riederer confirmed its use in ancient Egyptian New Kingdom (1570–1070 BCE) ceramics[54] in 1974. In 2001, its use was pushed back another thousand years to the Old Kingdom (2686–2181 BCE) by Heywood's analysis of some objects in New York's Metropolitan Museum of Art.[55] How did we miss that?

Table 4.1 Selected list of Egyptian pigments used from ancient times to the Roman period. Adapted from ref. 51 with permission from John Wiley & Sons, Copyright © 2004 John Wiley & Sons, Ltd, and ref. 52 with permission from Taylor & Francis, Copyright 2016. (Items marked with * are discussed in the text.).

Color	Pigment	Chemical formula	Approximate dates of use	Remarks
White	Calcite	$CaCO_3$	2494 BCE – Roman period	Very common
	Gypsum	$CaSO_4 \cdot 2H_2O$	2494 BCE – Roman period	Very common
	Anhydrite	$CaSO_4$	2494 BCE – Roman period	Not common
	Lead white	$2PbCO_3 \cdot Pb(OH)_2$	512 BCE – Roman period	Common later
	*Huntite	$Mg_3Ca(CO_3)_4$	2600 BCE–200 CE	Less common later periods
Black	Carbon black	C	3500 BCE – Roman period, possibly earlier	Charcoal
	Pyrolusite	MnO_2	3000 BCE–1292 BCE	Rarely used
	Galena	PbS	3100 BCE – Roman period	Cosmetic
	Soot	C	1187 BCE – Roman period	Lamp black
Red	Hematite	Fe_2O_3	4000 BCE – Roman period	Most common
	Cinnabar	HgS	747 BCE – Roman period	Rare early
	*Minium	Pb_3O_4	30 BCE – Roman period	Rare early
	*Realgar	As_4S_4	2494 BCE – Unknown	Red-orange
	Madder	$C_{14}H_8O_4$	1500 BCE – Roman period	Used as dye only prior to the Roman period
Yellow	Goethite	$FeOOH$	4000 BCE – Roman period	Yellow ochre
	Orpiment	As_2S_3	1991 BCE – Roman period	Degrades
	Jarosite	$KFe_3(SO_4)_2(OH)_6$	2494 BCE–30 BCE	Yellow brown
	*Lead antimonate yellow	$Pb_2Sb_2O_7$	1450 BCE – Unknown	Naples yellow
Blue	*Azurite	$2CuCO_3 \cdot Cu(OH)_2$	2613 BCE–1334 BCE	Single use
	*Ultramarine	$Na_7Al_6Si_6O_{24}S_3$	1550 BCE – Unknown	Single use
	*Cobalt blue	$CoO \cdot Al_2O_3$	2250 BCE – Unknown	Gap in usage
	*Egyptian blue	$CaCuSi_4O_{10}$	2639 BCE–395 CE	Used beyond end date
Green	Malachite	$CuCO_3 \cdot Cu(OH)_2$	4000 BCE – Roman period	Common
	Chrysocolla	$CuSiO_3 \cdot nH_2O$	1991 BCE–1292 BCE	Blue-green
	Green earth	$K(Mg,Fe^{2+})(Fe^{3+},Al)[Si_4O_{10}](OH)_2$	900 BCE – Roman period	Celadonite
	*Egyptian green	$CaSiO_3(2\%Cu)$, octahedral environment	2200 BCE–1069 BCE	Never used outside of Egypt

Heywood suggests it had "probably" been misidentified as magnesite ($MgCO_3$) or dolomite ($CaCO_3 \cdot MgCO_3$) which have similar chemical compositions. Huntite is a very soft, beautifully white, dense and chalky material with a talc-like feel that makes it an excellent pigment. It was widely available to the ancient Egyptians but it took over 4000 years for modern chemists to discover it again.

4.6.2 Minium

Minium, red lead, Pb_3O_4, is the colorant used to paint seven early Roman period (*ca.* 30 BCE–200 CE) Egyptian mummies from head to toe. This unusual collection, known as the "red shroud mummies," is now dispersed in several museums including the J. Paul Getty Museum and the Brooklyn Museum. A 2009 analytical project found the minium covering all the mummies came from a single geological source, the Rio Tinto site in Spain's pyrite belt. Some chemical detective work on the chemically complex minium further indicated it was produced from litharge, PbO, formed at high temperatures and subsequently re-fired at lower temperatures. Its presence on these mummies suggests that Spain and Egypt traded either in pigments or raw litharge during this period.[56] Minor amounts of another pigment, lead–tin yellow, Pb_2SnO_4, were found mixed in with the minium. Probably adventitious in this context, lead–tin yellow has been detected in a relief fragment from the Palace of Apries, lower Egypt (*ca.* 600 BCE).[57] Otherwise, it was unknown as a pigment until the fourteenth century CE.[58]

4.6.3 Realgar

The realgar you look upon today may not be the color applied by the ancient Egyptian artists centuries ago. Red realgar, As_4S_4 (Figure 4.4, left), tends to degrade very slowly upon exposure to light into yellow orange pararealgar, As_4S_4 (Figure 4.4, right). In the realgar structure, each arsenic atom is bonded to three nearest neighbors, two sulfur atoms and one other arsenic atom. Presumably, light energy breaks the As–As bonds, which are 30% weaker than the As–S bonds, allowing for the formation of free As that destabilizes the realgar structure, and causes the change in the reflected wavelength in the visible region (**Chapter 1.6**).[59]

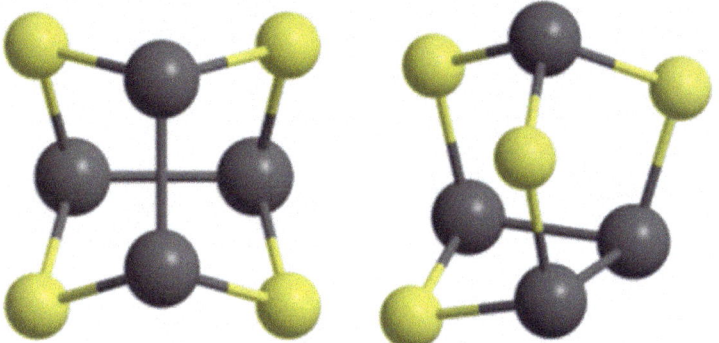

Figure 4.4   The two forms of As_4S_4 (Yellow = Sulfur; Gray = Arsenic). Realgar, left, is bright red; pararealgar, right, is yellow-orange. Figure courtesy of Liliana Mammino, University of Venda.

The process can be reversed upon heating. Eventual oxidation and loss of sulfur can lead to the formation of white arsenolite, As_2O_3. Once an artifact is removed from its safe resting place in, for example, an undisturbed tomb, environmental circumstances can greatly accelerate the process.

4.6.4   Lead Antimonate Yellow

This compound, $Pb_2Sb_2O_7$, is known by many names, among them, Naples yellow, giallolino, giallorino, zalulino, and jalloline. Its first documentation in the literature is a 1559 recipe by Cipriano Piccolpasso in his *Arte del Vasaio*.[60] Notable uses begin with Matthias Stomer's (1600–1650) *Arrest of Christ* (1630–1632) and comes to an apparent end with Carl Max Schultheiss's (1885–1961) *Flight into Egypt* (1933). However, the years from the 1600s to the 1930s marked a rediscovery of the pigment. It was originally used as early as 1450 BCE to color glass in a Theban palace. It enjoyed continued use in Egypt and the Middle East until about 400 BCE.[57] This, along with the huntite case, is another example of the Egyptians being the first to use a pigment but not receiving credit for its discovery until millennia later.

4.6.5   Azurite

As a readily available, beautiful blue basic copper carbonate mineral, $2CuCO_3 \cdot Cu(OH)_2$, azurite is oddly absent from the Egyptian palette. The only instance of its occurrence is the painted necklace and armlets of a female figure on a leather fragment tentatively dated to *ca.* 1550 BCE. No other instances of its use have been reported in Egyptian artifacts.[61]

4.6.6   Natural Ultramarine

The first recorded use of natural ultramarine was in the wall paintings of Mansur Depe, Turkmenistan, Central Asia, in the second century BCE.[62] Not surprisingly, considering its proximity to the source, it was also identified on the famous 6th century CE Bāmiyān Buddhas of Afghanistan that were destroyed in 2001.[63] However, the first experimental confirmation of ultramarine usage as a pigment predates the first of these reports by 1300 years – the Egyptians got the jump on everyone else. The object in question is the fragmented statue of a queen from Thebes dated stylistically to the latter part of the Seventeenth Dynasty (1571–1540 BCE).[61] Ultramarine use as a pigment, plus its origins, its various names, and how it is produced will all be treated in **Chapter 8.8**.

4.6.7   Cobalt Blue

Art historians attribute a fifteenth century CE start date to glassy cobalt oxide, CoO,[64] but the Egyptians used it as a colorant in glass from at least 1480 BCE. Its use then fell into decline after about 1300 BCE, whereas cobalt blue pigment derived from cobalt-bearing alum was continuously popular for painting on faience.[65,66]

4.6.8   Egyptian Blue

To the art history fan, mere mention of Egyptian blue conjures up visions of mystery, magic, ancient tombs, colorful temples and endless desert sands. To the scientist, its very existence is an astonishing technological advance achieved with little more than fire and simple tools. Its origins lost in the shadows of prehistoric time, it is the oldest synthetic pigment known. It is also one of the first pigments to be subjected to analytical scrutiny in modern times. Thus, it is a bridge from the past to the present, from the ancient world of gods and goddesses to the modern world of scientific investigation.

Although technical recipes for the production of Egyptian blue, found on Babylonian clay tablets, can be dated to the 17th century BCE, its presence has been identified on Egyptian artifacts from the Old Kingdom (2575–2134 BCE) that were manufactured eight centuries before.[67] It has been detected in many other places in the ancient world,[68] signifying its position as the most important blue pigment, and perhaps the only available blue pigment, for many thousands of years. This fact also testifies to its extraordinary stability, reasons for which can be found in its chemical constitution.

We now know the major ingredients in Egyptian blue manufacture are copper (as the metal, alloyed with other metals, or in oxidized form), silica in the form of sand (SiO_2), limestone ($CaCO_3$) and a flux, natron, which helps lower the melting point of the mix. Roman author and engineer, Vitruvius (d. 15 BCE), left us the recipe, but curiously omitted limestone, a key ingredient.[69] Even more curious, despite the abundance of written documentation on ancient Egyptian papyri about almost every conceivable subject, there is no information on production, trade or use of Egyptian blue.[70] When the specified mixture is heated to about 950 °C for a long period of time, preferably overnight, Egyptian blue is formed with the established chemical formula of $CaCuSi_4O_{10}$, which corresponds to the relatively rare mineral, cuprorivaite. The formula was established by **X-ray diffraction (XRD)** analysis.[71,72] Delamare[73] provides the details for this remarkable work that experimentally linked ancient Egyptian blue to the naturally occurring mineral, cuprorivaite. In actual fact, Egyptian blue is never produced in a pure form since the unreacted silica and other reactants added in excess are also present. When this mixture is ground up for use as a pigment, it contains a great deal of colorless, glassy material – often termed a frit. Tite and co-workers have analyzed numerous samples of Egyptian blue from sites in Egypt, Amarna, Nimrud, Nineveh, and Rome and found some of the material from Nimrud and Nineveh most closely approximated cuprorivaite's formulaic, or stoichiometric, composition, though some other localities never even came close.[74] The reason for this discrepancy is that their manufacturing technique involved a series of production steps in which the glassy mass had more copper and calcium compounds mixed in and reheated each time so that by the last step, the content was about 10% Cu and 10% Ca. However, not many samples of this high quality Egyptian blue survived in Mesopotamia because of degradation due to high humidity. There was also a gradual changeover to the use of lapis lazuli (**Chapter 8.8.3**) due to an Islamic influence, *i.e.*, opening of more trade with the source of lapis in Afghanistan.[75] Figure 4.5[76] is an example of Egyptian blue's use on an Egyptian coffin panel.

Because the actual recipe for producing Egyptian blue was a closely guarded secret in ancient times, there seem to have been relatively few manufactories – users relied on importation of the pigment from a few major sites.[73] Despite ignorance of the major role played by limestone in Egyptian blue synthesis, the Roman production centers at Pozzuoli, in the shadow of Mount Vesuvius,[¶] were very successful in producing it in very good yield. It is quite possible they succeeded because the sand in the vicinity had the required limestone content built right in. This hypothesis seems to be borne out because very few other groups succeeded in reproducing the recipe, possibly accounting for its quick decline in medieval and Renaissance art. There is another good chemical reason as well: the correct phase is very important. Solid phase synthesis of the correct ingredients invariably yields Egyptian blue; if the starting materials are first put into the molten phase, azurite is formed, which eventually degrades to malachite at room temperature.[77]

Figure 4.5 Egyptian (Thebes) coffin panel, *ca.* 1070–945 BCE. The Walters Art Museum. Reproduced from ref. 78.

[¶]Interestingly, Egyptian Blue was found as a naturally occurring material in Vesuvian lava.

4.6.8.1 Enter Han Blue. Egyptian blue's close relative, Han blue, sometimes referred to as Chinese blue, was a Far Eastern development that seems not to have overlapped technologies with the Middle Eastern variety. Its distinguishing characteristic is that a different alkaline earth element, barium, is used in its formation. The barium minerals available were witherite ($BaCO_3$) and barite ($BaSO_4$); use of the latter would impose great difficulty in the formation of Han blue because of the higher processing temperatures it would require.[||]

The generalized chemical reactions for the formation of Egyptian blue and Han blue are as follows:

$$Cu_2CO_3(OH)_2 + 8SiO_2 + 2CaCO_3 \xrightarrow{\text{Flux at } 800-900\,°C}$$

$$\underset{\text{Egyptian Blue}}{2CaCuSi_4O_{10}} + 3CO_2 + H_2O \qquad (4.1)$$

where under the reactants: Malachite, Sand, Lime.

$$CCu_2CO_3(OH)_2 + 8SiO_2 + 2BaCO_3 \xrightarrow{\text{Lead additive at } 900-1000\,°C}$$

$$\underset{\text{Han Blue}}{2BaCuSi_4O_{10}} + 3CO_2 + H_2O \qquad (4.2)$$

where under the reactants: Malachite, Sand, Witherite.

In these two similar reactions, the only difference in the products is the identity of Ca^{2+} in Egyptian blue and of Ba^{2+} in Han blue. These species are positively charged ions called counterions. They are present only to balance the charge of the anion, $CuSi_4O_{10}{}^{2-}$, which is responsible for the color of the pigment.

In a tutorial review, Heinz Berke[78] proposes an ingenious rationale for overcoming the barite problem based upon the known fact that the Chinese added lead salts as part of their production technique. At approximately 1000 °C, lead sulfate decomposes to lead oxide, which then reacts with barium sulfate to produce barium oxide. Since the energy required to decompose barium oxide is not as high as what is needed for barium sulfate, the reaction can then proceed to form Han blue

---

[||]Barite's Gibbs free energy of formation is over 200 kJ mol⁻¹, more negative than that of witherite, requiring that much more energy to break the chemical bonds necessary for forming new compounds from barite.

with much less expenditure of energy. Han blue's naturally occurring analog is called effenbergerite. Found in the Kalahari Desert in South Africa, it was first described in the literature in 1994.[79]

4.6.8.2   Egyptian Blue Production Conditions.   In a 2009 study of the production conditions of Egyptian blue, Kakoulli[80] mixed the following ingredients: 64.6 g silica, SiO_2; 7.2 g synthetic natron; 15.4 g malachite, $Cu_2CO_3(OH)_2$; 12.8 g $CaCO_3$. She then treated the mixtures according to the following conditions:

1. At 850 °C for 45 minutes, no Egyptian blue was formed.
2. At 950 °C for 24 hours, Egyptian blue was formed.
3. At 850 °C for 45 minutes with added $PbCO_3$, Egyptian blue was formed.
4. At 850 °C for six hours, pellets ground, reformed and heated for six more hours at 850 °C, Egyptian blue was formed.

From these trials, we learn that Egyptian blue formation depends on maintaining the ingredients at a very high temperature (about 1000 °C) for a very long time, a technological feat of the highest order in ancient times. It likely required the use of twin bellows like those used in iron works. Archaeological evidence for their use comes from a wall painting at Thebes showing the operation of twin bellows,[81] but the technology was probably known as early as 4000 BCE. The addition of $PbCO_3$ in trial 3, an ingenious trick that Berke had noted previously, was used by the Chinese in the production of Han blue: it possibly acted as a flux (to lower the melting point) as well as a possible catalyst. It certainly was effective in this case for the production of Egyptian blue as well.

Analysis of Egyptian blue from various sites, particularly of unused stores of the pigment, has produced an elemental profile of trace elements such as zinc, lead and iron that indicates a variety of starting materials. One thing is certain from the analyses: since the ratio of copper to calcium varies very little from sample to sample, there must have been a prescribed method of manufacture the ancient production sites were privy to.[73]

4.6.8.3 Origin of the Blue Color. The **chromophore** for both Egyptian blue and Han blue is the CuO_4^{6-}, the dark blue in Nefertiti's headdress in Figure 4.6, left;[82] it is the background blue in this Eastern Han mural painting, Figure 4.6, right.[83] Given the low probability of light absorption in the copper, it is not a very effective chromophore** and a great deal of colorant is necessary to achieve intense coloration.[78,84]

Spectral analyses of both Egyptian blue and Han blue have noted the colors they exhibit are anomalous in that they absorb virtually nothing in the blue region of the spectrum, in contrast to the behavior of the CuO_4^{6-} group in other environments.[84,85] Investigators have found that the presence of silicate in the compound greatly aids an almost complete absorption of radiation in the red, yellow, and green regions of the spectrum.[86] Indeed, the need for sand in the recipe (to provide the silicate environment) and, in fact, large excesses of it in the form of quartz, was

Figure 4.6   Left: The iconic headdress of Queen Nefertiti (*ca.* 1330 BCE) is painted with Egyptian blue, Altes Museum, Berlin. Reproduced from ref. 82, public domain image. Right: The blue background of this detail from an Eastern Han Dynasty tomb mural painting at Luoyang (Henan) depicting liubo players is Han blue (25–220 CE). Reproduced from ref. 83, public domain image.

**Quantum mechanically speaking, the energy level transitions in the copper ion are of the "d–d" forbidden type which typically produce pale colors because the transition events have little probability of occurring.

recognized in practice in the samples excavated in Pompeii, Cumae, and Liternum in the first century CE.[73]

4.6.9   A New Life for Egyptian Blue

In visiting museums containing examples of ancient marble statuary, the most striking visual perception is their gleaming white, almost shimmering, reality. Resurrected from excavations, covered with the grime of centuries, these pieces were often scrubbed to pristine cleanliness by restorers who were ignorant of the fact that they were originally polychrome pieces (**Chapter 7.5**). However, by probing the folds and corners of these works, the careful observer can discern the minutest traces of the original colors. These clues prompted cultural heritage scientist Giovanni Verri to probe further with a battery of instruments. He was able to demonstrate what every museum scientist now knows: the fragile pigments once decorating ancient marbles had detached or degraded over the centuries.

What Verri was stunned to discover is that Egyptian blue is brightly luminescent when viewed in infrared light. This phenomenon, visible **induced luminescence** (VIL), is especially useful in determining tiny amounts of a pigment that may not be otherwise visible.[87] When he digitally photographed the pigment, he said it gleamed like ice crystals.[88] This property is a form of photoluminescence, light emission resulting from excited electronic states following absorption of electromagnetic radiation. This phenomenon was first noted by George Gabriel Stokes (1819–1903) in 1852 when he described the blue glow of a quinine solution subjected to ultraviolet radiation.[89] From his observations, he formulated what came to be known as Stokes's Law, *i.e.*, the wavelengths of luminescent light were always longer than those of the incident light. Understanding the mechanism of photoluminescence[††] had to wait until the advent of quantum chemistry. Figure 4.7 [90] displays the reflectance spectrum (**Chapter 1.6**) of Egyptian blue that shows a

---

[††]Luminescence in all its forms has been the subject of much study and the names of various forms of luminescence arise from their source: photoluminescence, chemiluminescence, bioluminescence, triboluminescence, *etc.* Included in photoluminescence are the phenomena of fluorescence and phosphorescence, attributed to different relaxation mechanisms from excited states. Triboluminescence is light generated by friction.

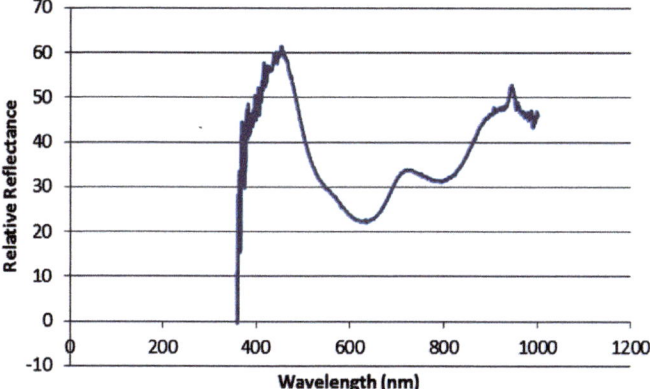

Figure 4.7   Reflectance spectrum of Egyptian blue powder. Reproduced from ref. 90 with permission.

large increase in reflectance (luminescence) beyond the limit of the visible region at 700 nm, peaking at 946 nm in the infrared region. The 60% reflectance on the left at 459 nm is responsible for Egyptian blue's blue color which is, happily, visible to the human eye.

Verri sees VIL[91] as a novel technology for the non-invasive detection of materials exhibiting this behavior, such as Egyptian blue, Han blue, and many other well-known pigments. Using this technique, he was able to prove the presence of Egyptian blue on some artifacts in the British Museum and on the wall paintings in the Tomb of Tutankhamen (see Box 4.2), among other sites.[92,93]

Due to VIL, some scientists are proposing how Egyptian blue's career as a pigment may expand into many other areas. For example, since human tissue is more transparent to infrared radiation than other wavelength ranges such as visible or ultraviolet light, Egyptian blue could be used as a coloring material for new medical imaging techniques. Its basic structure might also be a launching point for the development of novel types of security ink to thwart counterfeiting and forgeries. It could also substitute for normal fingerprint dusting powder on luminous surfaces, where it could be more easily seen.[96,97] Who knows what the future holds for Egyptian blue? It seems to be very bright indeed.

Box 4.2   King Tut's Tomb

Even though this chapter emphasizes pigments, it would be remiss to omit mention of King Tutankhamen (*ca.* 1342–1325 BCE), another modern icon of ancient Egypt. When archaeologist Howard Carter (1874–1939) unsealed his nearly intact tomb in 1923, he, King Tut, and Egyptology were immediately catapulted to the world's center stage. Over the course of the following decades, millions of tourists swarmed *en masse* to the site, about the size of a normal 20th century living room, and millions more have visited the traveling exhibit, "The Treasures of Tutankhamen." This author had the privilege of working part-time in the Objects Conservation Department at New York's Metropolitan Museum of Art preparatory to the show. Today, thanks to Factum Arte (https://www.factum-arte.com/), a digital mediation workshop based in Madrid, you can forego a trip to Luxor and simply visit the digital facsimile on their website.[94] But do not forget the mummy's (or Pharaoh's) curse: "Death or misfortune shall befall any who disturb Pharaoh." The financial backer of the King Tut excavation, Lord Carnarvon (1866–1923) traveled to Egypt in February of 1923 to view the official opening of the inner burial chamber. He never returned home. He died in a Cairo hotel room, victim of a lethal mosquito bite, on 5 April 1923.[95]

4.6.9.1   Expanded Chronology of Egyptian Blue.   The archaeological record shows Egyptian blue was known to every civilization West of the Caucasus, and Han blue, its barium counterpart, was well known in the Far East, but there is no evidence that their technological or artistic paths ever crossed. Egyptian blue's presence has been verified in a number of civilizations, including Egyptian necropoli,[48] Armenian wall paintings from the 2nd millennium BCE,[98] Etruscan tombs from the 7th century BCE (Figure 4.8),[99] Roman Age wall paintings,[100] and excavations from 1st century CE Pompeii.[101]

Of special note is the appearance of Egyptian blue in frescoes dated to 860 CE in the monastery of Müstair, Switzerland.[102] Its appearance at that late date suggests that the pigment was applied from a previously existing cache, or that it was synthesized on the spot. If the latter were the case, then of course, the "lost" recipe was not lost after all. There is some evidence that perhaps the original recipe was lost, but reinvented using an unusual starting material in the mix – zinc-containing brass. The Egyptian blue pigment in frescoes of the church of Santa Maria *foris portas* in Castelseprio, Lombardy, dated to 850–950 CE, is indeed rich in zinc. The research team responsible hypothesizes the zinc may have been used as a flux since there was a natron shortage at the time of its synthesis.[103] A similar instance of a poor quality Egyptian blue sample

Figure 4.8   Etruscan tomb painting, Tarquinia, Italy, showing the use of Egyptian blue. *ca.* 700 BCE. Reproduced from ref. 99 with permission from American Chemical Society, Copyright 2015.

dated to about the same time (*ca.* 845 CE) in a fresco of the lower church of the Basilica of San Clemente, Rome argues for the latter: it indicates a less sophisticated technology than that employed in Roman times.[104] An even more striking discovery was a report by Bredal-Jørgensen and co-workers[105] in 2011. They confirmed – beyond a doubt – the presence of Egyptian blue in a 16th century painting by Giovanni Battista Benvenuto (aka L'Ortolano ferrarese, *ca.* 1480 – *ca.* 1525) by optical microscopy, VIL and **Raman spectroscopy**, remarking it was available in a period in which it is normally considered not to exist. There is a sprinkling of reports for its use during the several centuries in between (950–1500).[106] Further "sightings" were reported by Chiari in 2018[107] and by a conservation team at the Universidade NOVA de Lisboa reporting its use in manuscripts as late as the 12th century.[108]

Using the data given above plus a chronology supplied by Berke,[78] we can construct a timeline for Egyptian blue's recent history (Figure 4.9).[90]

A similar chart could be constructed for Han blue which would mark its arrival in the Berlin patent office (1900), its synthesis and structure determination (1959), and the first evidence of its presence in ancient artifacts (1983).

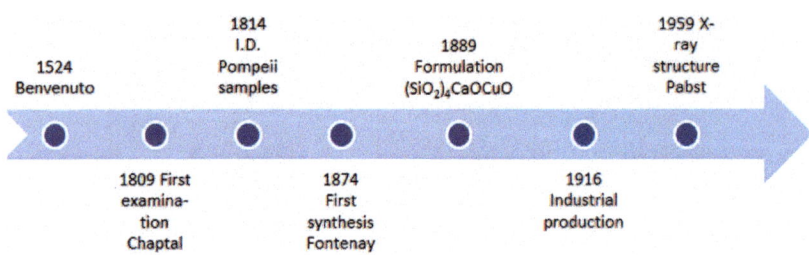

Figure 4.9 Milestones in the recent history of Egyptian blue. Reproduced from ref. 90 with permission.

This foray into the history of one of the world's most celebrated and unusual pigments, and its close relative, Han blue, has been one of surprising rediscovery. If we liken Egyptian blue to a river, we can discern its trickling source possibly in Mesopotamia, its full flow into the land of the Nile, its slow disappearance into the underground cavern of the Middle Ages, and then its own re-emergence along with the Renaissance. Its entry into the modern world of chemistry is marked by its formulation and structure determination, and finally it has achieved celebrity status as a recognized addition to the industrial scene and the artists' palette, as well as a possible candidate for medical use due to its VIL in the near infrared.

We would be remiss if we did not mention its adoption as a poster child in chemistry education as well. A team at Aix Marseille University selected the synthesis of Egyptian blue as the link between culture, history, cognitive development and science.[109] The group was particularly interested in applying the "particle nature of matter" ideas of Dorothy Gabel[110] as codified by Alex Johnstone's "triangle," *viz.* the mind's passage from macroscopic observation to sub-microscopic (molecular) understanding to symbolic representation.[111,112] A group at the Thacher School in Ojai, California, picked up on this idea and adapted it for use across a variety of curricular settings.[113]

We may be certain the story is not over yet – modern chemical research being what it is, Egyptian blue and Han blue will surely figure in future studies of the characteristics of pigments and how they can be modified.

4.6.10   Egyptian Green

This little-known and less-used turquoise blue-green pigment is a kind of stepchild with respect to Egyptian blue. Questions about its very existence arose when some researchers thought it was just a dusty, poorly preserved, weathered sampling of the blue. Others felt it was just a "misfired" Egyptian blue. A comprehensive study of about 50 archaeological samples of Egyptian green plus synthesis of experimental turquoise samples at different firing temperatures showed that: (1) Egyptian green is a pigment in its own right; (2) the color is determined by the CuO concentration, which increases with increasing firing temperature to the limit of 1150 °C; both Egyptian blue and Egyptian green have their own characteristic colors resulting from the same metallic ion in the same oxidation state but in different molecular geometries.[114,115]

4.7   THINGS ARE SELDOM WHAT THEY SEEM: PIGMENT DEGRADATION

Although ancient Egyptian pigments, with few exceptions, were mineral pigments and expected to remain unaltered over long periods of time, when artifacts are removed from their original home and placed in museums and other settings, air pollution, humidity, and large temperature fluctuations can cause serious degradation. Some major examples are the following:[116]

Orpiment, As_2S_3, a yellow pigment found on a large range of artifacts, will slowly transform into arsenolite, As_2O_3, which is white, on exposure to light. Release of sulfur from the orpiment can potentially interact with and damage other pigments nearby.

Red realgar, As_4S_4, readily transforms into yellow orange pararealgar, As_4S_4, which in time will oxidize to form white arsenolite, As_2O_3 (**Section 4.6.3**).

Green pigments, in general, with Cu as their **chromophore** tend to degrade with time into green atacamite, $Cu_2Cl(OH)_3$, one of the three forms of copper(II) trihydroxychloride.[117]

Red lead, Pb_3O_4, will blacken if exposed to sulfide, S^{2-}, gases, forming black PbS.

Red vermilion, HgS, will darken irreversibly on exposure to light; red cinnabarite transforms into gray-black metacinnabarite.

Lead white, $2PbCO_3 \cdot Pb(OH)_2$, will darken when exposed to sulfides in a similar manner to red lead.

Egyptian blue and Egyptian green present the most alarming prospect of all. Thought to be unalterable and stable due to their glassy structure, it is precisely the glass that makes them vulnerable. The problem first came to light in 1972 when Riederer reported the presence of atacamite, a copper-containing mineral, in eleventh dynasty tomb artifacts from Deir el-Bahari.[54] He considered it a pigment, but did not exclude the possibility it might be a degradation product. That gloomy prospect has now been confirmed through further work by Schiegl *et al.* in Heidelberg. They found atacamite changes the structure of faience artifacts into a hydrous gel rather than a glass. The process is of ongoing devitrification, decomposition of the residual glass, formation of the atacamite, and ultimately final destruction of the glaze. In some cases, the deterioration was so far advanced that all of the glass fragments were completely replaced by a fluffy, fine-grained network of atacamite, but still retaining the basic outlines of the original glass. This so-called "copper chloride cancer" has left entire walls barren of any blue or green paint. It changes Egyptian blue to green, Egyptian green to pale green, and renders the entire surface so fragile that it could crumble at the slightest tremor. Schiegl's final remark is "the impact of this discovery on the interpretation of the colour symbols of the ancient Egyptians is yet to come." The same may be said of all the other color alterations noted above.[118]

4.8   EPILOG

As we review the unparalleled contributions of the ancient Egyptian artists to the palette of civilization, Egyptian blue stands out as such a major contribution right down to the present day even if this were the only gift to us, it would be more than enough. But there are other remarkable, but little celebrated and little-known instances:

- discovery of huntite 4600 years before it was "rediscovered" in modern times
- discovery of lead–tin yellow, if only inadvertently
- discovery of lead antimonate yellow
- first experimental confirmation of use of ultramarine as a pigment
- long-term use of cobalt blue millennia before cobalt's discovery

Even before all the achievements of the ancient Egyptians cataloged above were known, Champollion himself expressed great appreciation of their civilization:

> *All that remains to add, to sum up, is that we Europeans are merely men of Lilliput, and no people ancient or modern has conceived of art or architecture on so sublime a scale, so broad, so grandiose, as did the Egyptians of old.*‡‡

The Egyptians gave us many of the "firsts" in pigment usage, most gathered from local surface sources. Although these ingenious people operated some mines as well, they were a peripheral activity because their supply chains were largely local. Mines only came to be the major means of extracting Earth's buried treasures when community needs outgrew what surface sourcing could supply. Some mines became the stuff of legend and mining developed into the mainstay industry of many a people as the next chapter will document.

REFERENCES

1. J. Milton, *On Shakespeare*, 1630.
2. R. David, Mummification, in *Ancient Egyptian Materials and Technology*, ed. P. T. Nicholson and I. Shaw, Cambridge University Press, Cambridge, 2000, pp. 372–389.
3. W. R. Dawson, *Proc. R. Soc. Med.*, 1927, **21**, 34–39.
4. D. C. Benjamin, *Stones and Stories: An Introduction to Archaeology and the Bible*, Fortress Press, Philadelphia, 2009, p. 33.
5. J. Ray, *The Rosetta Stone: And the Rebirth of Ancient Egypt*, Profile Books, London, 2008.
6. R. H. Fritze, *Egyptomania: A History of Fascination, Obsession and Fantasy*, Reaktion Books, London, 2016.
7. British Museum Collection, 2006, https://commons.wikimedia.org/wiki/File:Place_des_ecritures_Figeac.jpg, accessed June 2021.
8. D. Rankine, *Heka – the Practices of Ancient Egyptian Ritual and Magic*. Avalonia, London, 2006.
9. R. Sugg, *Lancet*, 2008, **371**(9630), 2078–2079.

‡‡Jean-François Champollion, letter from Egypt to his elder brother, 24 November 1828, as quoted in ref. 5, p. 56.

10. R. David, *Discovering Ancient Egypt*, Facts on File, New York, 1994, p. 16.
11. J. Pipe, *Egyptian Mummies: A Very Peculiar History*, The Salariya Book Co., Brighton, 2012.
12. *Mummies Around the World: An Encyclopedia of Mummies in History, Religion and Popular Culture*, ed. M. Cardin, ABC-CLIO, Oxford, 2015, p. 226.
13. R. Sugg, *Mummies, Cannibals and Vampires: The History of Corpse Medicine from the Renaissance to the Victorians*, Routledge, New York, 2nd edn, 2016, p. 108.
14. P. Schwyzer, Mummy is become merchandise: literature and the Anglo-Egyptian mummy trade in the seventeenth century, in *Re-Orienting the Renaissance*, ed. G. MacLean, Palgrave Macmillan, London, 2005.
15. T. Browne, *Fragment on Mummies*, University of Chicago, 1682, https://penelope.uchicago.edu/misctracts/mummies.html, accessed September 2020.
16. Tube of mummy brown in a coffin, https://commons.wikimedia.org/w/index.php?title=File:Tube_of_mummy_brown_in_a_coffin.JPG&oldid=354363348, accessed June 2021.
17. H. Pringle, *The Mummy Congress: Science, Obsession and the Everlasting Dead*, Hyperion Books, New York, 2001, pp. 194–202.
18. A. H. Church, *The Chemistry of Paints and Painting*, Seely, Service & Co., London, 3rd edn, 1901, p. 236.
19. *Field's Chromatography or Treatise on Colours and Pigments as Used by Artists*, Winsor & Newton, London, ed. T. Salter, 1869, p. 253, https://www.gutenberg.org/files/20915/20915-h/20915-h.htm, accessed July 2020.
20. R. Kipling, *Something of Myself and Other Autobiographical Writings*, Cambridge University Press, Cambridge, 1990, pp. 9–10.
21. P. McCouat, *J. Art Soc.*, 2019, http://www.artinsociety.com/the-life-and-death-of-mummy-brown.html, accessed August 2020.
22. R. Eveleth, *Smithsonian Magazine*, 2 April 2014, https://www.smithsonianmag.com/smart-news/ground-mummies-were-once-ingredient-paint-180950350/, accessed July 2020.
23. G. B. Armenini, *De Veri Precetti della Pittura, lib. 3.*, Francesco Tebaldini, Ravenna, 1587, p. 124.

24. *Artists' Techniques in Golden Age Spain: Six Treatises in Translation*, ed. Z. Veliz, Cambridge University Press, Cambridge, 1986.
25. C. I. Bothe, Asphalt, in *Artists' Pigments: A Handbook of their History and Characteristics*, ed. B. H. Berrie, National Gallery of Art, Washington, D.C., 2007, vol. 4, pp. 111–149; 115–116.
26. R. J. Gettens and G. L. Stout, *Painting Materials: A Short Encyclopedia*, Dover Publications, New York, 1966, p. 132.
27. R. Siddall, *Minerals*, 2018, **8**, 201–236.
28. K. Kelleher, *The Awl. Caput Mortuum, an Earthy Brown Made of Bodies or Minerals*, https://www.theawl.com/2017/11/caput-mortuum-an-earthy-brown-made-of-bodies-or-minerals/, accessed July 2020.
29. B. Demarchi, R. Boano, A. Ceron, F. Dal Bello, S. E. Favero-Longo and S. Fiddyment, *et al.*, *J. Archaeol. Chem.*, 2020, **119**, 105145, DOI: 10.1016/j.jas.2020.105145.
30. T. Watson, *Nat. News*, 2017, **546**, 17.
31. V. J. Scheunemann, A. Peltzer, B. Welte, W. P. van Pelt, M. Molak and C.-C. Wang, *et al.*, *Nat. Commun.*, 2017, **8**, 15694.
32. Herodotus, *Euterpe*, trans. W. Beloe, Edward Earle, Philadelphia, 1814, vol. 1, LXXXII, 381.
33. L. E. Warren, *J. Chem. Educ.*, 1934, **11**, 146–153.
34. P. Loyson, *J. Chem. Educ.*, 2011, **88**, 146–150.
35. M. Hartwig, Method in ancient Egyptian painting, in *Artists and Colour in Ancient Egypt*, ed. V. Angenot and F. Tiradritti, Montepulciano, Italy, 22–24 August 2008, Studi Poliziani di Egittologia 1, Montepulciano: missione archeologica Italiana a Luxor. ISBN 978-88-908083-0-2, pp. 28–56.
36. E. R. Russmann, The Egyptian character of certain Egyptian painting techniques, in *The Wall Paintings of Thera*, ed. S. Sherrat, The Thera Foundation, Athens, 2000, vol. I, pp. 71–76.
37. Anonymous Egyptian Tomb Artists, https://commons.wiki-media.org/wiki/File:Egyptian_harvest.jpg, accessed June 2021.
38. S. Colinart, Analysis of Inorganic Yellow Colour in Ancient Egyptian Painting, in *Colour and Painting in Ancient Egypt*, ed. W. V. Davies, The British Museum Press, London, 2001, pp. 1–4.
39. A. Hallmann-Mikołajczak, *Arch. Hist. Filoz. Med.*, 2004, **67**, 5–14.

40. G. Lefebvre, *J. Egypt. Archaeol.*, 1949, **35**, 72–76.

41. G. Pinch, *Red Things: The Symbolism of Colour in Magic, Colour and Painting in Ancient Egypt*, ed. W. V. Davies, The British Museum Press, London, 2001, pp. 182–185.

42. J. Baines, *Am. Anthropol.*, 1985, **87**, 282–297.

43. G. G. Singer, Color in ancient Egypt, *Terrae Antiquae – Arqueologia – Historia*, 2010, https://terraeantiqvae.com/profiles/blogs/color-in-ancient-egypt, accessed July 2020.

44. F. Steinmann, *Z. Ägypt. Sprache Alt.*, 1982, **109**, 149–156.

45. E. R. Caley, *J. Chem. Educ.*, 1927, **4**, 979–1002.

46. E. R. Caley, *J. Chem. Educ.*, 1927, **3**, 1149–1166.

47. K. R. Williams, *J. Chem. Educ.*, 2000, 77, 300–301.

48. P. Vandenabeele, R. García-Moreno, F. Mathis, K. Leterme, E. Van Elslande and F.-P. Hocquet, *et al.*, *Spectrochim. Acta, Part A*, 2009, **73**, 546–552.

49. B. M. Bryan, The ABCs of painting in mid-eighteenth dynasty and social meaning, in *Essays for the Library of Seshat*, ed. R. K. Ritner, Oriental Institute of the University of Chicago, Chicago, 2017, pp. 1–27.

50. L. Lee and S. Quirke, Painting materials, in *Ancient Egyptian Materials and Technology*, ed. P. T. Nicholson and I. Shaw, Cambridge University Press, Cambridge, 2000, pp. 104–120.

51. H. G. M. Edwards, S. E. Jorge Villar and K. A. Eremin, *J. Raman Spectrosc.*, 2004, **35**, 786–795.

52. D. A. Scott, *Stud. Conserv.*, 2016, **61**, 185–202.

53. G. T. Faust, *Am. Mineral.*, 1953, **38**, 4–23.

54. J. Riederer, *Archaeometry*, 1974, **16**, 102–109.

55. A. Heywood, The use of huntite as a white pigment in ancient Egypt, in *Colour and Painting in Ancient Egypt*, ed. W. V. Davies, The British Museum Press, London, 2001, pp. 5–9.

56. M. S. Walton and K. Trentelman, *Archaeometry*, 2009, **51**, 845–860.

57. S. B. Hedegaard, T. Delbey, C. Brøns and K. L. Rasmussen, *Heritage Sci.*, 2019, **7**, 54–86.

58. H. Kühn, Lead-Tin Yellow, in *Artists' Pigments: A Handbook of their History and Characteristics*, ed. A. Roy, National Gallery of Art, Washington, D.C., 1993, vol. 2, pp. 108–121.

59. D. L. Douglass and C. Shing, *Am. Mineral.*, 1992, 77, 1266–1274.

60. C. Piccolpasso, *Li tre libri dell'arte del vasaio*, trans. B. Rackham and A. Van de Put, Victoria and Albert Museum, London, 1934.
61. A. Heywood, *Metropolitan Museum Studies in Art, Science and Technology*, 2010, vol. 1, pp. 73–81.
62. N. Lapierre, *Arts asiatiques*, 1990, vol. 45, p. 28.
63. C. Blänsdorf, S. Pfeffer and E. Melzl, The polychromy of the giant Buddha statues of Bāmiyān, in *The Giant Buddhas of Bāmiyān: Safeguarding the Remains. Monuments and Sites XIX*, ed. M. Petzet, International Council on Monuments and Sites, Paris, 2009, pp. 237–264.
64. J. Riederer, *Dtsch. Farben-Z.*, 1968, **22**, 387–389.
65. Y. Abe, R. Harimoto, T. Kikigawa, K. Yazawa, A. Nishisaka and N. Kawai, *et al.*, *J. Archaeol. Sci.*, 2012, **39**, 1793–1808.
66. A. P. Laurie, *The Materials of the Painter's Craft*, T.N. Foulis, London, 1910, p. 19.
67. H. Berke and H. G. Wiedemann, *EASTM*, 2000, vol. 17, pp. 94–120.
68. A. Brysbaert, *Stud. Conserv.*, 2006, **51**, 252–266.
69. M. V. Pollio, *The Ten Books on Architecture*, trans. M. H. Morgan, Harvard University Press, Cambridge, 1914, Book VII, ch. XI, pp. 218–219.
70. J. Riederer, Egyptian Blue, in *Artists' Pigments: A Handbook of Their History and Characteristics*, ed. E. W. FitzHugh, National Gallery of Art, Washington, D.C., 1997, vol. 3, pp. 23–45.
71. A. Pabst, *Acta Crystallogr.*, 1959, **12**, 733–739.
72. F. Mazzi and A. Pabst, *Am. Mineral.*, 1962, **47**, 409–410.
73. F. Delamare, *Blue Pigments: 5000 Years of Art and Industry*, Archetype Publications, London, 2013.
74. M. S. Tite, M. Bimson and M. R. Cowell, Technological examination of Egyptian blue, in *Archaeological Chemistry IIIii*, ed. J. B. Lambert, American Chemical Society, Washington, D.C., Advances in Chemistry Series 205, 1984, pp. 215–242.
75. D. Ullrich, *PACT: Revue du Groupe européen d'études pour les techniques physiques, chimiques et mathématiques appliquées à l'archéologie*, 1987, vol. 17, pp. 323–332.
76. Coffin Panel with Paintings of Funerary Scenes, Walters Art Museum, https://commons.wikimedia.org/w/index.php?title=File:Egyptian_-_Coffin_Panel_with_Paintings_of_Funerary_Scenes_-_Walters_622_-_Detail_B.jpg&oldid=457127065, accessed June 2021.

77. G. A. Mazzochin, D. Rudello, C. Bragato and F. Agnoli, *J. Cult. Herit.*, 2004, **5**, 129–133.
78. H. Berke, *Chem. Soc. Rev.*, 2007, **36**, 15–30.
79. G. Giester and B. Rieck, *Mineral. Mag.*, 1994, **58**(393), 663–670.
80. I. Kakoulli, *Greek Painting Techniques and Materials from the Fourth to the First Century BC*, Archetype Publications, London, 2009, pp. 41–42; 138–140.
81. C. J. Davey, *Levant*, 1979, **11**, 101–111.
82. Nefertiti, https://commons.wikimedia.org/w/index.php?title=File:Nefertiti_berlin.jpg&oldid=454678123, accessed June 2021.
83. Eastern Han Dynasty tomb mural painting at Luoyang, https://commons.wikimedia.org/w/index.php?title=File:Eastern_Han_Luoyang_Mural_of_Liubo_players.jpg&%20oldid=285268525, accessed June 2021.
84. P. García-Fernández, M. Moreno and J. A. Aramburu, *Inorg. Chem.*, 2015, **54**, 192–199.
85. R. J. Ford and M. A. Hitchman, *Inorg. Chim. Acta*, 1979, **33**, L167–L170.
86. P. García-Fernández, M. Moreno and J. A. Aramburu, *J. Chem. Educ.*, 2016, **93**, 111–117.
87. G. Verri, The application of visible-induced luminescence imaging to the examination of museum objects, in *SPIE Europe Optical Metrology. O3A: Optics for Arts, Architecture and Archaeology II, Proceedings Volume 739105*, ed. L. Pezzati and R. Salimberi, SPIE, Munich, 14–18 June 2009, pp. 739105–739112.
88. M. Talbot, *The New Yorker*, 29 October 2018, pp. 44–51.
89. G. G. Stokes, *Philos. Trans. R. Soc. London*, 1852, **142**, 463–562.
90. M. V. Orna, Archaeological Blue Pigments: Problem Children from the Get-Go, in *Archaeological Chemistry: A Multidisciplinary Analysis of the Past*, ed. M. V. Orna and S. C. Rasmussen, Cambridge Scholars Publishing, Newcastle upon Tyne, 2020, pp. 333–396.
91. J. Dyer, G. Verri and J. Cupitt, Multispectral imaging in reflectance and photo-induced luminescence modes: a user manual, *European CHARISMA Project*, published online as version 1.0, October 2013, https://www.academia.edu/7276130/J_Dyer_G_Verri_and_J_Cupitt_Multispectral_Imaging_

in_Reflectance_and_Photo_induced_Luminescence_ modes_a_User_Manual_European_CHARISMA_Project, accessed July 2020.

92. G. Verri, The Courtauld Institute of Art, https://courtauld. pure.elsevier.com/en/persons/giovanni-verri, accessed July 2020.

93. G. Accorsi, G. Verri, M. Bolognesi, N. Armaroli, C. Clementi, C. Miliani and A. Romani, *Chem. Commun.*, 2009, 3392–3394.

94. D. Zalewski, *The New Yorker*, 28 November 2016, pp. 66–79.

95. Quoted in T. Wilkinson, *A World beneath the Sands*, W.W. Norton, New York, 2020, p. 426.

96. SciShow, *Egyptian Blue: How an Ancient Color Can Save Lives*, https://www.youtube.com/watch?v=K2Bwmdl61Sw&vl=en, accessed July 2020.

97. B. W. Pogue, F. Leblond, V. Krishnaswamy and K. D. Paulsen, *Am. J. Roentgenol.*, 2010, **195**, 321–332.

98. K. Hovhannissian, *The Wall-Paintings of Erebooni*, Armenian SSR Academy of Sciences, Yerevan, 1973.

99. M. V. Orna, Historic Mineral Pigments: Colorful Benchmarks of Ancient Civilizations, in *Chemical Technology in Antiquity*, ed. S. C. Rasmussen, American Chemical Society, Washington, DC, 2015, pp. 17–69, Figure 12.

100. G. A. Mazzocchin, F. Agnoli and M. Salvadori, *Talanta*, 2004, **64**, 732–741.

101. R. Piovesan, R. Siddall, C. Mazzoli and L. Nodari, *J. Archaeol. Sci.*, 2011, **38**, 2633–2643.

102. *Die Mittelalterlichen Wandmalereien im Kloster Müstair, Grundlagen zur Konservierung und Pflege*, ed. A. Wyss, H. Rutishauser and M. A. Nay, Hochschulverlag AG an der ETH Zürich, Zürich, 2002.

103. M. Nicola, L. M. Seymour, M. Aceto, E. Priola, R. Gobbetto and A. Masic, *Archaeol. Anthropol. Sci.*, 2019, **11**, 5377–5792.

104. L. Lazzarini, *Stud. Conserv.*, 1982, **27**(2), 84–86.

105. J. Bredal-Jørgensen, J. Sanyova, V. Rask, M. L. Sargent and R. Hoberg Therkildsen, *Anal. Bioanal. Chem.*, 2011, **401**, 1433–1439.

106. M. C. Gaetani, U. Santamaria and C. Seccaroni, *Stud. Conserv.*, 2004, **49**, 13–22.

107. M. Nicola, M. Aceto, V. Gheroldi, R. Gobbetto and G. Chiari, *J. Archaeol. Sci.*, 2018, **19**, 465–475.

108. M. J. Melo, P. Nabais, R. Araújo and T. Vitorino, *Phys. Sci. Rev.*, 2019, DOI: 10.1515/psr-2018-0017.
109. O. Morizot, E. Audureau, J.-Y. Briend, G. Hagel and F. Boulc'h, *J. Chem. Educ.*, 2015, **92**, 74–78.
110. D. Gabel, *J. Chem. Educ.*, 1993, **70**, 193–194.
111. A. H. Johnstone, *Int. J. Chem. Educ.*, 1991, **36**, 7–10.
112. R. J. Petillion and W. S. McNeil, *J. Chem. Educ.*, 2020, **97**, 1536–1542.
113. C. R. Vyhnal, E. H. R. Mahoney, Y. Lin, R. Radpour and H. Wadsworth, *J. Chem. Educ.*, 2020, **97**, 1272–1282.
114. S. Pagès-Camagna, S. Colinart and C. Coupry, *J. Raman Spectrosc.*, 1999, **30**, 313–317.
115. S. Pagès-Camagna and S. Colinart, *Archaeometry*, 2003, **45**, 637–658.
116. L. Green, Colour transformations of ancient Egyptian pigments, in *Colour and Painting in Ancient Egypt*, ed. W. V. Davies, The British Museum Press, London, 2001, pp. 43–48.
117. J. B. Sharkey and S. Z. Lewin, *Am. Mineral.*, 1971, **56**, 179–192.
118. S. Schiegl, K. L. Weiner and A. El Goresy, *Naturwissenschaften*, 1989, **76**, 393–400.

CHAPTER 5

Buried Treasure: The Earth Yields Up its Secrets[†]

The meek shall inherit the Earth, but not its mineral rights.[1]

J. Paul Getty

5.1 INTRODUCTION

Among the oldest legacies of Earth's once semi-liquid crust, ores and mineral deposits continued to yield up astonishing surprises that transformed the world we live in. The pigments they provided have shape-shifted into international drivers of hostilities and diplomacy in some very unexpected ways.

5.2 HEIGH HO, HEIGH HO, IT'S OFF TO WORK WE GO

If you have ever followed the adventures of Snow White in her skirmishes with her wicked stepmother, you are no doubt familiar with her pals, Doc, Grumpy, Sneezy, and company of Disney fame, at least if you subscribe to the Disney version. However, under different names, these little guys were well known in folk

---

[†]Glossary entries and chapter cross-references are in boldface.

March of the Pigments: Color History, Science and Impact
By Mary Virginia Orna
© Mary Virginia Orna 2022
Published by the Royal Society of Chemistry, www.rsc.org

stories for centuries before (Figure 5.1),[2] mostly for their mining inactivity as cataloged by the doyen of all mining literature, Agricola:[3]

> *...there are...weak demons, which some of the Germans...call hobgoblins, because they are imitators of humans. For they passionately ridicule joy: and they seem to do many things, but they do nothing completely. Some call them mountain devils, since they are commonly noticeable in height, since they are certainly ¾ as tall as a dwarf. Moreover they appear as old men clothed in the manner of miners, that is, clothed with a shirt and dressed with a piece of cloth hanging from their loins (Figure 5.1)...These are not accustomed to do damage to miners, but they wander in wells and mines and although they do nothing, they seem to train themselves in every habit of laborers, now they dig cavities, now they pour what is dug out into vessels, and now they maneuver the hauling machine. Although in fact the dirt sometimes irritates the workers, it rarely harms them. The hobgoblin never harms them unless it is first provoked by loud laughter or insults.*

Figure 5.1   The Dwarves Warn Snow White about her Stepmother. Franz Jüttner (1865–1925): Reproduced from ref. 2.

Thus wrote the early modern skeptic, Georgius Agricola, in one of his more credulous moments. Agricola, aka Georgius Bauer (1491–1555), gifted us with a comprehensive treatise on ancient mining practice, *De Re Metallica*.[4] This work laid the groundwork for our modern disciplines of geography, geology and geochemistry.

Miners, in general, accepted without question such creatures as hobgoblins, and even today, faith in their existence has not entirely disappeared. This is not to be wondered at since there is no place on Earth that evokes evidence of the supernatural as does the underground mine. It is here that death awaits in the invisible form of toxic gases. It is here that miners' lamps distort every movement and every shape into a grotesque mockery of reality. It is here that the selfsame substance being mined consumes the hands and feet of the miner. It is here that the very earth groans and creaks as restless rocks uncannily move from place to place. It is here in the dank darkness that death itself imprints its being on every sound and motion.

It is no wonder, then, that mines, especially the more fabled ones, have long held an air of mystery for fiction fans and history buffs. The number one best-seller of the Victorian era was H. Rider Haggard's *King Solomon's Mines*,[5] mapped out by a dying scribe in his own blood. On a more prosaic note, mines came to be the means of extracting Earth's buried treasures, no matter what that treasure happened to be. Salt, slate, coal, precious stones, metals, minerals and even some uninteresting-looking rocks were found to contain exotic elements and other surprises. But before we unpack these wonders, we have to look at how these mineral deposits got here in the first place.

5.3 ALL THE WORLD'S A STAGE

What's the world coming to? Shakespeare said that "all the world's a stage"[6] – perhaps he should have said "staging area" because the world is coming to something that it is not yet. The world, even if it seems permanent to us, has been in constant flux for the past 4.5 billion years. Change's orderly process involves first, building; then, balancing, and finally, chemistry. Forming the world involves building blocks, elements like silicon, aluminum, iron, all derived ultimately from the simplest element, hydrogen. Balance involves motion, interchange with the environment, pumping stuff in and out. Chemistry involves controlling the rate of the change: slowing things down, speeding them up, and supplying what's needed.[7] In a seismic shift that completely changed the game 3.7 billion years ago, the first life appeared: microbial mats sucking the sun's energy into their hydrothermal anaerobic environment. Countless eons later, about 2.33 billion years ago, tiny little cyanobacteria began the second great seismic shift: they started spewing free oxygen into the atmosphere – the Great Oxygen Event (GOE). From then on, both life and Earth evolved together in a process called co-evolution.[8] Before the GOE, water was responsible for forming about a thousand different minerals on the Earth's surface. After GOE, thanks to oxygen's rather aggressive appetite, about 3500 more mineral species now enrich the mineralogist's catalog – an utter impossibility in a non-living world.[9] When we gaze upon the oxygen-rich colorful mineral pigments like green malachite ($CuCO_3 \cdot Cu(OH)_2$), deep blue azurite ($2CuCO_3 \cdot Cu(OH)_2$), and pale Egyptian blue, cuprorivaite ($CaCuSi_4O_{10}$), they are clear-cut signs of the action of life itself on the mineral environment. Meanwhile, the staging area geared up for a grand production of the minerals we now value and procure.

5.4   MINES, ALL MINES

Nature has liberally sprinkled this rich mineral diversity around the entire planet in deposits sometimes close to the surface, and at other times, miles below in deeply entrenched veins, seams and domes. They may occur in areas close to human commerce, or in virtually inaccessible regions like the Hindu Kush mountain range.[10] How these treasures came to be discovered and wrested from their original resting places comprises the entire history of mining.[‡] Discovery often involved prospecting, which is an art, a science, or witchcraft, depending on the seeker. For example, despite his reputation as the "skeptical chymist," Robert Boyle (1626–1691) firmly believed in the efficacy of the divining rod. Mining centers more on metallurgy than it does on the other

---

[‡]Mindat.com is a site that shows localities for the minerals in their database, and so much more.

commodities like pigments, salt, coal, *etc.*, but there can be no doubt that revenues from mines used to be the economic backbone of many a nation.

5.4.1 First Use of Ores

Extractive activity of the Paleolithic period was driven by color. Ancient peoples recognized their earth colors by their ores, the earliest exploitation known being of ochres, the yellow and red iron oxides so familiar in prehistoric art. By the end of the Upper Paleolithic (*ca.* 12 000 BCE), even green-colored stones (chrysocolla, malachite and turquoise) were recorded for the first time as being part of the self-respecting cave household.[11] These metal and metal-bearing ores were utilized for many thousands of years along with the rare metals sometimes found in their native states.[12] Thermal decomposition of substances by strong heating and their conversion to oxides by heating in air was well known. How, when and where the ancient crafts workers transitioned to extracting metals from their ores, *i.e.*, smelting, is a matter of debate.[13] What is not in question is the fact that this was a paradigm shift, "an intellectual and technical landmark" in the history of human development.[14] From the post-smelting event onward, the history of extractive metallurgy and the history of pigment extraction run parallel to each other.

Mining began with simple resource gathering that gradually became more systematic as use demanded. For example, Paleolithic artists had clear ideas regarding the availability of their pigments and they went to great lengths, literally, to obtain them. As more useful pigments were discovered and exploited, it soon became necessary to move into "industrial scale" extraction; at first, the open-pit variety, and later on, the burrowing of horizontal tunnels called adits that would extend for some distance under a mountain. And it was usually a mountain – all the literature tells us that ores were to be found in the mountains – and that is quite so. Gradually, vertical and sloping mine shafts became more necessary, leading to the types of mines that we are used to today: vertical shafts to a certain depth, and then a grouping of small, narrow horizontal galleries radiating out from the vertical shaft. With this development, problems of lighting, ventilation and drainage had to be dealt with – and this gave rise to what today we call mining engineering.

5.4.2   Sources of Mining Information

Two great fonts of mining lore arose almost simultaneously in the sixteenth century, Vannoccio Biringuccio (1480–1539) from Italy, and Agricola, from whom we heard earlier. Although both of their treatises center on metals, both of them also treat pigments since they are often the compounds from which metals can be extracted. Biringuccio, in his posthumously published *Pirotechnia*,[15] treats of the deposit locations, processing, and properties of many important pigments. In his description of orpiment (As_2S_3), a yellow ore of arsenic, he tells us that "it comes from very deep mines in Hellespont and Cappadocia, for it is a material that Nature hides from us, teaching us to leave it alone as harmful." Later he gives the recipe for a depilatory that would not meet today's safety standards: "Orpiment mixed with lye and lime shaves every hairy spot without cutting," but later cautions, "I advise you not to use them except by force of necessity."[15] Two centuries later, Biringuccio's advice went unheeded when Carl Wilhelm Scheele (1742–1786) discovered a vibrant green pigment when he heated copper sulfate and arsenic trioxide together to produce copper arsenite, $CuHAsO_3$. Introducing it in 1778, he had second thoughts because of its toxicity, but his need to claim priority won out. It went on to fame in its use as wall hangings and wallpaper, and gained infamy as the invisible killer in candy, clothing, medicine and the wall hangings as well. Its counterpart, Emerald Green, copper acetoarsenite, came along two decades later. Both were extensively used by artists such as J. M. W. Turner (1775–1851), Édouard Manet (1832–1883) and Paul Gauguin (1848–1903).[16]

Concerning antimony, which to Biringuccio always meant the sulfide, Sb_2S_3, he praises its yellow color for painting earthenware vases and for tinting enamels and glasses. He tells us that it is imported from Germany to Venice for another purpose, "for the use of those masters who make bells, because they find that by mixing a certain part of it with the metal the sound is greatly increased."[15]

In Agricola's *De Re Metallica*,[3] also published posthumously in 1556, we have the work of a true Renaissance scholar. Though Agricola was northern European by birth, he studied medicine and spent some years in Venice editing Greek works and other texts.

He was fully immersed in humanist culture and steeped in the systematic observation of Nature.[17] Consequently, his great work was received as the *ne plus ultra* of its time, published in ten editions and three languages. A very large percentage of the work was technologically new; much else was the fruit of Agricola's own personal observations and experiences. When he was not working at his "day job" as a physician, he spent almost all of his spare time in smelters and mines. With his great work, he moved geology, mineralogy and mining engineering out of the dark ages. The twelve books, in broad outline, deal with prospecting, mining methodology, assaying, processing, and smelting of ores, and the separation of one metal from another. It became the standard reference work on mining for 200 years. The entire work is interspersed with intriguing woodcuts, the preparation of which considerably delayed the first publication (Figure 5.2).[18]

What set Agricola apart from his peers was his willingness to speculate on the origins of things in spite of what the Bible and other writers had pronounced on previously. H. C. and L. H. Hoover, the translators of the 1912 version, say quite definitely:

If we strip his theory of the necessary influence of the state of knowledge of his time, and of his own deep classical learning, we find two propositions original with Agricola, which still

Figure 5.2 Rectangular Assay Furnace. This furnace is good for assaying ores; the method of its use is described in Book VII of *De Re Metallica*:[4] Reproduced from ref. 18.

to-day are fundamentals: (1) That ore channels were of origin subsequent to their containing rocks: (2) That ores were deposited from solutions circulating in these openings. A scientist's work must be judged by the advancement he gave to his science, and with this gauge one can say unhesitatingly that the theory which we have set out above represents a much greater step from what had gone before than that of almost any single observer since. Moreover, apart from any tangible proposition laid down, the deduction of these views from actual observation instead of from fruitless speculation was a contribution to the very foundation of natural science.[4]

5.5   THE FABLED MINERAL/PIGMENT MINES AND THEIR TREASURES

Many volumes are needed to discuss even the most famous of the world's mines. Included here are those that are most closely related to some important historic pigments, and some that opened up new horizons in the search for material color.

5.5.1   Ngwenya – the Oldest Mine in the World

At a site called Lion Cavern, in present-day Eswatini (formerly Swaziland), ancient miners had cut into a 500-foot-high cliff face to shape a 30-foot deep workspace. A mountain shaped like a threatening recumbent crocodile gave its name to this, the oldest mine in the world – "ngwenya" means crocodile in siSwati – but its great age, over 43 000 years, was no match for 20th century commercialization. Red hematite, iron oxide, Fe_2O_3, is the most abundant red pigment in the world. It, and specularite, a particularly sparkling variety of the ore, were priceless commodities – presumably not so much for art, but more for ritual and medicine – right through the Middle Stone Age that flourished here until about 20 000 years ago.[19] In 1964, the Anglo-American company's excavators and bulldozers invaded to gouge the crocodile to extinction by open cast mining techniques. It removed 28 million tons of ore in 13 years, most of it to form the chassis of Toyotas and Hondas, before economic productivity fell too low.[20] A brief spate of renewed activity from 2011–2014 produced almost 3 million metric tons before the mine's final closure.[21] Today, the Ngwenya Iron Ore Mine Museum's only occupants are abandoned rusting backhoes, cranes and dump trucks. Specularite still has great press in the modern world: it is omnipresent on sites like Etsy and Chakra meditation practices. As the most ancient mine, we would place this site as no. 1 on your geotourism list.

There are other ancient hematite mines in Africa, notably those in Zambia and the ancient mines of Meroe of the Kingdom of Kush (present-day Sudan). Archaeological excavations at these sites place mining activities there from about 2700 years before present, nothing near the antiquity of Ngwenya.[22] Considering the fact that iron is the fourth most abundant element in the Earth's crust at 5.6%, it is not surprising that there are many hematite mines worldwide, some of them of great antiquity. And element abundance has nothing to do with chance, but everything to do with atomic structure and energy. Iron, according to nuclear chemists, has an extremely stable nucleus and, as such, was one of the most likely to be produced in the elemental crapshoot when the universe began.

5.5.1.1 Hormuz. If you really want to see some spectacular red, then make your way to the Islamic Republic of Iran, where Hormoz (var. Hormuz) Island, 60 miles offshore in the Persian Gulf, awaits your inspection. Here you will find a bright red beach contrasting with a deep blue sea. The Hormoz ochre mine is small with a total hematite deposit of 390 000 tons but it is unique in terms of its high concentration of Fe_2O_3. The natives call the red soil of this area "gelak," which they use as a spice in various kinds of food. One particular recipe, called "suragh," involves steeping sardines in the salty Hormoz red soil mixed with sour orange peel and serving with bread (Figure 5.3).[23] Marco Polo described the island in his travelogue after his visit of 1290 CE. It fell under the domination of the Portuguese Empire in 1507 and remained so until 1622.

5.5.1.2 El Cóndor Mine, Atacama Desert. In what is now northern Chile, the El Cóndor mine was exploited as an important hematite source from approximately 300–1500 CE. Beginning as a small-scale craft activity, it rose in importance during the Inca expansion period (*ca.* 1400 CE) to become a large complex

Figure 5.3   View of the red salty hematite soil valued by Hormuz natives for making a specialty dish called "suragh." Reproduced from ref. 23, http://dx.doi.org/10.4236/oje.2014.411060, under the terms of the CC BY 4.0 license, https://creativecommons.org/licenses/by/4.0/.

with a specialized infrastructure and a build-up of wide distribution and consumption networks. If your geo-tour includes the New World, this might be a good stopping off place on your way home.[24] Or you might consider a stopover in Cancún, Mexico in Quintana Roo province. There along the coast, archaeologists have uncovered an even older ochre mine that disappeared with rising seas 7000 years ago. Prior to that event, it was mined for extraordinarily high-quality ochre for 2000 years.[25]

5.5.1.3   The Gamut of Ochres.   The variety of colors in red ochres is enormous. They can range from almost purple-black to light pink depending upon the mix of minerals other than the original **chromophore**, iron oxide. A charming salmon pink from Pozzuoli, Italy, called *terra rosa* and a rich red earth from Sinope, on the Black Sea coast, were favorites of early medieval artists. In ancient Greece and Rome, the latter was sold in sealed packets stamped with their source to guard against counterfeits. These red shades found great use in wall paintings over the course of centuries, but they were never bright and brilliant enough to attract late medieval painters.[26] You will hardly ever find it in medieval illuminated manuscripts. It is worth mentioning that up until the end of the 18th century, red ochre was the only red pigment that could be used in ceramics – no other available red would withstand the high firing temperatures.[27] However, the bright blue pigment extracted from our next mine is a different story.

5.5.2 Sar-e-Sang Mine

Ultramarine blue draws its name from the geographic region from which it has been mined for the past 7000 years: the land that lies beyond (ultra) the sea (mare), namely beyond the Caspian Sea, beyond the High Pamirs, in Afghanistan's Hindu Kush. Marco Polo wrote of it when he sojourned in the area for about a year in 1272, though he never actually visited the mine: "There are mountains likewise in which are found veins of lapis lazuli, the stone which yields the azure color, here the finest in the world."[28] The principal mine, Sar-e-Sang, is surrounded by numerous other excavations distributed at walking distances of several hours from one another around the town of Badakhshan. They mostly consist of near-horizontal shafts about 9 feet high and wide, and perhaps a thousand feet deep. Despite the tales of romance and mystery surrounding the source of the fabled blue stone, the reality is quite different. The area is very poor, the men work long hours for a pittance, and in case of illness or injury, it is a two-hour jeep ride, if there is one running, to the nearest medical help. And all the miners suffer from chronic bronchitis since they do not wear masks.

When renowned international journalist Victoria Finlay visited Sar-e-Sang in 2001, she learned that she was the first woman to do so. She also chose to go when Afghanistan was one of the hardest places to visit due to the political situation. In fact, she never made it to the mine on the first try. She had to return in the following year, making a journey fraught with peril and frustration. But in her own delightful and imaginative way, as she traversed the first 1000 feet of Mine No. 1 she noted where the extracted stones would have gone. First to the Egyptian tombs, a little farther on, into the blue of the Bamiyan halos, and a bit farther on, to the place where Armenian monasteries would have gotten the blue for their illuminated Gospel Books. A few steps further, it was Titian's turn, then Michelangelo's, then "Hogarth's, Ruben's and Poussin's: a whole art history in one little pathway."[29] Once extracted, the miners themselves carry the blocks of lapis, sometimes as heavy as 100 kg, back to camp on their backs; thence it is transported 400 km by donkey or by truck to Kabul lapidaries, sometimes taking nine days. Smuggling operations across the Afghan Pakistan border continue to this day.

5.5.2.1 Lapis Lazuli, the Mineral. The raw mineral is also known as lapis lazuli, or simply lapis, literally "blue stone," (from the Latin "lapis" = stone; Persian "lazhward" = blue); it consists of the blue material intimately mixed with a high proportion of colorless material, mostly silicates and aluminates. Well-formed single crystals of any size are only found at Sar-e-Sang (Figure 5.4). There are other lapis deposits around the world, notably in the Lake Baikal Region, Siberia, as well as in China, Chile, Myanmar, and even in San Bernardino County, California. However, the

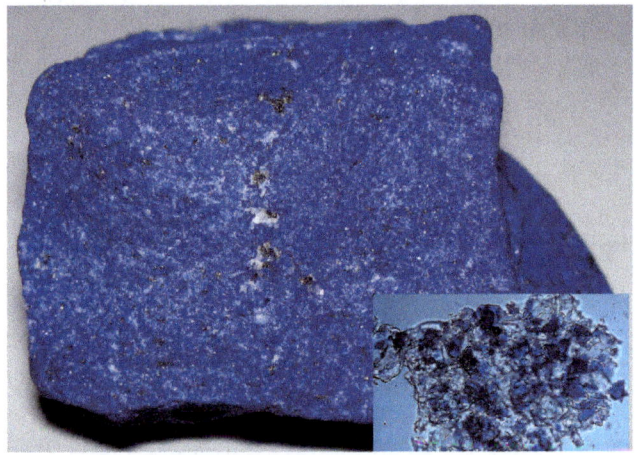

Figure 5.4   A fine example of lapis lazuli (lazuritic metamorphite) from the Sar-e-Sang Deposit, Precambrian Sakhi Formation, 4.5 cm width; Photo: James St John; https://flickr.com/photos/47445767@N05/33761908008. Reproduced from ref. 34, https://flickr.com/photos/47445767@N05/33761908008, under the terms of the CC BY 2.0 license https://creativecommons.org/licenses/by/2.0/. Inset: A sample of raw lazurite (photographed at 50× magnification). Note the large proportion of colorless material surrounding the blue portions signifying that this sample was never transformed into the more highly blue-concentrated pigment by the wax-ball process. Reproduced from ref. 34 with permission from American Chemical Society, Copyright 2013.

quality is mediocre, at best, and the veins are very small, some only millimeters thick. In effect, Sar-e-Sang seemed to be the only option for high-quality lapis.[30] However, in 2004, an international team of scientists (Germany and Hungary) collected lapis samples from the four major sources (out of 13 known deposits) and analyzed them for principal components and trace elements. They were able to determine that the Afghanistan and Lake Baikal samples were quite similar, whereas those from Chile and the Ural Mountains were quite different. These data can now enable sample provenance for these four sites.[31]

5.5.2.2 Lazurite, the Pigment. Laborious grinding of the lapis lazuli yields a useable pigment that can be termed "lazurite," but there has always been some confusion about terminology and the form of the pigment.[32]

5.5.2.3 Ultramarine. Transformation of the lazurite into the fine pigment we would term "ultramarine" involves a more labor-intensive process that starts, again, with grinding it to a fine powder followed by incorporation into a dough-like, hydrophobic mass of waxes, resins, and oils. This mix is kneaded repeatedly while submerged in an alkaline solution made from potash. The blue ultramarine particles, being hydrophilic, or attracted to water, fall to the bottom of the solution, whence they are collected in fractions and dried. Most of the colorless portion remains behind in the hydrophobic dough.[33,34]

Due to Badakhshan's harsh climate, the mines were virtually inaccessible, only open to the outside world for four months out of the year. Only a small fraction of pigment could be wrested from the bulk material in the purification process. Adding to these factors the numerous price markups by middle merchants along the way, it is no surprise that ultramarine was the most expensive colorant ever known. Weight for weight, it was more expensive than the purest gold, making it necessary for the patron to pay for it separately when commissioning a work of art.[33]

5.5.2.4 Chemical Structure. Chemically, the pigment is an alumino-silicate cage-like structure in which other elements are embedded. Inorganic pigments usually require the presence of a metal-based **chromophore**, but ultramarine is unique: the unit responsible for the deep blue color is S_3^-, the trisulfide radical anion, which is unstable in the free state (Figure 5.5).[35]

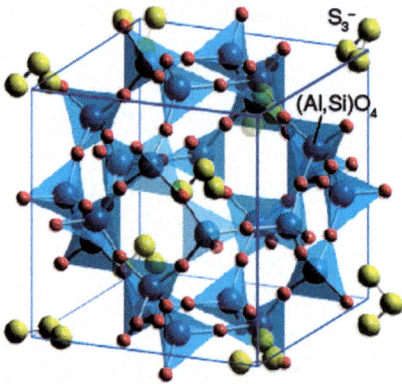

Figure 5.5   Ultramarine structure: trisulfide radical anions (shown as three yellow balls) in a host cage-like structure. Reproduced from ref. 35 under the terms of the CC BY-NC-SA 2.0 UK license, https://creativecommons.org/licenses/by-nc-sa/2.0/uk/, Copyright © 2008–2021 The University of Liverpool.

The nature of this entity was the subject of speculation for centuries until F. Albert Cotton and co-corkers finally identified it in 1976.[36] They showed that the cause of the blue color was a strong absorption band centered at around 625 nm that literally sopped up the entire red and green portions of the visible spectrum, leaving your eye to feast on the blue. The greater the amount of sulfur in the pigment, the deeper the blue. But if sulfur is replaced by the element selenium, the crystal color becomes blood red.

In future chapters of this volume, we will participate in the journey that natural ultramarine has made through art history. However, there is a synthetic variety that needs our brief attention.

5.5.2.5   Synthetic Ultramarine. Cloak and a Virtual Dagger.   The story of artificial ultramarine is a bit less straightforward because there was a great deal of money at stake. Apparent "sightings" of the compound in lime and soda kilns had been reported as early as 1787. In 1814, Nicolas Vauquelin (1763–1829), the distinguished French chemist and discoverer of beryllium and chromium, analyzed one such sample and found that its chemical composition was identical to that of natural ultramarine hard won from lapis lazuli. This find eventually caused the *Société d'Encouragement pour l'Industrie Nationale* to offer a substantial prize to the individual who came up with a feasible method of manufacture. The prize was eventually awarded to Jean-Baptiste

Guimet (1795–1871) in 1828. This was not without a heated challenge from Christian Gottlieb Gmelin (1792–1860), Professor of Chemistry at the University of Tübingen.[37] Guimet, a Frenchman, may have won the verdict from the French *Société*, but the shadow of Gmelin's plagiarism accusation stalked his career. Gmelin made up for the slight by developing a thriving market for his product. He was able to supply his ultramarine at a cost of 10% of the natural product, which sent the latter's use into rapid decline. As supplies of the natural variety dried up and became more difficult to obtain, being offered by only a handful of commercial pigment suppliers, the price ratio of natural to synthetic ultramarine jumped to 80 to 1.

Appropriately enough, synonyms for artificial ultramarine are French ultramarine, French blue, and Guimet's blue. Artificial ultramarine has been identified in a number of paintings by the French Impressionists and in many 19th century German paintings.[30] There are several methods of manufacture, and they seem to be relatively simple: the major ingredients, sodium sulfate, sulfur, silica, and iron-free clay (to provide the aluminates) are mixed well and heated to red heat (about 500 °C) for several hours in the absence of air. The product's particles are of regular size and shape, which distinguishes it from the natural material, which has, in general, irregularly shaped and larger particles mixed with fairly large amounts of crystalline mineral impurities. One big ecological drawback is the emission of SO_2 gas in great amounts during manufacture.

However, when ultramarine was first synthesized, its method of manufacture was shrouded in secrecy because prize money was at stake. Success rates for reproducing it have been spotty ever since, largely because most of the recipes omit a key ingredient or other piece of essential information (a common ploy adopted from antiquity). A group at the University of Belfast undertook to study and experimentally carry out the various literature recipes for producing ultramarine, and they also devised three trials of their own. They report on all of these trials along with an evaluation of the presence of impurities that could impinge on the observed color, such as the S_2^- **chromophore**,[§] and the optimal factors for success. This labor of love is available online as an open access article.[38]

[§]The S_2^- **chromophore** is yellow, thus giving an undesirable green cast to blue ultramarine.

On our tour of the world's pigment mines, the next most ancient brings us to Spain and the production of red again – a red almost as highly prized as ultramarine blue.

5.5.3   The Ancient Mine of Almadén

The Almadén cinnabar mine was an economic mainstay of the Roman Empire for centuries, eventually morphing into the major means of making New Spain a profitable venture for the Spanish regime. It is presently a UNESCO World Heritage Site. Cinnabar is a compound of mercury and sulfur, HgS. Its synthetic counterpart, vermilion, depends on the availability of elemental mercury for its manufacture.

5.5.3.1   Elemental Mercury.   If you are old enough to remember sealing several mL of mercury into a glass tube as part of your high school chemistry experience, welcome to the club! It was the most fun and fascinating thing we did in the laboratory all year, and we also had a souvenir to bring home to share with the family. Those days are gone forever, but they speak volumes regarding two facts we know about mercury: its silvery, snaky behavior is fascinating, captivating and mysterious. However, its toxicity is now being recognized and dealt with in the academic sector. Our experience with elemental mercury is common to our chemical ancestors of antiquity. They, too, knew about its toxicity, and they, too, were enthralled by its striking properties, some of which may be interpreted by ancient names, including "living water," "flying eagle," "tail of the dragon," "white smoke," "runaway slave," and "the whole secret."[39] Quicksilver in English, *Quecksilber* in German, and *hydrargyrum* (Hg) in Latin, the element takes its name from Mercury, the ancient Roman God who was fleet-of-foot. Found mainly in its compounds as the mineral cinnabar, elemental mercury was perhaps not unknown to the ancient Egyptians. Heinrich Schliemann (1822–1890) reported finding a small flask of it in a tomb dated to 1500–1600 BCE at Kurna (Thebes).[40] However, a much more recent report calls this claim into question and concludes that we cannot give credence to an appearance earlier than the "liquid silver" of Aristotle (385–323 BCE)[41] in the late 4th century BCE.[42] Vitruvius has a lot to say about elemental mercury, including its capacity to form amalgams and its great density (only gold is denser among the known substances of ancient times).[43] Apparently, its chief use in Roman times was for gilding and other decorative purposes.[44] Mercury's most common chemical compound is bright red cinnabar (from Ancient Greek, *kinnabari*).

5.5.3.2 Formation of Cinnabar Deposits. Uncounted eons ago, volcanic activity gave rise to an extraordinary geochemical anomaly. It opened up tectonic fissures in the Earth's crust through which cinnabar ascended to fill the cracks and solidify not far from the Earth's surface, forming the rare deposits that have been exploited for many centuries. Spain's Almadén mine (Al-Mahedan is Arabic for "the mine"), the world's largest mercury mine, has supplied approximately one-third of the total world's historic mercury production, some 260 000 tons.[45,46] Historically, the Iberian peninsula was one of the regions of the Roman Empire most fully exploited for its rich mineral resources. Though the archaeological record for these mining operations is scarce because of modern activity which obliterated much of the evidence, there are some literature sources that have helped identify where and when mines were fully operative. For example, early Christian writers discussed the condemnation of Christians to hard labor in many areas throughout the empire.[47] To be condemned to Almadén was a virtual death sentence because of the toxicity of the cinnabar powder which would have been ubiquitous in the mine atmosphere. Mercury toxicity has been a documented occupational hazard for centuries (See Box 5.1).[48]

Box 5.1 Mercury Toxicity

Since mercury chiefly affects the central nervous system, neuromuscular problems, such as uncontrollable shaking, are one of the symptoms. A classic story of an occupational hazard involving mercury involves what happened in a small city in northwestern Connecticut. Danbury, in the 19th century, became the center of the hat industry. In the felting process, the fur that would become the basis of the hat had to be harshly treated with an alkaline material – typically urine. Some of the hatters took to using their own urine, and some among them were found to be producing superior felt. These individuals were being treated for syphilis with a medication called calomel, chemically mercury(I) chloride, Hg_2Cl_2. The felting industry quickly modified its protocol to include mercury(II) nitrate as the felting agent of choice, leading to massive mercury poisoning symptoms among the workers that became known as the "Danbury Shakes." Due to the pioneering industrial toxicology research of Alice Hamilton, MD, Connecticut's governor finally banned the use of mercury in hat production in 1941. The expression "mad as a hatter" is still in common use, though not many people are aware of its origin: Lewis Carroll's *Alice in Wonderland*.

5.5.3.3 Historic Uses of Cinnabar: Amalgamation. The earliest documented use of cinnabar is at the Neolithic sites of Kfar HaHoresh in Israel[49] and Çatalhöyük in Anatolia (7400–6000 BCE).[50] Almost the entire cinnabar supply that made its way into ancient Rome came from the mines in Almadén. Mercury's special ability to join with other metals, thus concentrating them, to form amalgams (Box 5.2) was well known to the Romans; the Roman Emperor strictly controlled mercury production because of this property.[51,52] The Arabic alchemists' interest in cinnabar, from which they could produce elemental mercury, or "quicksilver," lay in its property of amalgamation. Geber (8th century) in his *Book of Mercury* and Rhazes (865–925?) and his *Book of Secrets* wrote of it extensively. The alchemists found that they could transform free metals, such as copper, into "gold" using this process.[53] In the Christian world, Roger Bacon (1214–1294) accepted the mercury-sulfur theory of the early alchemists Avicenna (980–1037) and Albertus Magnus (1199?–1280) that the two elements

---

Box 5.2   Alloys and Amalgams

The Bronze Age in human history owes its name to a substance called an alloy, bronze, which is a mixture of copper and tin. The amount of tin can vary, but the most useful bronze contained about 10–12% tin by weight. An alloy is not a chemical compound – the metals mixed together do not chemically bind to one another, but their presence together changes the entire crystal structure and therefore changes the physical properties as well. For example, bronze is stronger than either copper or tin alone; it also has a lower melting point and is easier to work. To amalgamate something, in common parlance, we mean to join, so "Amalgamated Transit" would signify the union of several transportation companies. Chemically speaking, the word has a special meaning: the joining of two metals with one another to form a special kind of alloy, those of mercury and gallium. Mercury is a particularly good choice for purifying silver in this manner. Silver ores have to be crushed into very small particles; the smaller the particles, the easier it is for the mercury to come into contact with the silver atoms. Because mercury is a liquid at room temperature, it can penetrate into an ore and "bond" with the silver in it. Metals form chemical compounds by donating electrons to a non-metal; for example, iron donates electrons to oxygen to form iron oxide. But if metals mix and bond with other metals, such as mercury and silver, they simply mix their electrons in a large, mobile electron sea to form what chemists call a solid solution. Mercury dissolved in another metal is a solution analogous to a solution of a salt dissolved in water. Once the solid solution is formed, it is easy to separate it from the rest of the ore material, and then simply heating the amalgam will drive off the mercury to leave highly purified silver behind, and the collected mercury can be recycled.

were compounds of different materials, containing both fixed and unfixed components.[54] Albertus Magnus refers on various occasions to obtaining silver by quicksilver.[55]

Although Almadén's cinnabar production continued right through the fall of the Roman Empire and beyond, though somewhat reduced because of the lack of a military garrison to protect the miners and the resource, it found new life with the discovery of the New World. In the sixteenth century, silver was discovered in New Spain (Mexico) and other Spanish territories, but the ores, typically containing less than 1% silver, had to be processed. At the time, the most efficient, and in fact, virtually the only way to do this was by forming mercury amalgams. Without mercury, there would be no silver; without silver, the economy of New Spain would collapse. So, Almadén's importance rose enormously over the next few centuries, becoming the Spanish Crown's economic mainstay. Recognizing the need for more education in this area, the King of Spain founded the Mining Academy at Almadén in 1777, the fourth such institution worldwide.

5.5.3.4 Almadén's Final Days and Reconfiguration. When the Függer banking family from Germany took over the lease of Almadén in 1525, they reintroduced the penal aspect of the work, and for the following 250 years, almost 2000 men were sent to "redeem their guilt" as *"forzados"* (forced laborers) in the mines of Almadén. It has been estimated that about 25% of those forced laborers succumbed to mercury poisoning before their sentences expired.[56] Another sociological reality revealed by the very dimensions of the mine galleries is the fact that they were, in places, so narrow and small that they could not have been worked except by children or dwarves.[57] In fact, the archaeological record reveals that in some mining sites, about half of the shoes discovered could only have been worn by children, and who knows how many workers went barefoot? A heartrending little tombstone to a four-year-old "niño minero" named Quartulus, found in Baños, Spain, is now in the collection of the National Archaeological Museum in Madrid (Figure 5.6).[58,59] No one can even hazard a guess regarding the number of children who died.

Although many modern improvements over the centuries made the mine more productive, environmental pressures in the late 1980s coupled with changing technology in the chemical industry

Figure 5.6   Quartulus, el "Niño minero;" Museo Arqueológico Nacional. Inv. 16744. Photo: Miguel Ángel Otero. Reproduced from ref. 58 with Permission from Museo arqueologico Madrid.

caused a sharp decline in demand for mercury. Progressive recession led to the mine's final closure in 2003. A prime site for the study of industrial archaeology as well as the ethnography and heritage of the region led to promotion of the mine's environs for ecotourism.[45] Almadén is now a major part of this project, having achieved, together with the mine of Idrija, Slovenia, the distinction of designation as a UNESCO World Heritage Site. Erasmus Darwin (1731–1802) immortalizes both in his verse:[60]

On vermeil beds in Idria's mighty caves

The living silver rolls its ponderous waves.

An enduring monument to Almadén's historic importance is Alexander Calder's spectacular mercury fountain. Commissioned in 1937 by the Spanish Republican Government as a memorial to

Figure 5.7   Alexander Calder. Mercury Fountain. 1937. Joan Miró Museum, Barcelona. The artist took advantage of the fact that mercury is the only metal which is liquid at room temperature. Reproduced from ref. 61 under the terms of the CC BY-SA 4.0 license, https://creativecommons.org/licenses/by-sa/4.0/deed.en.

the siege of Almadén by Franco's troops, it now resides at the Joan Miró Museum in Barcelona (Figure 5.7).[61]

Our mining journey continues northward to Germany and Scandinavia, regions of Europe plentifully blessed with exotic minerals and rare elements, some of which remained undiscovered until the 20th century.

5.5.4   Discovery of Cadmium

Göttingen, Germany, a university town situated near the mining areas in the Harz Mountains, holds a special place in art history. It is the discovery site of the element cadmium that has given us, for over two centuries, beautiful reds, oranges and yellows as artists' pigments. The Harz region was the center of zinc mining, though zinc itself, a by-product of silver mining, was deemed boring, colorless, and good only for laundry tubs.[62] However, zinc compounds found their way into the medicine cabinet as early as the 18th century. In 1817, when a pharmacy near Göttingen reported an unnerving discoloration in its zinc medicines,¶ it called upon the regional Inspector of Pharmacies for help. Since the inspector, Friedrich Stromeyer (1776–1835), moonlighted at the University of Göttingen as a chemistry professor, he used his chemistry skills to isolate a yellow impurity from the zinc that turned out to be the sulfide of an unknown element.[63] Because of the element's close association with zinc,‖ which used to be smelted in furnaces called "cadmia fornacum," Stromeyer dubbed his find "cadmium."[64]

5.5.4.1 Cadmium Yellows, Oranges and Reds. Very soon after his discovery, Stromeyer suggested cadmium's use as a pigment,[65] but there was so little of the metal available at the time that it did not become popular until it was commercially exploited in the 1840s.[66] Cadmium sulfide, CdS, is a naturally occurring mineral called greenockite that is found as a yellow coating on zinc ores, but its occurrence is quite rare. It is readily prepared in the laboratory or factory by either heating cadmium metal with sulfur in the absence of air (dry process) or by mixing solutions containing cadmium ions and sulfide ions to form insoluble yellow CdS (wet process). E. A. Parnell, as early as 1846, mentions a very clever way of dyeing a textile, such as silk, with cadmium sulfide by immersing the fabric in a solution of cadmium chloride and then passing it into a solution of a soluble sulfide to form CdS directly on the material.[67] Though CdS is a bright yellow solid, its color can vary depending on its particle size. The color can also be changed by replacing some of the sulfur atoms in its crystal structure with selenium, a slightly larger close relative of sulfur, lying just below it in the periodic table. Depending on the degree of selenium substitution, colors of these cadmium sulfoselenides can range from orange (100 S/12 Se) to orange-red (100 S/15 Se) to dark red (100 S/30 Se) to dark maroon (100 S/40 Se).[68] Some formulations of cadmium-based pigments with other compounds are marketed as artists' pigments as well. Among

---

¶From the time of the alchemists, color was used as a criterion for distinguishing between a pure and an impure substance. In this case, noting an improper color, because zinc compounds were indeed colorless, led directly to discovery of a new element. See ref. 37, p. 72.

‖Cadmium is a member of the "zinc family," lying just below zinc in the periodic table.

these are the lithopones (a mix with barium sulfate, $BaSO_4$) and cadmium/mercury pigments, solid solutions with variable Cd/Hg ratios.[69] Cadmium pigments are offered by a number of art suppliers as oil colors, watercolors and pastels. Early complaints about their non-permanence and tendency to fade have been overcome by care in preparation: to prevent fading, it is essential that no free sulfur be present in the pigment since it can react in moist air to initially form sulfurous acid which can in turn react with the sulfide in the pigment to form colorless sulfate.[70] Prolonged use of cadmium pigments could cause serious damage to health by exposure through inhalation and if swallowed. They are listed as a cancer suspect agent.

5.5.4.2 Cadmium's Career in Pigments. Cadmium-based pigments were favored by artists due to their warm colors, good hiding power, and lightfastness early on. Cadmium sulfide first found limited use in oil painting in France and Germany in 1829.[71] It was first shown in public at the Great 1851 Exhibition in London[72] and rapidly took its place as one of the most important artists' pigments by the turn of the century. Among its eager users were Vincent van Gogh (1853–1890) in his *Self-Portrait* of 1886–1887, Henri Matisse (1869–1954) in his *Bathers by a River* of 1916–1919 and numerous works by Claude Monet (1840–1926) spanning 35 years of his long career from 1873 to at least 1908 (Figure 5.8).[64,73]

Figure 5.8   Claude Monet. Haystacks, 1891. A painting dripping with cadmium reds, yellows and oranges. Reproduced from ref. 73.

5.5.5 Discovery of Cobalt and Cerium

It is no accident that Sweden heads the list of countries where numerous elemental discoveries were made. It sits directly on the Baltic shield, a geological phenomenon where, eons ago, igneous and hydrothermal processes gave rise to hundreds of different minerals, many with unusual properties. Sweden, likewise, possessed the intellectual infrastructure in the 18th and 19th centuries that enabled her chemists to brilliantly probe and unravel the complex mixtures emerging from her mines.[74]

5.5.5.1 Cobalt and the Pellugruvan Mine of Riddarhyttan. We saw, earlier in this chapter, that not only did Agricola but also virtually anyone associated with mines believe in the existence of gnomes and goblins that inhabited them. The miners often blamed these instigators of mischief, called *kobelts*, for work-related problems. When they encountered a particularly corrosive material that injured their hands and feet, they used the same epithet – *kobelt*. Hence, the substance, which was probably a cobalt-bearing arsenical ore, comes down to us as the name of the metallic element it contains.[3] However, although there was infinite confusion among zinc, cobalt, arsenic and bismuth minerals, Agricola at least recognized that "the slag of bismuth,... when melted make[s] a kind of glass [that] will tint glass and earthenware vessels blue."[75]

It was Georg Brandt, Jr., Assay Master of the Stockholm Mint, who dispelled all the confusion surrounding cobalt by isolating it from a sample of cobaltite, $CoAsS$, sourced from the Pellugruvan Mine in Riddarhyttan, 150 km northwest of Stockholm. He was able to show that minute amounts of this element were responsible for the blue color in a variety of minerals. His investigations took place over a period of about 15 years from 1730 to 1745, so no exact date can be given for cobalt's discovery. Brandt found that it could produce a glassy blue pigment called "smalt" that was well known to glassmakers in Europe at the time,[76] but possibly had its origins in such exotic places as Inner Mongolia and China.[77] Cobalt occurs in Nature as a single **isotope** with an atomic weight of 59, ^{59}Co. The Co-60 in common parlance is a radioactive isotope of cobalt produced in a cyclotron and is useful as a radiation source in cancer chemotherapy.

5.5.5.2 Cobalt before its "Discovery". Georg Brandt actually arrived very late on the cobalt scene – it had been "discovered" and used many centuries before, but never isolated as an element as such. Its use in ceramics can be traced to Egypt and the Middle East from the 16th to the 1st century BCE. Strangely, it fell out of use for almost a thousand years, reappearing in the 10th century CE in Persia, in the 11th in Constantinople, and the 12th century through the Italian Renaissance in Italy. Cobalt came to Italy from the Erzgebirge (Ore Mountains) in Saxony, and with it, the della Robbia family "wrote one of the most illustrious pages in the history of that vibrant artistic period."[78] No one who has ever visited Florence can forget the distinctive white-upon-brilliant blue rondos, tiles, tombs and altarpieces created with secret recipes for almost a century – from 1442 to 1527 – until betrayal of the secret by a family maid. In the following centuries, cobalt exports from Saxony took on industrial proportions and its compounds soon entered the standard pigment catalog.

5.5.5.3 The Cobalt Blues: Smalt, Cobalt Blue (Thénard's Blue), and Cerulean Blue. These three blue pigments, as their collective name implies, have the element cobalt in the form of Co(II) oxide, or cobaltous oxide, in common. In smalt, the ground up oxide is mixed with glass;[79] cerulean blue is $CoO \cdot nSnO_2$, cobaltous stannate;[80] cobalt blue, or Thénard's Blue, is $CoO \cdot Al_2O_3$, cobaltous aluminate.[81] The cobalt blues cannot be easily distinguished visually; instrumental methods are absolutely necessary.

5.5.5.3.1 Smalt. Although smalt, glassy CoO, has been used as a colorant for glass for millennia, it did not come into its own as an artists' pigment until the mid-1550s. Because of its glassy nature, it was difficult to handle, but it could provide a beautiful blue at a small fraction of the cost of ultramarine. Because smalt is a glass, we would expect a fair degree of permanence, but when used in various media, it does tend to discolor, as documented in an oil painting ascribed to Bartolomé Esteban Murillo (1617–1682) that hangs in the National Gallery, London.[79] Unfortunately, smalt has betrayed many other painters over time. The blue skies of Paolo Veronese (1528–1588) have turned gray, the blue robe of Tintoretto's (1518–1594) *Madonna of the Stars* is now brownish, and the backgrounds and curtains of many of Rembrandt's (1606–1669) paintings are now brownish.[82] This shortcoming could be attributed to the fact that as it degrades, the Co(II) ion that imparts the strong blue color to the pigment tends to gradually assume a different configuration.

5.5.5.3.2 Cobalt Blue. Soon after Brandt discovered cobalt and identified it as the cause of the blue color in its compounds, other researchers took up the challenge of further syntheses. But the compound did not enter the world of the artist until Louis Jacques Thénard (1777–1857) found it in his search for a less expensive substitute for ultramarine. Forever afterwards, it has also been called after the re-discoverer, "Thénard's Blue." A start date for the use of cobalt blue is usually given as 1802 to 1805, although Thénard only published his *opus magnum* a decade later.[83] The pigment he found is a powerful bright blue that is exceptionally permanent and chemically inert. It works well in various media and does not interact adversely with other pigments. It would seem to be the ideal pigment except for two other qualities: it is expensive but also highly toxic.[37]

5.5.5.3.3 Cerulean Blue. In 1805, by roasting tin and cobalt oxides together to form cobaltous stannate, $CoO \cdot nSnO_2$, Andreas Höpfner, a German chemist, succeeded in first synthesizing cerulean blue, Pigment Blue 35.** It did not enter the artists' palette, however, until 1860 when it was marketed in England by George Rowney & Company. The National Gallery has documented the use of Rowney's "Ceruleum" in the sky of S. Giorgio Maggiore from the Dogana, 1859, and in some subsequent paintings.[84]

Cerulean Blue is bright but lacks the deep blue shade of cobalt blue, so it works best for painting blue skies, as its name implies (*caeruleus* in Latin means "dark blue" or "blue"). It is lightfast, inert to chemicals, and mixes well with other colors. However, it is not as opaque as cobalt blue and turns "chalky" when used as a watercolor. It, too, is very expensive. However, it became a staple among artists, particularly the Impressionists, and can be found in works by Berthe Morisot (1841–1895) and Claude Monet

---

**It is often necessary to specify the *Colour Index* number for cerulean blue because other pigments with different properties are often sold under this name. Artists who wish to manage their colors so that they always know their chemical identities are always counseled to purchase them by using the C.I. numbers as a guide. A handy website for this purpose is http://www.artiscreation.com/Color_index_names.html#.X0GBcH4pCM8 (accessed August 2020).

(1840–1926).[85,86] It has been said that Paul Signac (1863–1935) squeezed countless tubes of the pigment dry, as did many of his fellow artists.[87]

5.5.5.4 Cobalt Yellow.   There is, indeed, a cobalt yellow, called aureolin, first synthesized by N. W. Fischer in 1831,[88] but not introduced to the art world until 1851 by Gillot Saint-Èvre (1791-1858).[89] Chemically, it is potassium cobaltinitrite, $K_3[Co(NO_2)_6]$. Because the pigment has only moderate tinting strength and brightness, as well as dubious permanence, it has not been popular as an artists' pigment except as an occasional watercolor.[90]

Sweden remained at the forefront of new element and new pigment discoveries for the next century until the center of activity of this exciting enterprise takes us to France, but for an entirely different reason.

5.5.5.5 Cerium and the Bastnäs Mine of Riddarhyttan.   In 1803, a wealthy Swedish amateur mineralogist, Wilhelm Hisinger (1766–1852) engaged an up-and-coming young chemist, Jöns Jacob Berzelius (1779–1848) in the task of examining a strange dense mineral that he had found in his iron mine, the Bastnäs. Surprisingly, their analysis revealed a totally new oxide which they named cerium after the mineral in which it was found, cerite, named in turn after the asteroid, Ceres, that had been discovered two years earlier in 1801 by Giuseppe Piazzi (1746–1826).[††]

5.5.5.5.1 Cerium Pigments.   Cerium remained in the backwater, as far as pigments are concerned, for two centuries until the cerium sulfides entered the market in the early 1990s. They belong to a new group of what are called "high performance pigments" that have many desirable properties for use in high stress situations such as automotive coatings. They have a pleasing range of color depth from light orange to dark red and burgundy. They are thermally stable, highly opaque, lightfast and much brighter than iron oxide. They seem to be thoroughly benign both ecologically and toxicologically and show promise as a replacement for the cadmium oranges and reds.[91] They exist in three different crystalline forms, α-Ce_2S_3, β-Ce_2S_3 and γ-Ce_2S_3, that exhibit

---

[††]Cerium was discovered simultaneously by Martin Klaproth in Berlin; today both groups share credit for the discovery.

different color properties. The α-form is of little interest because of its brownish-black color; the β-form is very versatile since its burgundy color can be fine-tuned to lighter shades by substituting oxygens for the sulfur atoms in the crystal. The γ-form is dark red but it can only be produced at temperatures higher than 1100 °C, a difficult requirement, industrially speaking. We will probably be seeing more of this new type of pigment on the market in the future.[92]

But long before its debut in pigments, cerium had another very productive life as the principal element in cerite, the ore from which a whole parade of new elements emerged like a series of Russian dolls. It all began on a little-known island that leaped into prominence when the Ytterby Mine, in Sweden, began to yield up new element after new element in a seemingly unending sequence.

5.5.6   Ytterby Mine, Resarö Island, Sweden

Ask a chemist what the most famous mine in the world is and you would likely hear, "Ytterby, of course." Now, Ytterby is not necessarily a household word, but a glance at the "southern" portion of the periodic table reveals the strange names of three unfamiliar elements: terbium (Tb, no. 65), erbium (Er, no. 68) and ytterbium (Yb, no. 70). Combined with yttrium, Y, element number 39, these four elements constitute a group named after the Ytterby Mine (Figure 5.9),[93] from which they were all extracted. The mine, now closed but with some signage indicating its historic importance, once supplied high-quality quartz and feldspar to the British porcelain industry.[94] Along with cerium (Ce, no. 58), these elements are part of a larger group of 17 called collectively the "rare earth elements," or REEs. They leaped into prominence in the late 20th and early 21st century due not only to their colorful properties but also due to their magnetic properties that have become indispensable to the electronics industry. There are many other important mines in Scandinavia, including the Loos Cobalt Mine and the Falun Mine.

5.5.6.1   REEs.   Cerium occurs in the ore cerite, the parent ore from which are extracted the so-called lighter REEs, lanthanum, praseodymium, neodymium, europium and gadolinium.

Figure 5.9   Ytterby Mine Entrance, Plaque and Street Sign. The plaque reads: ASM International (formerly American Society for Metals) has designated Ytterby Mine an historical landmark. Four periodic elements – yttrium, terbium, erbium and ytterbium – were isolated from the black stone gadolinite mined here, and were named after the Ytterby Mine. 1989. Reproduced from ref. 93 with permission from James L. Marshall, "Rediscovery of the Elements".

Yttrium occurs in the ore ytterbite, from which are extracted the heavier REEs terbium, dysprosium, holmium, erbium, thulium, ytterbium and lutetium. While these separations sound simple enough NOW, they entailed difficult, monotonous, time-consuming work full of traps and snares that chronologically overlapped three centuries: from 1794 when Johan Gadolin (1760–1852) discovered yttrium, to 1907 when Georges Urbain (1872–1938) isolated lutetium.[95,96] Furthermore, REEs are by no means rare! They are, in fact, quite ubiquitous: for example, yttrium is more abundant in the Earth's crust than cadmium and mercury. However, they are so scattered around the Earth, so embedded in their mineral substrates, and so chemically similar to one another that they are difficult to process even today.[97] While there are substantial deposits of REEs in the continental United States, Chinese mines have dominated the REE market beginning in the 1980s. The Bayan Obo mine in Inner Mongolia harbors the largest REE deposit in the world. In 1957, the Chinese began open-pit mining here, extracting the estimated 6% of REE oxides from a matrix of other ores, including iron.[98]

5.5.6.2 REEs and Pigments. We have seen that the REE cerium is a maroon/red/orange high performance pigment source. Other REEs have not yet been exploited for pigment use because they are so valued for their magnetic properties. The star of the color show, however, is europium, utilized primarily for its unique luminescent behavior. Ultraviolet radiation can excite europium atoms to higher energy states, which the europium in turn emits as visible radiation. Europium serves as the basis for several commercial blue phosphors used for color TV, computer screens and fluorescent lamps. In energy efficient fluorescent lighting, europium provides not only the necessary red but also the blue.[99] As an aside to their use as pigments, we must make sure you also realize that wind turbines, hybrid automobiles and mobile telephones, among other things, cannot operate without their fair complement of REEs.

Heading southward back down the continent, we encounter the last, and possibly the most significant, of the historic mines in Europe. What can be more significant than the discovery of almost the whole bottom half of the periodic table? Read on – and soon you will agree that you never know when throwaway stuff should be saved.

5.5.7   St Joachimsthal Mine, Bohemia

The St Joachimsthal Mine is situated in the famous Erzgebirge (ore mountains) region of Bohemia, the most productive metal producing region in Europe. It closed down at the time of the fall of the Roman Empire. In 770 CE, Charlemagne (742–814) reopened the mines, and soon the gold, silver and lead extracted from them were enriching the Holy Roman Empire. Later on, the mines started yielding up uranium and cobalt to be used as blue and pink colorants for glass and ceramic glazes.[100] Even after its radioactive properties were discovered, uranium salts such as sodium uranate, $Na_2U_2O_7$, had very little commercial value.

5.5.7.1 Radioactivity. In 1896, Henri Becquerel (1852–1908) reported that uranium[‡‡] emitted radiation capable of exposing light-sensitive silver salts by penetrating light-opaque paper. Becquerel's great insight was that the radiation emitted by the uranium originated from the substance itself and not from any type of excitation producing phosphorescence or fluorescence. The young doctoral student at Paris's Sorbonne, Marie Skłodowska Curie (1867–1934) seized upon this discovery and set about trying to determine the nature of this mysterious phenomenon and whether other substances exhibited it as well. That she would succeed beyond all expectations is now a matter of history: she is the most famous woman scientist in the world.[101] In the course of her research, Marie found that thorium also emitted these strange rays and she eventually described the new phenomenon by the term "radioactivity." Her crucial next step was to quantitatively test museum mineral samples for radioactivity and, to her surprise, she found that several ores, including uranium ores, exhibited a degree of radioactivity many times greater than uranium and thorium alone. For example, when she tested the mineral pitchblende, about 80% of which is U_3O_8, she found that its radioactivity was several times that of the same weight of pure uranium. This observation led her to propose a very daring idea: that the ores could be hiding a hitherto unknown element.[102,103]

At this point, uranium took a back seat in terms of radioactivity research. Yes, it was the first substance to be identified as radioactive, but there seemed to be much more intense radiation emanating from ores that contained little or no uranium at all. So, uranium fell back into its normal role as a colorant for ceramic glazes, principally being the glaze for the famous orange-red FiestaWare pottery that teachers love to expose to a Geiger

---

[‡‡]Martin Klaproth (1743–1817) detected uranium, U, element no. 92, in 1789 in a sample of pitchblende and succeeded in isolating the oxide. See M. E. Weeks, *The Discovery of the Elements*, Easton, PA, *Journal of Chemical Education*, 1968, 7th edn, pp. 268–269. It was the first member of the actinide elements (no. 89–103) to be discovered.

counter in front of incredulous students.[§§] Uranium would remain in the backwater for the next 40 years – only to shapeshift into an astonishing form that no one could have anticipated.

Meanwhile, Marie and her husband Pierre (1859–1906), who immediately dropped his own research to team up with her, began to put the pitchblende ore into a systematic separation process. At this stage in their research, the St Joachimsthal Mine in Bohemia (now Jáchymov in the Czech Republic) came into play. In order to purify a measurable amount of radium, they needed a great deal more of the expensive pitchblende. Through the good offices of the Vienna Academy of Sciences, the Austrian Government gifted them with one ton of waste pitchblende tailings from which the uranium had already been extracted for glassmaking. The ore came from the St Joachimsthal Mine which, at that time, was under Austrian control. After months of intense labor in an unheated, poorly equipped shed, they were able to announce the discovery of a new element, polonium, Po, in July 1898. Continuing the process, five months later they announced the identity of another new element, radium, Ra.

Over the next two and a half years, they labored with yet more tons of pitchblende residues until, in July 1902, they were able to exhibit 0.1 g of purified radium to the world.[104]

5.5.7.2 Nuclear Consequences.   Radioactivity, as evinced by uranium, was to dominate the scientific, political, economic, and social scenes of the first half of the 20th century. During that century, all the rest of the actinides, and most of the transactinides (element no. 104–118), were isolated or synthesized. This led to spectacular discoveries, overturned assumptions and theories, and gave glimpses of a Nature full of unexpected surprises. But the biggest surprise of all happened in 1938 when Otto Hahn (1879–1968) and Lise Meitner (1878–1968) realized that they had discovered the second of uranium's peculiar properties – fission of its nucleus, releasing potentially the most destructive or the most life-giving energy force in Nature.[105] It was certainly a twisted path from uranium's casual ability to decorate earthenware cups and saucers to becoming the most sought-after commodity in any market, legal or illegal, in the world.

[§§]Modern FiestaWare no longer contains radioactive salts.

5.6 CONCLUSION

Mines, especially the more fabled ones, have long held an air of mystery, even romance, for the armchair traveler and history buff. Over the course of millennia, people extracted every possible buried commodity from deliberately constructed mines: salt, slate, coal, precious stones, iron and other metalliferous ores, minerals and later, some uninteresting-looking rocks that were found to contain exotic elements and other surprises. Although mineral pigments were not high on the list of major commercial resources, the same technologies were used to extract and process them. One of these pigments, cinnabar, when put to a different use, became the economic mainstay of the Spanish colonies of the New World. Another element, uranium, discovered in 1789, at first found no use except for coloring glass and ceramics. Two of its peculiar properties, discovered much later, launched the world into the nuclear age. It is a lesson to be learned: you never know what the Earth's crust can yield up or what surprises you may experience by simply trying to find a suitable coloring material. An RSC resource designed to spark the interest of schoolchildren in the relationship between minerals and their transformation into paints can be found at this site.[106]

Digging deeper into the earth's crust supplied colorful commodities that enabled civilizations to flourish. In much the same way, delving into the seas provided a much sought-after color that conferred a royal aura on those who wore it. As the next chapter will surely demonstrate, there is no shade that can match the color purple.

REFERENCES

1. Cited in M. H. Manser, *The Facts on File Dictionary of Proverbs*, Infobase Publishing, NY, 2nd edn, 2007, p. 186.
2. J. Grimm and W. Grimm, *Sneewittchen*/Illustrated by Franz Jüttner, Scholz, Mainz, 1905, p. 12, https://doi.org/10.24355/dbbs.084-200510210200-683, accessed June 2021.
3. G. Agricola, *De Animantibus Subterraneis*, H. Froben, Basel, 1549; as translated in M. L. Aldrich, A. E. Leviton and L. L. Sears, *Proc. Calif. Acad. Sci.*, Fourth Series, 2009, **60**(9), §502.

4. G. Agricola, *De re metallica*, translated from the first Latin edition of 1556 by H. C. Hoover and L. H. Hoover, Salisbury House, London 1912.
5. H. R. Haggard, *King Solomon's Mines*, Cassell & Co., London, 1885.
6. W. Shakespeare, *As You Like It*, II, VII.
7. R. J. P. Williams and R. Rickaby, *Evolution's Destiny: Co-evolving Chemistry of the Environment and Life*, RSC Publishing, London, 2012.
8. R. Hazen, *The Story of Earth: The First 4.5 Billion Years, from Stardust to Living Planet*, Penguin Random House, New York, 2013.
9. B. McFarland, *A World from Dust*, Oxford University Press, New York, 2016. pp. 173–174.
10. Mineral Database. www.mindat.org, accessed August 2020.
11. D. E. Bar-Yosef Mayer and N. Porat, *Proc. Natl. Acad. Sci. U. S. A.*, 2008, **105**(25), 8548–8551.
12. L. Weeks, Metallurgy, in *A Companion to the Archaeology of the Ancient Near East*, ed. D. T. Potts, Blackwell Publishing, Oxford, 2012, pp. 295–316.
13. M. Radivojević, Th. Rehren, E. Pernicka, D. Šljivar, M. Brauns and D. Borić, *J. Archaeol. Sci.*, 2010, **37**, 2775–2787.
14. J. B. Lambert, *Traces of the Past: Unraveling the Secrets of Archaeology through Chemistry*, Addison-Wesley, Reading, MA, 1997, p. 173.
15. V. Biringuccio, *Pirotechnia: The Classic 16th Century Treatise on Metals and Metallurgy, Pub. 1540*, trans. C. S. Smith and M. T. Gnudi, Dover Publications, Mineola, NY, 1990.
16. I. Fiedler and M. A. Bayard, Emerald green and Scheele's green, in *Artists' Pigments: A Handbook of Their History and Characteristics*, ed. E. W. FitzHugh, National Gallery of Art, Washington, DC, 1997, vol. 3, pp. 219–271.
17. B. Varani, *Geol. J.*, 1994, **32**(2), 151–160.
18. G. Agricola, *De re metallica libri XII*, Wellcome Images, London, 1494–1555, https://wellcomecollection.org/works/zauqw76v, accessed June 2021.
19. R. A. Dart and P. Beaumont, *Curr. Anthropol. Res. Rep.*, 1969, **10**(1), 127–128.
20. Ngwenya Mines. UNESCO World Heritage Application, https://whc.unesco.org/en/tentativelists/5421/, accessed August 2020.

21. J. J. Barry, The Mineral Industry of Swaziland, *U.S. Geological Survey Minerals Yearbook*, USGS, United States Printing Office, Washington, DC, 2017, pp. 42.1–42.3.
22. J. Humphris, R. Bussert, F. Alshishani and T. Scheibner, *Azania: Archaeol. Res. Afr.*, 2018, **53**(3), 291–311.
23. A. Yazdi, M. A. Arian and M. M. R. Tabari, *Open J. Ecol.*, 2014, **4**, 703–714.
24. M. Sepúlveda, F. Gallardo, B. Ballester, G. Cabello and E. Vidal, *J. Anthropol. Archaeol.*, 2019, **53**, 325–341.
25. M. Price, Underwater Caves in Mexico Preserve One of the World's Oldest Ochre Mines, *ScienceMag*, July 2020, https://www.sciencemag.org/news/2020/07/underwater-caves-mexico-preserve-one-world-s-oldest-ochre-mines, accessed August 2020.
26. D. V. Thompson, *The Materials and Techniques of Medieval Painting*, Dover Publications, New York, 1956, pp. 98–99.
27. N. Zumbulyadis, R. Fuchs II and E. Uffelman, *Keramos*, 2020, **247**, 15–34.
28. *The Travels of Marco Polo, the Venetian*, ed. T. Wright, George Bell & Sons, London, 1886, p. 84.
29. V. Finlay, *Color: A Natural History of the Palette*, Random House, New York, 2004, p. 310.
30. J. Wyart, P. Bariand and J. Filippi, *Rev. Geogr. Phys. Geol. Dyn., 2 Ser.*, 1972, **14**(4), 443–448. Republished in *Gems & Gemology* (Winter, 1981), 184–190.
31. J. Zöldföldi, S. Richter, Zs. Kasztovszky and J. Mihály, Where Does Lapis Lazuli Come from? Non-Destructive Provenance Analysis by PGAA, *34th International Symposium on Archaeometry*, Zaragoza, Spain, 3–7 May 2004.
32. G. Frison and G. Brun, *J. Am. Inst. Conserv.*, 2016, **16**, 41–55.
33. J. Plesters, Ultramarine blue, natural and artificial, in *Artists' Pigments: A Handbook of Their History and Characteristics*, ed. A. Roy, National Gallery of Art, Washington, DC, 1993, vol. 2, pp. 37–54.
34. M. V. Orna, Artists' Pigments in Medieval Illuminated Manuscripts: Tracing Artistic Influences and Connections – A Review, in *Archaeological Chemistry VIII*, ed. R. A. Armitage and J. H. Burton, American Chemical Society, Washington, DC, 2013, pp. 3–18, https://pubs.acs.org/doi/abs/10.1021/bk-2013-1147.ch001.

35. N. Greeves, University of Liverpool, https://www.chemtube3d. com/ss-ultramarine/, accessed June 2021.

36. F. A. Cotton, J. B. Harmon and R. M. Hedges, *J. Am. Chem. Soc.*, 1976, **98**(10), 1417–1424.

37. F. Delamare, *Blue Pigments: 5000 Years of Art and Industry*, Archetype Publications, London, 2013.

38. I. Hamerton, L. Tedaldi and N. Eastaugh, *PLoS One*, 2013, **8**(2), e50364, https://doi.org/10.1371/journal.pone.0050364.

39. M. P. Crosland, *Historical Studies in the Language of Chemistry*, Harvard University Press, Cambridge, MA, 1962, p. 36.

40. E. O. von Lippmann, *Entstehung und Ausbreitung der Alchemie: Ein Beitrag zur Kulturgeschichte*, Julius Springer, Berlin, 1919, p. 601.

41. Aristotle, *Meteorology IV*, 8, trans. E.W. Webster, http:// classics.mit.edu/Aristotle/meteorology.4.iv.html, accessed August 2020.

42. R. Krauss, *Jahrbuch Preussischer Kulturbesitz für, 1985*, 1986, vol. 22, pp. 171–183.

43. Vitruvius, *De Architectura.* VII, 8, trans. M. H. Morgan, Harvard University Press, Cambridge, MA, 1914, https:// www.gutenberg.org/files/20239/20239-h/20239-h. htm#Page_215, accessed August 2020.

44. E. R. Caley, *J. Chem. Educ.*, 1926, **3**(1), 1149–1166.

45. P. L. Higueras Higueras, L. Mansilla Plaza, S. L. Álvarez and J. M. Esbrí Victor. The Almadén mercury mining district, in *History of Research in Mineral Resources*, ed. J. E. Ortíz, O. Puche, I. Rábano and L. F. Mazadiego, Cuadernos del Museo Geominero, 2011, vol. 13, pp. 75–86.

46. O. Puche Riart, Heritage values of Almadén and its setting, in *International Congress: Mining and Industrial Heritage: Its Impact on Major Cultural Routes of Universal Value. The Mine of Almadén and Other Mining Sites Linked to the Intercontinental Spanish Royal Road through The Mercury Route*, ICOMOS-Spain, Madrid – Almadén, 12–18 November 2006.

47. J. C. Edmondson, *J. Rom. Stud.*, 1989, **79**, 84–102.

48. J. Sloane, Mercury: Element of the ancients, *Dartmouth Toxic Metals*, https://sites.dartmouth.edu/toxmetal/mercury/mercury-element-of-the-ancients/, accessed August 2020.

49. A. N. Goring-Morris and L. K. Horwitz, *Antiquity*, 2007, **81**, 902–919.
50. D. S. Çamurcuoğlu, The wall paintings of Çatalhöyük (Turkey): materials, technologies and artists, PhD Thesis, Institute of Archaeology, University College London, London, 2015.
51. G. Chic García, Estrabón y la práctica de la amalgama en el marco de la minería sudhispánica: Un texto mal interpretado, in *La Bética en su problemática histórica*, ed. C. González Román, Universidad de Granada, Granada, 1991.
52. M. Castillo Martos, Use of Mercury (Quicksilver) from Almadén in the Amalgamation of Precious Metals, in *International Congress: Mining and Industrial Heritage: Its Impact on Major Cultural Routes of Universal Value. The mine of Almadén and Other Mining Sites Linked to the Intercontinental Spanish Royal Road through The Mercury Route*, ICOMOS-Spain, Madrid – Almadén, 12–18 November 2006.
53. M. Bargalló, *La minería y la metalurgia en la América española durante la época colonial*, Fondo de Cultura Económica, México D.F., 1955.
54. W. R. Newman, *Ambix*, 2014, **61**(4), 327–344.
55. M. Castillo Martos, Alberto Magno: precursor de la ciencia renacentista, in *La ciencia de los filósofos*, "Themata" no. 17, 1996.
56. J. A. Prior Cabanillas, *La Pena de Minas: Los Forzados de Almadén, 1646-1699*, Universidad de Castilla La Mancha, Almadén, 2003.
57. Heritage of Mercury: Almadén, Idrija, 2011. Nomination File, UNESCO, https://whc.unesco.org/en/list/1313/documents/, accessed July 2020.
58. Quartulus, el "Niño minero;" Museo Arqueológico Nacional. Inv. 16744. Photo: Miguel Ángel Otero, http://www.historiaclasica.com/2007/05/quartulus-el-nio-minero.html, accessed June 2021.
59. Th. Stöllner, Mining and economy. a discussion of spatial organisations and structures of early raw material exploitation, in *Man and Mining. Studies in Honour of Gerd Weisgerber*, ed. Th. Stöllner, G. Körlin, G. Steffens and J. Cierny, Ruhr-Universität, Bochum, 2003, pp. 415–446.

60. E. Darwin, *The Botanic Garden*, Jones & Co., London, 1825, Canto II, L, pp. 405–406. As quoted in C. C. Gaither and A. E. Cavazos-Gaither, *Chemically Speaking*, Institute of Physics Publishing, Philadelphia, 2002.

61. Font de Mercuri, https://commons.wikimedia.org/w/index.php?title=File:Font_de_Mercuri.JPG&oldid=416478756, accessed June 2021.

62. P. Levi, *The Periodic Table*, trans. R. Rosenthal, Schocken Books, New York, 1984, p. 36.

63. J. L. Marshall and V. R. Marshall, *The Hexagon of Alpha Chi Sigma*, 2012, vol. 103((1),), pp. 4–9.

64. I. Fiedler and M. A. Bayard, Cadmium yellows, oranges and reds, in *Artists' Pigments: A Handbook of Their History and Characteristics*, ed. R. L. Feller, National Gallery of Art, Washington, DC, 1986, vol. 1, ch. 5, pp. 65–108.

65. F. Stromeyer, *Ann. Philos.*, 1819, **14**, 269–274.

66. G. H. Bachhoffner, *Chemistry as Applied to the Fine Arts*, J. Carpenter, London, 1837, p. 117.

67. E. A. Parnell, *A Practical Treatise on Dyeing and Calico-Printing,* Harper, New York, 1846, p. 429.

68. P. J. Curtis and R. B. Wright, *J. Oil Colour Chem. Assoc.*, 1954, **37**, 26–43.

69. E. L. Moore, Cadmium/mercury sulfides, in *Pigment Handbook*, ed. T. C. Patton, Interscience, New York, 1973, vol. 1, pp. 395–399.

70. J. N. Friend, *An Introduction to the Chemistry of Paints*, Longmans, Green, London 1910, pp. 38–39.

71. W. K. Kelley, Cadmium colors in coatings industry, *Am. Paint J. Conv. Dly.*, 1970, **55**(15), 22–23.

72. A. P. Laurie, *The Pigments and Mediums of The Old Masters*, Macmillan, London, 1914, p. 16.

73. C. Monet, Meules, https://commons.wikimedia.org/w/index.php?title=File:Monet_grainstacks_W1273.jpg&oldid=355118603, accessed June 2021.

74. J. L. Marshall and V. R. Marshall, *The Hexagon of Chi Alpha Sigma*, 2003, vol. 94(1), pp. 3–8.

75. G. Agricola, *De Natura Fossilium*, Forben Press, Basel, 1546, p. 347.

76. M. E. Weeks, *J. Chem. Educ.*, 1932, **9**, 22–30.

77. R. J. Gettens and G. L. Stout, *Painting Materials: A Short Encyclopedia*, Dover Publications, New York, 1966, p. 159.

78. A. Zucchiatti, A. Bouquillon, I. Katona and A. D'Alessandro, The 'Della Robbia Blue': A case study for the use of cobalt pigments in ceramics during the Italian Renaissance, *Archaeometry*, 2006, **48**, 131–152.

79. B. Mühlethaler and J. Thissen, Smalt, in *Artists' Pigments: A Handbook of Their History and Characteristics*, ed. A. Roy, National Gallery of Art, Washington, DC, 1993, vol. 2, pp. 113–130.

80. M. Bacci, D. Magrini, M. Picollo and M. Vervat, *J. Cult. Herit.*, 2009, **10**, 275–280.

81. A. Roy, Cobalt blue, in *Artists' Pigments: A Handbook of Their History and Characteristics*, ed. B. H. Berrie, National Gallery of Art, Washington, DC, 2007, vol. 4, pp. 151–177.

82. B. Berrie, Mining for color: new blues, yellows and translucent paint, in early science and medicine, *Early Sci. Med.*, 2015, **20**, 308–334.

83. R. Harley, *Artists' Pigments: 1600–1835*, Butterworths, London, 1970, p. 53.

84. National Portrait Gallery: British Artists' Suppliers, https://www.npg.org.uk/research/programmes/directory-of-suppliers/r.php?searched=tim+smit&advsearch=all-words&highlight=ajaxSearch_highlight+ajaxSearch_highlight1+ajaxSearch_highlight2, 1650–1950, accessed August 2020.

85. E. Taggart, The history of the color blue: from ancient Egypt to the latest scientific discoveries, https://mymodernmet.com/shades-of-blue-color-history/, accessed August 2020.

86. Winsor and Newton, Cerulean Blue, http://www.winsornewton.com/na/discover/articles-and-inspiration/spotlight-on-colour-cerulean-blue, accessed August 2020.

87. K. St Clair, *The Secret Lives of Color*, Penguin Books, New York, 2017, p. 205.

88. N. W. Fischer, *Ann. Phys.*, 1831, **21**, 160–163, series ii.

89. R. Mayer, *The Artist's Handbook of Materials and Techniques*, Viking Press, New York, 3rd edn, 1970, p. 49.

90. M. Cornman, Cobalt yellow (aureolin) in *Artists' Pigments: A Handbook of Their History and Characteristics*, ed. R. L. Feller, National Gallery of Art, Washington, DC, 1986, vol. 1, pp. 37–46.

91. M. Comstock, *J. Surf. Coat. Aust.*, 2016, December, 10–29.

92. J.-N. Berte, Cerium pigments, in *High Performance Pigments*, ed. E. B. Faulkner and R. J. Schwartz, Wiley-VCH, Weinheim, 2009, pp. 27–40.

93. J. L. Marshall and V. R. Marshall, *Rediscovery of the Elements*, ISBN 978-0-615-30793-0.

94. J. L. Marshall and V. R. Marshall, Northern Scandinavia: an elemental treasure trove, in *Science History, a Traveler's Guide*, ed. M. V. Orna, American Chemical Society, Washington, DC, 2014, pp. 209–257.

95. M. Fontani, M. Costa and M. V. Orna, *The Lost Elements: The Periodic Table's Shadow Side*, Oxford University Press, New York 2015, p. 171.

96. *Episodes from the History of the Rare Earth Elements*, ed. C. H. Evans, Springer, Heidelberg, 1996, pp. 37–65.

97. R. Adunka and M. V. Orna, *Carl Auer von Welsbach: Chemist, Inventor, Entrepreneur*, Springer, Cham, 2018, p. 17 ff.

98. Y. Xie, Z. Hou, R. J. Goldfarb, X. Guo and L. Wang, *Rev. Econ. Geol.*, 2016, **18**, 115–136.

99. A. Stwertka, *A Guide to the Elements*, Oxford University Press, New York, 1996, p. 156.

100. H. Aldersey-Williams, *Periodic Tales*, Penguin Books, London 2011, pp. 160–171.

101. M. V. Orna, Paris, a scientific theme park, in *Science History, a Traveler's Guide*, ed. M. V. Orna, American Chemical Society, Washington, DC, 2014, pp. 135–158.

102. M. Curie, *C. R. Acad. Sci., Ser. II: Mec., Phys., Chim., Sci. Terre Univers*, 1898, **126**, 1101–1103.

103. M. Rayner-Canham and G. Rayner-Canham, *Women in Chemistry: Their Changing Roles from Alchemical Times to the Mid-Twentieth Century*, American Chemical Society, Washington, DC, 1998, pp. 99–100.

104. A. J. Ihde, *The Development of Modern Chemistry*, Dover Publications, New York, 1984, pp. 487–489.

105. M. V. Orna and M. Fontani, Discovery of the actinide and transactinide elements, in *The Heaviest Metals: Science and Technology of the Actinides and Beyond*, ed. W. J. Williams and T. P. Hanusa, John Wiley and Sons, New York, 2019, pp. 3–30.

106. Making Paint with Minerals, https://edu.rsc.org/resources/making-paint-with-minerals/1590.article, accessed October 2020.

CHAPTER 6

Purveyors of Purple: The Oceans' Gift to the World of Color[†]

Purple puts us in touch with the part of ourselves that is regal.[1]

Byllye Avery

6.1 INTRODUCTION

The world's first major industry was founded on the minuscule drops of a purple secretion found in a lowly snail. The color soon became the stuff of myth, magic and mystery. Identified with royalty, it was the most coveted color in history. For many, it still is.

6.2 PURPLE IN MYTH, MAGIC AND MYSTERY

No other color in history has been so coveted or so condemned as the color purple. The Roman troops marched lockstep under purple banners, while at the same time, thousands of slaves toiled in stench-filled vats covered in slime. Fine Senators outfitted with purple-trimmed togas strutted to the Forum; the dye works themselves were a foul-smelling blight upon the

---

[†]Glossary entries and chapter cross-references are in boldface.

March of the Pigments: Color History, Science and Impact
By Mary Virginia Orna
© Mary Virginia Orna 2022
Published by the Royal Society of Chemistry, www.rsc.org

landscape. Purple dye was the status symbol *par excellence*, and it was also the two-edged sword that cut through the essence of a society and laid bare its injustices and inadequacies. Yet the aura of myth, magic and mystery has lingered through the centuries: purple holds sway even today.

6.2.1   Purple in Myth

To the ancient Greeks, purple was the color of power, purpose, and personality. A purple garment was like a suit of armor and a badge of nobility, even of divinity in many cases, like those of Perseus and Theseus.[2] When Danaë and her infant son, Perseus, were tossed exiled into the sea, she hoped for rescue by the infant's father, Zeus, who would recognize Perseus by the purple cloth that covered him. Similarly, when Theseus had to prove his divine origin as son of Neptune to Minos, King of Crete, he did so swathed in a mantle of purple.[3] Why was purple so special in ancient times? Perhaps because it has an amazing story, starting life in a lowly snail that somehow became the most luxurious and coveted shade in history.

The primal tale of a past mythic time has Hercules, or Melqart, strolling along the beach near Tyre with his dog, Cerberus, newly ransomed from Hades. Cerberus, literally the dog from hell, scarfed up a bunch of sea snails that abounded along the coast, staining his mouth with a brilliant purple color (Figure 6.1).[4] History-debunker Jacob Bryant[5] attributes this story to the ancient practice of crediting every useful invention to the gods. For example, Ceres bestowed bread and corn on humanity. Every deity was looked up to as the cause of some blessing. The Tyrians and Sidonians were famous for the manufacture of purple, a very exquisite dye. Hercules, their favorite son, was deemed the discoverer, but some insisted that the dog did the discovering. Bryant debunks this idea too. He says that anyone who knows anything about these snails knows the "shell has strong and sharp protuberances with which a dog would hardly engage," concluding, "this story is founded upon the same misconception, of which so many instances have been produced in mythology and elsewhere."

Because of its devilishly difficult and costly production, the dye/pigment extracted from these purple-bearing snails has

Figure 6.1   The Discovery of Purple. Theodoor van Thulden (1606–1669). Museo del Prado. Reproduced from ref. 4.

always enjoyed great social and economic value. Known variously by such grandiose names as Tyrian purple, Royal Purple, and Imperial Purple, it was also highly esteemed because of its iridescent shine and its fastness to light, washing, and rubbing.[6] For this reason, its manufacture was strictly regulated both by law and/or economics. The first instance of such a law in ancient Rome, the 215 BCE *Lex Oppia*, was designed to regulate the Romans' "frantic passion for purple."[7] Instead, it had the opposite effect: it simply raised purple's prestige and the general lust to possess it.[8] Let's take a closer look at the color purple, how this little chemical factory we call a snail produces it, and how human beings process and use it. An incredible journey!

6.2.2   The Magic of Purple

The human eye is a marvelous instrument. It can simultaneously transmit numerous wavelengths of light to the brain, which then integrates these data to interpret from them a single reported color. For example, simultaneous red, orange, yellow and green light waves entering the eye report out as yellow. When it comes to the turn of red and blue, they report out as purple. Artists and colorists have long known that to make purple, they need only combine a red pigment and a blue pigment in the proper proportions. The problem, of course, is that the new combined pigment is duller than either of its components because both pigments absorb two-thirds of the visible spectrum (**Chapter 1.5**). Very little light is being reflected. No wonder that the purple pigment mix is dull, whereas a true purple pigment, one that reflects both the red and the blue regions and only absorbs the green region is much, much brighter because lots of reflection and little absorption is taking place. Up until the mid-19th century CE, only the purple derived from shellfish could fill that role.

Please note that throughout this chapter, this purple is referred to as either a dye or a pigment. The term depends upon its use. A dye is usually a water-soluble material capable of being absorbed into the interiors of fibers to impart color; a pigment is a finely divided water-insoluble solid material that adheres to a surface through the action of a binder in which it is insoluble: this purple has seen life in both of these capacities. "Colorant" is the more general term that embraces both of these designations. In addition, as a pigment, except for the blue, plant-derived colorant called woad (**Chapter 3.8**), this purple pigment is the first we have encountered extracted from a living species and consists mainly of carbon atoms arranged in a particular structure; such chemical compounds are termed "organic." We will learn more about how the atoms are put together to make this pigment in **Chapter 12.5.2**.

6.2.2.1   Source of the Color.   The color itself resides as a colorless precursor in shellfish of several genera and almost 40 species designated at one time, for simplicity, under the collective term *murex*. The major purple dye mollusks were *Hexaplex trunculus*, once also called *Murex trunculus* or the banded dye murex, *Stramonita haemastoma* (formerly *Thais haemostoma*) and *Bolinus brandaris* (formerly *Murex brandaris*) – now collectively known as muricids, or historically, murex.[9,10] Although taxonomically, these mollusks are referred to as muricids, the older common and popular name for them is murex, a term we will use throughout the rest of this chapter. They are residents of the coastlines of the Mediterranean Sea, Atlantic Ocean, Caribbean Sea, the Sea of Japan and other tropical to semi-tropical seas around the world.

Among his many firsts, Aristotle described the anatomy and habits of the "purple murex."[11] To extract the dye from the shellfish, it was necessary to harvest each one individually by smashing the shell and dissecting the hypobranchial gland (Figure 6.2)[12] to remove the yellowish mucus inside. When placed in the sun, this fluid would magically change color from light green to sea green to dark green to bluish and finally, after an hour or two, to the desired purple color.

Following a multi-century hiatus of murex purple knowledge (Section 6.4), the first person to describe this sequence was William Cole of Bristol in 1681 when he noticed the color change in a mollusk native to British shores. Full of glee, Cole announced his discovery to the world.[13] Then in 1832, Bartolomeo Bizio (1791–1862), a Venetian chemist, re-discovered the colorant's marine source and realized that it was a brominated form of the plant colorant, **indigo**.[14] Bizio's 30 year-long ground-breaking studies were casually overlooked by the scientific community, possibly because he published in a little-known northern Italian journal. His re-discovery included: (a) the recognition of the source of Tyrian purple from two marine gastropod species, *Hexaplex trunculus* and *Bolinus brandaris*; (b) light plus oxygen activated the formation of purple from the mucus of the hypobranchial gland, but heat had no effect; (c) a foul-smelling sulfur-based compound emitted during the process; (d) a species-specific color difference between the two snails; (e) large amounts of copper in snail ashes, leading eventually to the discovery of haemocyanin, the first-known copper protein (Box 6.1).

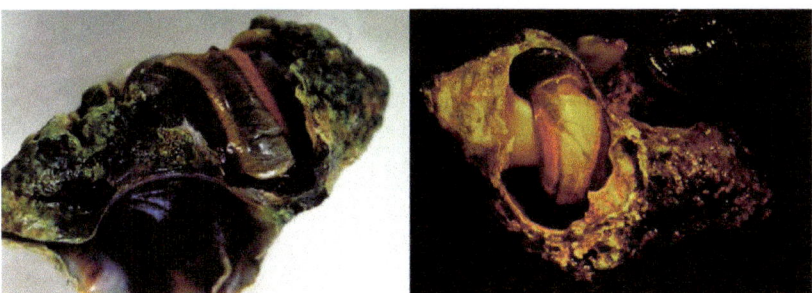

Figure 6.2   Two *Hexaplex trunculus* mollusks with their shells cracked open to expose their hypobranchial glands: (left) gray vein, (right) off-white vein. Reproduced from ref. 12 with permission from American Chemical Society, Copyright 2013.

Box 6.1   An Ecological Approach to Shellfish Dyeing

Edward Schunck, a gentleman-chemist of the 19th century, regaled members of the Chemical Society with a traveler's narrative of the purple dye works in Nicaragua: "The colour most valued is the Tyrian purple, obtained from the murex shellfish, which is found on the Pacific coast of Nicaragua... The process of dyeing the thread illustrates the patient assiduity of the Indians... Each shell is taken up singly, and a slight pressure upon the valve which closes its mouth forces out a few drops of the colouring fluid, which is then almost destitute of colour. In this each thread is dipped singly, and after absorbing enough of the precious liquid, is carefully drawn out between the thumb and finger, and laid aside to dry. Whole days and nights are spent in this tedious process... The fish is not destroyed by the operation, but is returned to the sea where it lays in a new stock of colouring matter for a future occasion."[15] A model of ecological responsibility: No shells, no waste, no stench, preservation of a renewable resource.

6.2.2.2   The Magic Transformation.   The little gland contains water-soluble forms of a bromine-containing substance called tyrindoxyl sulfate. The bromine has been selectively concentrated from seawater despite the almost 300-fold predominance of chlorine by weight; why and how this is done remains part of the mystery of this marvelous pigment. It was once thought that bromine contributed to the toxic effects of the secretions used in predation, but it is now recognized that this is not the case.[16] Alternatively, it has been suggested that the bromine's function is hormonal, analogous to the known role of iodine in human thyroidal activity.[17] The biological purposes of many of the other exotic compounds produced by the hypobranchial are also very poorly understood. Some hypotheses include anti-predation, anesthetizing, detoxification and pheromonal functions.[18]

Deliberately cracking the snail's shell and simultaneously piercing the hypobranchial gland initiates **purple dye formation**. This mechanism releases the **enzyme** aryl sulfatase (aka purpurase), which, by enzymatic action, ruptures the tyrindoxyl sulfate linkage and initiates a series of photochemical air oxidation steps that form intermediates leading to the greenish tyriverdin. At the same time, the process liberates the sulfur component as odoriferous compounds, such as dimethyl disulfide, $(CH_3)_2S$, related by the presence of sulfur to compounds that cause "bouquet of skunk."[12,19] The end results are compounds related to the blue **indigo** (IND) pigment synthesized by the *Indigofera* genus of plants.

The only difference is that they may contain one or two bromine atoms, as first determined by Paul Friedländer,[20,21] who identified the 6,6'-dibromoindigo (DBI) colorant by preparing about 1.4 g of it from 12 000 snails. He also suspected the presence of a more soluble colorant;[22] 70 years later, a more soluble red-blue component, 6-monobromoindigo (MBI), was found.[23] Hence, departing from IND (which is blue), we have MBI, and DBI with bluish-purple and reddish-purple hues. The bromine atoms' presence shifts the color from blue toward purple.[24] These are the main processes of color formation. There are several other colorants produced that vary from species to species of the shellfish, among them IND and indirubin, depending also upon the aquatic environment in which they are generated and develop. The overall final products confer the perceived color on the dye.[25,26]

6.2.2.3 Production of the Dye. Tyrian purple production was one of the mainstays of the Phoenician economy. Tyre's expeditions ended up founding such famous cities as Utica and Carthage where they set up factories and trading stations – initiating the world's first major industry about 1500 BCE.[27] However, since there is archaeological evidence of purple usage elsewhere in the Mediterranean since the middle Bronze Age,[28] many archaeologists think the Phoenicians merely brought the art to its technical perfection.[29,30] Zvi Koren has analyzed the archaeo-chemical evidence and posited that while the craftspersons in the Aegean and nearby areas discovered the purple pigment for use in wall paintings about four millennia ago, it was the Phoenicians who were the first to successfully chemically convert the pigment to a water-soluble dye for textile dyeing about five centuries later.[12]

Obviously, purple dye production was a harsh and gritty business, leaving thousands of discarded rotting mollusks piled in heaps along the shoreline. Described as hideous and abominable,[31] more than one Roman writer has compared the stench with the old shoes of a veteran soldier or from an amorous goat.‡ It is no wonder that sites with remains of the dye industry have all been found downwind from population settlements.[32] As Victoria Finlay trenchantly remarked, "the beauty of the color is in the effect, not the process."[33]

---

‡Common chemical names for three particularly unpleasant-smelling organic acids produced by goats in rut are caproic, caprylic and capric acids, from *caper*, Latin for goat.

6.2.3   The Mystery of Purple

How the purple was made was the purview of a select few; the only written recipe was that of Pliny the Elder, and even that was so vague that conjectures are still being made about what he meant. But the mystery deepens when, in the process of making purple, it was also possible to make many shades of it, as well as a blue-purple or violet color, *tekhelet*, and a red-purple, *argaman*, highly prized by Jewish color makers. The maddening part of the mystery was the irreproducibility of the color: sometimes it was reddish, sometimes bluish, all from the same type of shellfish. It took modern instrumental methods of chemical analysis to make a dent in this tough conundrum.

6.2.3.1   Processing Purple.   The simplest way to deal with purple is by direct coloring, namely by just rubbing the cloth to be colored with the mucus from the snail's hypobranchial gland.[34] Much better results come from the complicated practice of **vat dyeing**. The actual processing of the purple product was not simple. Details of production were a closely guarded secret, penetrated by Pliny, but not quite.[35] First, we must distinguish between the insoluble variety, the purple pigment, and the near-colorless, water-soluble, **reduction** product, called the *leuco*-form, suitable as a purple dye.

Vat dyeing is a carefully controlled process ensuring that all the components remain dissolved in the dye bath prepared to receive the textile to be dyed. The start involves stripping snail flesh from live large snails, along with the crucial hypobranchial gland, and steeping the mashed up bodies in a vat filled nearly to the brim with liquid, presumably seawater, to which, according to Pliny, some "salt" had been added. For small snails, everything went into the vat, shell and all. Since it is vitally necessary to keep the colorless precursors as the *leuco* forms, the "salt" should be one that keeps the whole mix alkaline – the precursors would precipitate out as solids in acid solution.[36] Common alkaline "salts" available in antiquity were lime and natron[§] (**Chapter 4, Box 4.1**). The **reducing agent** needed to maintain solubility may have been tin or even some of the odiferous sulfur compounds

---

[§]Natron was readily available as a common Egyptian export that had many lives: as a bleach, detergent, mouthwash, medicine, mummifier and dye-bath additive.

already present.[34] Another possibility, honey, was mentioned by Plutarch,[37] which prompted some modern experimental archaeologists to use glucose from grape juice instead, but only to prevent re-**oxidation**.[38] The vat contents should be kept closely covered and heated, but not boiled, for three days, a step necessary for the fermentation enabled by naturally occurring bacteria that function in an absence of air. Hence, the covering keeps out air as well as the sunlight that might hasten the production of the insoluble colorants. Mild heating could take place by piping very hot water or placing charcoal embers around the outside of the dye vat. An average-sized vat would have contained about 100 L.[17] Koren carried out this process numerous times in a laboratory equipped with the requisite instrumentation to monitor (and thereby confirm) the ancient process.[39] He found that the snails themselves were a self-contained, built-in, all-in-one system for the dyeing process; the only other necessities were a lukewarm temperature and a substance to maintain an alkaline dye bath.

6.2.3.2 Dyeing with Purple.   Once the dye vat is ready, about 6 to 9 days after initiation of the biochemical **reduction** process, a clean woolen fleece is typically dipped into the bath and allowed to steep for about 5 hours. In a second dip, the fleece will absorb almost all of the rest of the dye, and then the easiest part of the process takes place: hanging the fleece out to dry so that the purple color develops through air **oxidation**. Following the development of the purple, the fleece can be immersed a third time to try to exhaust the remainder of the leuco-dye in the dyebath. As far as we know, dyeing textiles with this process is the first industrial **oxidation–reduction** process in history.

Of particular interest is the production of the so-called "biblical" shades, red-purple *argaman*, now accepted to be chiefly DBI, and blue-purple *tekhelet*, chiefly IND, with a little DBI and MBI, both produced from *Hexaplex*-related shellfish. Since the end of the 19th century, investigators generally accepted that *Hexaplex trunculus* is the source of the ancient dye as discussed by Ziderman.[40]

In 1985, Otto Elsner and Ehud Spanier reported on their findings regarding the sex-dependence of the dyestuffs in specimens of *Stramonita haemastoma* taken from Haifa Bay, Israel.[38] They found that the male glands contained predominantly blue

dye-producing indoxyl dye precursors, while female glands pre-
dominated in bromoindoxyls, precursors for purple dyes, while
others have found the opposite to be the case. The dye mix
obtained will contain all three dyestuffs, IND, DBI, and MBI,
in varying proportions depending, among other things, on the
ratio of males to females in the shellfish catch used.[41] As if this
is not complicated enough, in a challenge to "Ripley's Believe It
or Not," male shellfish are latent hermaphrodites, undergoing
spontaneous sex changes, thus throwing the indigo/bromoin-
digo ratio more toward the purple side. Furthermore, females,
undergoing a biological process called imposex, can develop
the anatomic male sexual apparatus but continue to operate as
females. Compared with the glands of the shapeshifting females,
male glands seem to contain less of the completely unbromi-
nated indoxyl precursor, yielding upon air **oxidation** more blue
indigo and MBI.[42,43] The resulting perceived dye color on the textile can depend on several other factors including fiber affinity
and illumination. Another complicating factor is MBI's reaction
to heat: even mild heating to about 60 °C can change its color
from violet to blue,[44] not by expelling the bromine component,
but presumably by distorting the crystal structure. Is it any wonder that the ancient dyers had difficulty controlling the color
as well as explaining it? We, at least, have some biological and
chemical theory to rely on; the ancients had only experience and
tradition.

6.2.3.3   The Mystery Continues.   It turns out that not all murex,
not even of the very same genus and species, are alike. This fact
came to the fore in the so-called Jewish Snail Debate. Several
"schools of thought" arose regarding the production and true
color of *tekhelet*, including one group that subscribed to a *tekhelet*
that turned out to be identified as Prussian Blue (**Chapter 11.3**),
an 18th century synthetic pigment. Another school specified
the exact shade as a light blue, but ran into difficulty when con-
fronted with the problem of mandating any particular shade
derived from natural sources.[45] As Zvi Koren remarked, "when
performing natural dyeings, it is very difficult to obtain exact
reproducible results with natural materials. The fact that I claim
that the original tekhelet was blue-purple does not of course
indicate the exact shade of bluish-purple or purplish-blue."[46] The

scientific reason for this claim that also underscores the hopelessness of ever determining the "true" color as described in an ancient document is that every snail contains a different amount of the colorants and, indeed, a different number of colorants. For example, *H. trunculus* collected in Spain contained as many as nine pigments, whereas the same species collected in northern Israel contained only five. Furthermore, the different percentages of these pigments in the samples yield different perceived colors for the overall mixtures.[47] In another example, *H. trunculus* snails harvested from Dor beach, Israel, located about 50 km south of Akhziv contained the same four colorants as those in the Akhziv sample, but in entirely different proportions![48] It is no wonder that the contending groups cannot agree – the wonderful variety in Nature is against them. However, there is one incontrovertible fact that is definitive for a true murex purple dye: it must contain DBI, the signature component. Another important chemical marker is MBI, a compound that is only present in appreciable quantities in a single species, *Hexaplex trunculus*. This chromatic bio-marker is present in all archaeological purple artifacts analyzed to date, indicating that all murex purple dyeings done in antiquity used *H. trunculus*, sometimes alone and sometimes mixed with other species (Box 6.2).[49]

Box 6.2   Murex Chemistry 101

Marine gastropods are tiny little chemical factories that produce numerous molecules for the purposes of sexual attraction, feeding, warfare, and reproduction, among other activities. One very interesting example is the presence of the choline ester murexine, synthesized along with the precursors of the Tyrian purple color in the hypobranchial gland. When a murex snail sets out to feed on a bivalve, it injects its chemical warfare agent, murexine, a powerful muscle relaxant, into its prey, rendering it helpless. In the egg-laying stage of the murex, it transfers the murexine into the egg capsules where it tranquilizes the larvae and relaxes the reproductive tract. At the beginning of the reproductive season, these species, signaled by chemical pheromones, form mounds of hundreds of individuals that spawn together to produce an egg mass with a purple hue – these little snails are born to the purple! They are also predators themselves as "embryonic cannibals," and as adults, they use another biosynthesized chemical, hydrochloric acid, HCl – at the tip of the boring instrument, it is very effective in disintegrating other snails' shells, which are all made of calcium carbonate, $CaCO_3$.

6.2.4   All Those Shells!

Piles of murex shells, middens, are found all around the Mediterranean and, not surprisingly, there are heaps of them in the Beirut, Tyre and Sidon areas, often filling large holes cut in solid sandstone rock. It was William Wilde, an Irishman, a scientific traveler and a keen observer, who drew the world's attention to these archaeological remains in his memoir published in Dublin in 1840.[50] These shells, remains of the purple dye industry, are valuable relics that can reveal the kinds of murex used for this endeavor, and there is a lively research initiative in this area. Figure 6.3 is an image of the three most important of the murex dye shellfish,[51] *Bolinus brandaris,*[52] *Hexaplex trunculus*[53] and *Stramonita haemastoma.*[54] Many of the shells are also found in the vicinity of kilns, suggesting that lime production was a by-product of the dye industry.[27] Large quantities of shells were found at an inland site in Crete lying near a "stove," suggesting that they were transported inland for the purpose of lime manufacture.[55] Shells were also used as subfloor packing at some sites, presumably for drainage; this usage was common in Crete.[56] However, the mere existence of the shells does not necessarily mean that they were used exclusively for dye and lime; once the poisonous component, the hypobranchial gland, was removed, the snails would make a perfectly good food.[57]

Figure 6.3   The three most common purple-producing murex snails inhabiting the Mediterranean (from left to right): *Bolinus brandaris,* *Hexaplex trunculus* (also commonly known as *Murex trunculus*) and *Stramonita haemastoma.* Reproduced from ref. 52–54 under the terms of the CC BY-SA 3.0 license, https://creativecommons.org/licenses/by-sa/3.0/deed.en.

6.3   USES OF MUREX PURPLE

Obviously, the most important use of murex purple throughout history has been as a dye, However, regardless of the high price of manufacture, its first use seems to have been as a pigment, not as a dye.

6.3.1   Murex Purple Pigment

To prepare murex purple for use as a pigment, it is necessary to collect a significant number of shellfish and extract the dye precursor from the hypobranchial gland, a very laborious process. Unlike dye processing where the gland and the snail flesh and even the shells can be mixed up in the dye vat, that precious purple drop must be harvested individually from each snail and, as we know, from thousands of them to make up just a small amount of pigment. Once freed of shellfish debris, and washed, filtered, dried and ground up, it can be applied to any surface using a binder such as egg white or gum Arabic (but egg yolk and oil should be avoided because they will reduce the shade to a humdrum grayish-blue).[58] A fine mixture of the purple pigment with chalk or clay, referred to by Pliny as *purpurissum*, was also used as a paint pigment, but very rarely.

The earliest confirmed use of murex purple as a pigment comes from sample analysis of wall paintings at Akrotiri on the island of Santorini (Thera).[59] The frescoes at Akrotiri depict priestesses and noble ladies with red-purple and purple stripes on their dresses.[60] These wall paintings, dated to the 18th to 17th centuries BCE, may be the world's oldest known pieces of art with traces of the famous ancient purple. It has also been identified as the purple pigment on pottery found at Kastri on Kythera, another Greek island.[61] A bit of purple powder found in association with the Akrotiri wall paintings was rich in chalk-aragonite leading to the assumption that it was prepared by immersion of the chalk in a dye vat to prepare *purpurissum*.[27] However, there is, to date, no archaeological evidence that purple dyeing was ever done in the Aegean prior to the 14th century BCE. The only other instances, to date, of pigment use for Tyrian purple before the Christian era come from mainly wall paintings in Greece, for example, 4th century and 3rd century BCE Macedonia,[62] with one notable exception: a stone jar found in Israel.[63] Only isolated instances of

---

Box 6.3   Purple Mosaics

So, where did the magnificent violet color come from that is such an astounding feature of the mosaics in Imperial Roman times, especially those found in the basilicas and tombs of Ravenna? Fortunately, the collection of antiquities in the Martin von Wagner Museum in Würzburg holds about 250 samples of mosaic glass attributable to Roman origin. Analysis of these samples – a first for Roman mosaic glass – revealed that the violet component in every case was Mn^{3+} by the addition of Mn_2O_3. Depending on the content of Mn_2O_3, but also on the thickness of the glass, various shades of purple were observed, ranging from pink to violet and to dark purple with increasing Mn_2O_3 concentration: pink 1.7 wt%; violet 2.2 wt%; dark purple 3.6 wt%.[67]

---

murex purple use in manuscripts have been identified, notably in one in the Vatican Library called *Barb.Lat.* 570, British Isles, from the second half of the 8th century CE, on a single page in the ms. The likely source was the dog whelk (*Purpura lapillus*), a common shellfish on the coasts of Britain and Ireland.[64] Another instance was the purple in the 6th century Syrian manuscript, the *Vienna Genesis* (Codex Vindobonensis Theol.gr.31).[65] The only recorded use of the purple in easel painting is on the banners on either side of the Virgin Mary in the painting, *The Madonna and Child*, by the Master of St. Ceciila (1290–1295) at the J. Paul Getty Museum (Box 6.3).[66]

6.3.2   Use of Murex Purple as a Dye

Two archaeological sites in Syria, Tell Mishrife and Chagar Bazar, yielded up some of the oldest textile fragments dyed with murex purple, dated to the 18th to 16th centuries BCE. A more recent, but unexpected surprise came when, during the excavation of the Herodian fortress of Masada in Israel, a refuse dump contained the first murex purple dyed textile discovered in Israel. It is also one of the oldest fabrics ever found dyed with this colorant.[68] A personal description of this event and a color photograph of the sample appeared in 2006.[69] Purple-dyed fabrics gradually became widely available due to the technical expertise of the Phoenicians who had the market cornered for almost two millennia. And lest we think that these fabrics were used only for clothing, there is some evidence of purple's medicinal properties as well. For example, Marcellus Empiricus, physician of Bordeaux (4th – 5th

Century CE), in his *De Medicamentis*, says that the blood of a green lizard caught on wool and then wrapped in a purple cloth will be effective in preventing eye diseases.[70]

6.4   THE PURPLE DYE MARKET

Purple, mainly as dyed textiles, was a highly profitable and greatly valued trading commodity in the Mediterranean region. The very identity of the Phoenicians as the ultimate purple purveyors is found in their name: the Greek word "phoenix," (φοϊνιξ) means "red-purple." The market itself was highly complex since consumer selection depended on shade, price, and social value. Not only were purple-dyed materials or purple-pigmented objects on the market, but there was also a lively exchange or migration of the highly-skilled craftsmen who plied their trade either as freemen or as slaves, or were transferred between cities in order to expand influence. And, as with any other highly-valued goods throughout history, there were dupes and fakes galore. If, for example, a Roman could purchase an undyed fleece for 175 denarii per pound, whereas the same piece dyed with Tyrian purple would cost a staggering 16 000 denarii, what could stop an unscrupulous dealer from trying to sell a cheap substitute? Virtually nothing. Many, if not all, of the color-modifying variables were practically undetectable to even the expert consumer.[71]

Over time, the economic value of the fabric was superseded by its symbolic value or its use for political advantage. What mattered was the equation of the color with divinity or absolute power. By wearing the color alone, the owner became endowed with royalty, elevated from the masses. Its use conferred not only social status on the wearer but also protection and invulnerability. As the Roman Republic evolved into Empire, the government assumed ever greater control over the use (and abuse) of this very precious asset. By the time Justinian came to power in 527 CE, the wearing of the purple was restricted to the imperial family and purple dye workers became a hereditary labor caste regulated by the state. In the Late Roman Empire, creative use of plant-derived purple – and not from a shellfish source, but often orchil from various species of lichen – came to the fore, chiefly by dyeing specially prepared parchment for use as "prestige" manuscripts and Bibles (Figure 6.4).[72] A mythic dye was

Figure 6.4   The Codex Argenteus, a 6th century CE Gothic language manuscript. The pages are of high quality vellum stained with plant-based purple and inscribed with gold and silver lettering. Reproduced from ref. 72.

transformed into a mythic color, but never as bright as the true murex purple.[73,74]

6.5   THE DECLINE AND FALL OF MUREX PURPLE

After a long and successful history spanning two millennia, the murex purple dye industry in the Levant, the eastern Mediterranean landmass from southern Turkey to northern Egypt, fell victim to the Arab conquerors in 638 CE. Purple production and commerce continued in the Byzantine Empire, presumably aided by the purple dyers who may have fled into Asia Minor, which remained under Byzantine control.[39] Nevertheless, the great Jewish traveler, Benjamin of Tudela (*ca.* 1160) mentions observing red-purple dyeing activity in the vicinity of Tyre that seems to have survived into the 12th century.[75]

The final blow came on 29 May 1453 with the fall of Constantinople; on that day, the ancient art of purple dyeing became extinct, at least on an industrial scale. Any supplies that the West may have hoped for were completely cut off. Pope Paul II, in 1464, acknowledged this fact by decreeing that from thenceforth, cardinals' robes should be dyed with the red insect dye, kermes (*Kermes vermilio*).[76]

6.6 BORN-AGAIN PURPLE

Exactly 403 years after the demise of purple dyeing, it was miraculously reborn in a simple makeshift workroom by an 18 year-old student, William Henry Perkin. Emulating Isaac Newton who had discovered the key to the solar spectrum while on enforced vacation from school (it was a plague year), Perkin made his own discovery during Easter vacation in March, 1856. Perkin's genius was to derive from a crude black goo (coal tar, a by-product of the coke industry) a shimmering liquid that readily dyed silk the shade long lost to conquest: his immediate instinct was to name it "Tyrian purple." Continuing its life as "mauve," it went on to make history as the first of thousands of modern synthetic dyes, enabling the developments of such disciplines as pharmaceutical chemistry, medicinal chemistry, synthetic organic chemistry, and thermodynamics. Stay tuned to all of this in **Chapter 12**.

Meanwhile, many grams of DBI and its molecular relatives can now be synthesized in the modern laboratory at a small fraction of the time and effort that it took to obtain the natural product. However, if you thought that getting royal purple from rotten shellfish was a bit repulsive, how would you feel about mucking around with an ever-present inhabitant of human intestines and feces? Biochemists at Seoul National University have been able to coax *Escherichia coli*, that little bacterium that comfortably lives among us all, to do the devilishly hard job of manufacturing DBI, even getting the bromines into the right spots on the molecule. A little genetic engineering slotted three enzymes into the *E. coli*, enabling it to transform an essential amino acid, tryptophan, into DBI in three steps. By varying one of the enzymes, they could also orchestrate other distinctly colored pigments. A great advantage to this method is the lack of toxic by-products common to other synthetic colorants.[77,78]

But there has been little interest among artists in pursuing the synthetic track because there are so many other options now available to modern arts and crafts persons. Inge Boesken Kanold has been experimenting with Tyrian purple since 1978; it is now the keystone of her artistic career. Following the November 2000 (19th) meeting of *Dyes in History and Archaeology* (DHA) in Edinburgh and instructed by John Edmonds's book,[79] she reconstructed a purple fermentation vat using fresh murex from the Saturday market in Apt, France. She succeeded in dyeing some wool and some parchment, she learned how to "top off" the vat following a cooling down process, and she also warned of possible pitfalls using additives like salt and honey.[80]

A potentially useful property of DBI, visible **induced luminescence**, or VIL, promises to provide a non-invasive approach to the examination of historic artifacts containing indigoid colorants. Many museums have extensively used the photo-induced VIL property common to other important colors such as madder (**Chapter 10.3.6**) and Egyptian Blue (**Chapter 4.6.9**) to detect their presence even when they are not visible to the naked eye. In DBI's case, since as a pigment it would be mixed in not only with binders, but also with other indigoids, VIL may have the capability of easily and accurately picking it out of the crowd.[81]

Another exciting role for DBI lies in the organic electronics field. Research on natural or Nature-inspired materials for use in electronics applications such as RF-ID packaging tags and sensors embedded in commercial textiles (or human skin) has massive potential. Thin films of Tyrian purple have the ideal molecular orientation for forming semiconductor integrated circuits that would be completely biocompatible and biodegradable. As such, DBI is a prime candidate for air-stable organic electronic devices.[82]

6.7 CONCLUSION

If we have learned nothing else from this excursion into the life and times of a single color, it is to realize that an entire universe is contained in a simple biochemical process. Tyrian, or murex, purple was one of the most luxurious and coveted commodities of the ancient world. It meant life, death, and livelihood to many thousands of people for almost four millennia. Depending on supply, demand, politics, and the law, its symbolic value kept pace with and often outpaced its monetary value. But its real importance lay in its mysterious, almost mystical, attraction to the human imagination despite the fact that its origin was the lowly murex snail. Modern chemists have spotlighted it as a rich resource for studies in materials science and pharmacology; it has a permanent place in cutting-edge art historical and archae-ological research, and recent doctoral dissertations have focused on its socio-political, symbolic and religious connotations throughout history.

As influential as the Phoenicians were on Mediterranean life, it was the Romans who appropriated the entire region and imprinted on it their language, culture, and way of life. A cataclys-mic tragedy at the very high point of this civilization ensured that material evidence of its existence would be preserved for genera-tions to come. The next chapter reveals how this happened.

REFERENCES

1. B. Avery, *An Altar of Words*, Broadway Books, New York, 1998, p. 131.
2. W. Virtually, https://wandervirtually.com/the-story-of-perseus/, accessed August 2020.
3. G. W. Pieper, Conflict of character in Bacchylides' Ode 17, *TAPhA*, 1972, **103**, 395–404.
4. Theodoor van Thulden – The Discovery of Purple, https://commons.wikimedia.org/w/index.php?title=File:Theodoor_van_Thulden_-_The_Discovery_of_Purple.jpg&oldid=367321537, accessed June 2021.
5. J. Bryant, *A New System, or an Analysis of Ancient Mythology*, ed. J. Walker, *et al.*, Project Gutenberg, London, 3rd edn, 1807, https://www.gutenberg.org/files/19584/19584-h/19584-h.htm#footnote76, accessed August 2020.

6. J. Bridgeman, Purple dye in late antiquity and Byzantium, in *The Royal Purple and the Biblical Blue*, ed. E. Spanier, Keter Publishing House, Jerusalem, 1987, pp. 159–165.

7. Pliny the Elder, *The Project Gutenberg EBook of the Natural History of Pliny*, trans. J. Bostock and H. T. Riley, 2018, vol. 9, p. 60.

8. M. Pons, Tyrian purple: its evolution and reinterpretation as social status symbol during the Roman Empire in the West, MA thesis, Brandeis University, May 2016, p. 9.

9. I. I. Ziderman, *Rev. Prog. Color. Relat. Top.*, 1986, **16**, 46–52.

10. I. I. Ziderman, *Biblical Archaeol.*, 1990, **52**, 98–101.

11. Aristotle, *The History of Animals, Book IV*, trans. D. W. Thompson, http://classics.mit.edu/Aristotle/history_anim.4.iv.html, accessed August 2020.

12. Z. C. Koren, New chemical insights into the ancient molluskan purple dyeing process, in *Archaeological Chemistry VIII*, ed. R. A. Armitage and J. H. Burton, American Chemical Society, Washington, DC, 2013, pp. 43–67.

13. W. Cole, *Philos. Trans.*, 1685, **15**, 1278–1286.

14. F. Ghiretti, *Experientia*, 1994, **50**(9), 802–807.

15. E. G. Squier, *Nicaragua, its People, Scenery, Monuments*, vol. I, 1852, p. 286 as quoted in E. Schunck, J. Chem. Soc., 1880, 37, 613–617.

16. W. J. Clench, *Johnsonia*, 1947, **2**(23), 61–91.

17. L. Naegel and C. Cooksey, *J. Shellfish Res.*, 2002, **21**(1), 193–200.

18. P. E. McGovern and R. H. Michel, *The Spectrum*, 1991, **4**, 6–7.

19. J. Wisniak, *Indian J. Hist. Sci.*, 2004, **39**(1), 75–100.

20. P. Friedländer, Zur Kenntnis des Farbstoffes des antiken Purpurs aus Murex brandaris, *Monatsh. Chem.*, 1901, **30**, 247–253.

21. P. Friedländer, *Ber. Dtsch. Chem. Ges.*, 1909, **42**(1), 765–770.

22. P. Friedländer, *Ber. Dtsch. Chem. Ges.*, 1922, **55**, 1656–1658.

23. J. Wouters and A. Verhecken, *J. Soc. Dyers Colour.*, 1992, **108**, 404.

24. D. Rudd, M. Ronci, M. R. Johnston, T. Guinan, N. H. Voelcker and K. Benkendorff, *Sci. Rep.*, 2015, **5**, 13408.

25. C. J. Cooksey, *Molecules*, 2001, **6**, 736–769.

26. I. Karapanagiotis, *Sustainability*, 2019, **11**(13), 3595.

27. D. Reese, *Libyan Stud.*, 1979–1980, **11**, 79–93.

28. P. E. McGovern and R. H. Michel, *MASCA*, 1985, **3**, 66–70.
29. I. Herzog, Hebrew porphyrology, in *The Royal Purple and the Biblical Blue*, ed. E. Spanier, Keter Publishing House, Jerusalem 1987, pp. 18–131; 39.
30. E. Soriga, Mari(ne) Purple: Western Textile Technology in Middle Bronze Age Syria, in *Treasures from the Sea: Sea Silk and Shellfish Purple Dye in Antiquity*, ed. H. L. Enegren and F. Meo, Oxbow Books, Oxford and Philadelphia, 2017, pp. 79–95.
31. Martial (Marcus Valerius Martialis), *Epigrams*, Book 6, XCIII, Bohn's Classical Library, 1897.
32. N. Karmon and E. Spanier, Archaeological evidence of the purple dye industry from Israel, in *The Royal Purple and the Biblical Blue*, ed. E. Spanier, Keter Publishing House, Jerusalem, 1987, pp. 147–158.
33. V. Finlay, *Color: A Natural History of the Palette*, Random House, New York, 2004, p. 369.
34. P. E. McGovern and R. H. Michel, *Acc. Chem. Res.*, 1990, **23**, 152–158.
35. B. Marín-Aguilera, F. Iacono and M. Gleba, *J. Mediterr. Archaeol.*, 2018, **31**(2), 127–154.
36. F. Brunello, *The Art of Dyeing in the History of Mankind*, Neri Pozza, Vicenza, 1973, p. 45.
37. Plutarch, *Lives of the Noble Greeks*, ed. E. Fuller, Dell Publishing, New York, 1959, p. 309.
38. O. Elsner and E. Spanier, The Dyeing with Murex Extracts, an Unusual Dyeing Method of Wool to the Biblical Sky Blue, *Proceedings of the 7th International Wool Textile Research Conference, Tokyo*, 1985, vol. V, pp. 118–130.
39. Z. C. Koren, *Dyes Hist. Archaeol.*, 2005, **20**, 136–149.
40. I. I. Ziderman, *Color. Technol.*, 1986, **16**(1), 46–52.
41. L. Naegel and C. Cooksey, *J. Shellfish Res.*, 2002, **21**(1), 193–200; Table 196.
42. R. H. Michel, J. Lazar and P. E. McGovern, *J. Soc. Dyers Colour.*, 1992, **108**(3), 145–150.
43. K. S. Tan, *Mar. Pollut. Bull.*, 1997, **34**, 577–581.
44. I. I. Ziderman, Y. Sasaki, M. Sato and K. Sasaki, Thermochromic behaviour of 6-bromonindigotin: key to understanding purple dyeing with banded dye murex, in *Diversity of Dyes in History and Archaeology*, ed. J. Kirby, 2017, pp. 393–403.

45. G. Sagiv, *Contemp. Jew.*, 2015, **35**(3), 285–313.
46. Z. C. Koren's reply to the Stermans' response. Continuing the tekhelet debate, https://www.biblicalarchaeology.org/daily/biblical-artifacts/artifacts-and-the-bible/zvi-c-korens-reply-to-the-stermans-response/, accessed August 2020.
47. M. V. Orna, *The Chemical History of Color*, Springer, Heidelberg, 2013, p. 66.
48. Z. C. Koren, *Isr. J. Chem.*, 1995, **35**, 117–124.
49. Z. C. Koren, Archaeological shades of purple from flora and fauna from the ancient Near East, in *Archaeological Chemistry: A Multidisciplinary Analysis of the Past*, ed. M. V. Orna and S. C. Rasmussen, Cambridge Scholars Publishers, Newcastle-upon-Tyne, 2020, pp. 256–300.
50. W. R. Wilde, *Narrative of a Voyage to Madeira, Teneriffe, and along the Shores of the Mediterranean*, vol. II, William Curry, Jun, & Co., Dublin 1840, p. 149.
51. D. Reese, *Mediterr. Archaeol. Archaeom.*, 2009, **10**(1), 113–141.
52. Bolinus brandaris, https://commons.wikimedia.org/w/index.php?title=File:Bolinus_brandaris_2.jpg&oldid=385855670, accessed June 2021.
53. Hexaplex trunculus armigerus Settepassi, https://commons.wikimedia.org/wiki/File:Hexaplex_trunculus_armigerus_01.JPG, 1970, accessed June 2021.
54. Stramonita haemastoma, https://commons.wikimedia.org/w/index.php?title=File:Stramonita_haemastoma_01.JPG&oldid=439472071, accessed June 2021.
55. A. Carranante, Purple-Dye Industry Shell Waste Recycling in the Bronze Age Aegean? Stoves and Murex Shells at Minoan Monastiriki (Crete, Greece), in *Archaeomalacology Revisited*, ed. C. Çakirlar, Oxbow Books, Oxford, 2014, pp. 9–18.
56. D. S. Reese, personal communication, 12 August 2020.
57. E. Spanier and N. Karmon, Muricid snails and the ancient dye industries, in *The Royal Purple and the Biblical Blue*, ed. E. Spanier, Keter Publishing House, Jerusalem 1987, pp. 179–192.
58. I. Boesken Kanold, http://pourpre.inge.free.fr/, accessed August 2020.
59. E. Aloupi, Analysis of a purple material found at Akrotiri, in *Thera and the Aegean World III*, ed. D. A. Hardy, The Thera Foundation, London, 1990, vol. 1, pp. 488–490.

60. R. R. Stieglitz, The Minoan origin of Tyrian purple, *Biblical Archaeol.*, 1994, **57**(1), 46–54.

61. A. Brysbaert, *Stud. Conserv.*, 2006, **51**, 252–266.

62. H. Brekoulaki, *Revue Archéologique*, 2014, **1**(57), 1–36.

63. Z. C. Koren, *Microchim. Acta*, 2008, **162**(3–4), 381–392.

64. C. Porter, G. Chiari and A. Cavallo, The Analysis of Eight Manuscripts and Fragments from the Fifth/Sixth Century to the Twelfth Century, with Particular Reference to the Use of and Identification of "Real Purple" in Manuscripts, ART 2002, *7th International Conference on Non-destructive Testing and Microanalysis for the Diagnosis and Conservation of the Cultural and Environmental Heritage*, Antwerp, Belgium, June 2002.

65. M. Aceto, A. Agostino, G. Fenoglio, P. Baraldi, P. Zannini, C. Hofmann and E. Gamillscheg, *Spectrochim. Acta, Part A*, 2012, **95**, 235–245.

66. Y. Szafran and N. Khandekar, Varnish and Early Italian Paintings: Evidence and Implications, in *Early Italian Paintings: Approaches to Conservation: Proceedings of a Symposium at the Yale University Art Gallery, April 2002*, ed. P. S. Garland, Yale University Press, New Haven, 2003.

67. V. Gedzevičiūtė, N. Welter, U. Schüssler and C. Weiss, *Archaeol. Anthropol. Sci.*, 2009, **1**, 15–29.

68. Z. C. Koren, The Unprecedented Discovery of the Royal Purple Dye on the Two Thousand Year-Old Royal Masada Textile, in *The Textile Specialty Group Postprints*, ed. P. Ewer and B. McLaughlin, American Institute for Conservation of Historic and Artistic Works, Washington DC, 1997, vol. 7, pp. 23–34.

69. Z. C. Koren, Color my world, a personal scientific odyssey into the art of ancient dyes, in *For the Sake of Humanity: Essays in Honour of Clemens Nathan*, ed. A. Stephens and R. Walden, Martinus Nijhoff Publishers, Leiden, Netherlands, 2006, pp. 155–189.

70. L. Thorndike, *A History of Magic and Experimental Science*, Columbia University Press, New York, 1923, vol. 1, p. 590.

71. N. M. Sussman, *J. East. Mediterr. Archaeol. Herit. Stud.*, 2020, **8**(2), 159–173.

72. Wulfila Bible, https://commons.wikimedia.org/w/index.php?title=File:Wulfila_bibel.jpg&oldid=415188126, accessed June 2021.

73. M. Guckelsberger, The purple murex dye in antiquity, Essay for BA Degree, Latin, University of Iceland, Reykjavik, December 2013.

74. M. Aceto, E. Calà, A. Agostino, G. Fenoglio, M. Gulmini and A. Idone, *et al.*, *Spectrochim. Acta, Part A*, 2019, **215**, 133–141.

75. I. Herzog, Hebrew porphyrology, appendix B, in *The Royal Purple and the Biblical Blue*, ed. E. Spanier, Keter Publishing House, Jerusalem 1987, pp. 139–145.

76. C. P. Biggam, *Knowledge of Whelk Dyes and Pigments in Anglo-Saxon England. Anglo-Saxon England*, ed. M. Godden and S. Keynes, Cambridge University Press, Cambridge, vol. 35, 2006, pp. 23–55.

77. B. Halford, *Chem. Eng. News*, 2020, **98**(43), 11.

78. J. Lee, J. Kim, J. Song, W.-S. Song, Y.-G. Kim and H.-J. Jeong, *et al.*, *Nat. Chem. Biol.*, 2020, 1–9.

79. J. Edmonds, The mystery of imperial purple dye, *Historic Dye Series No. 7*, Little Chalfont, 2000.

80. I. Boesken Kanold, *Dyes Hist. Archaeol.*, 2005, **20**, 150–154.

81. G. Verri, C. Martin de Fonjaudran, A. Acocella, G. Accorsi, D. Comelli and C. D'Andrea, *et al.*, *Dyes Pigm.*, 2019, **160**, 879–889.

82. E. D. Glowacki, G. Voss, L. Leonat, M. Irimia-Vladu, S. Bauer and N. S. Sariciftci, *Isr. J. Chem.*, 2012, **52**, 540–551.

CHAPTER 7

In the Shadow of Vesuvius: A Window on the Ancient Palette[†]

Even as we stand on uncertain ground, the earth beneath
us moves quiet and wild, its boundaries shifting, its muscles
wavering.[1]

Alberto Rios

7.1 INTRODUCTION

When Vesuvius violently erupted in 79 CE, it created a time capsule of Pompeii and its surroundings that continues to yield up
the secrets of Roman wall painting that would otherwise have
been lost. Other spectacular finds have given us a whole new view
of the art of the ancient world.

[†]Glossary entries and chapter cross-references are in boldface.

March of the Pigments: Color History, Science and Impact
By Mary Virginia Orna
© Mary Virginia Orna 2022
Published by the Royal Society of Chemistry, www.rsc.org

7.2   VESUVIUS FLEXES HIS MUSCLES

In the dead of night on 24–25 August of 79 CE,[‡] a thunderous pyroclastic blast enveloped Pompeii and the adjacent towns of Herculaneum, Stabiae, Oplontis and Terzigno.[2] "...[T]he avenging angel summoned up the demon of Vesuvius and pointed out to him the city which is condemned to destruction.... At the voice of the angel the specter of destruction rose from the crater, revealing himself from the waist upwards..., one hand forking up lava with his hellish pitchfork, while with the other he scatters ash on the condemned city."[3] Such is the 19th century description that evokes the idea of a "just" punishment for a city of sin. But there is a double significance to the story of Pompeii's destruction: not only is it the remains of a society wiped off the face of the earth in an instant but also the occasion that reawakened an ancient culture. Archaeological investigation continues to unwrap the mortuary bandages of this extraordinary time capsule. There are no superlatives that can describe how important Pompeii is for the study of Greco-Roman art, the main topic of this chapter, due to the excellent state of preservation of the ruins.[4,5] It is the only extensive record left to us of the classic treatment of wall painting.[6]

After her long, seventeen-century sleep under a blanket of three meters of pumice and ash, Pompeii was roused in 1748, not by a prince's kiss, but by a king's excavation order.[§] As the digging proceeded, news of the archaeologists' spectacular finds captured the public's imagination, eventually giving rise to some brilliant works of art. Barely two months before his untimely death in December, 1791, Wolfgang Amadeus Mozart (1756–1791) produced his final opera, *The Magic Flute*, with the set modeled on the Temple of Isis in Pompeii. Thirty-four years later, in 1825, Giovanni Pacini (1796–1867) staged his *L'ultimo Giorno di Pompei* at the Naples San Carlo Theatre. Nine years after that, Edward Bulwer-Lytton (1803–1873) produced his eponymous (but in English) novel, which, in turn, inspired Karl Bryullov's masterpiece of the same name (Figure 7.1).[7]

[‡]There is some recent evidence showing that this date, noted by Pliny the Younger, may be incorrect and that the eruption occurred in the fall, perhaps mid- to late-October. Excavations have uncovered a market where autumn fruits and vegetables were displayed. Additional evidence regarding a numismatic inscription and a wall graffito are inconclusive.
[§]In 1763, Charles III, King of Naples and Sicily, ordered the first official excavation of the then unidentified site which was discovered to be the ancient city of Pompeii.

Figure 7.1   Karl Bryullov (1799–1852). The Last Day of Pompeii (1830–1833). State Russian Museum, Saint Petersburg. Reproduced from ref. 7.

While all of this artistic activity was transpiring, eventually spawning more plays and an entire genre of historical films on similar themes, a quiet experimental revolution that would prove to be far more informative and lasting was taking place in the laboratory of Jean-Antoine Chaptal (1756–1832) in France and in the portable laboratory of Sir Humphry Davy (1778–1829) in Italy.

7.3   THE EARLIEST-KNOWN ANALYSES OF ARTISTS' PIGMENTS

Though much has been written over the centuries regarding the content of artistic works, it took many centuries for art historians and scientists to begin to actually examine objects rather than accept what was "said" about them in the past.

7.3.1   What Pliny Said

Pliny the Elder (23–79 CE) is one of the most cited historic sources on the use of pigments in late Greek and early Roman times. Gaius Plinius Secundus, known almost universally as Pliny the Elder, composed his monumental treatise *Historia naturalis*[8] late in life, and fortunately it survived, a fate not granted to many of his other works. He was a busy Roman official who found the time to transcribe much of what was known in the ancient world. Books XXXIII through XXXVII describe the natural history of metals, stones, precious stones, paintings and colors. However, he could only describe; he could not analyze. Ironically, he perished in the eruption that buried Pompeii, but the full circumstances of his death are questionable.[9]

7.3.2 Early Analytical Work

The earliest chemical examination of pigments in historical painting seems to begin in 1780 with the work of Johann Friedrich Gmelin (1748–1804) who looked at the colors and binding medium on a mummy sarcophagus, with inconclusive results.[10] In 1800, the physician John Haslam (1764–1844) provided the first definitive results on the pigments on some medieval wall paintings in London prior to their demolition.[11]

7.3.2.1 Jean-Antoine Chaptal. Chaptal, a former academic advisor to Antoine Laurent Lavoisier (1743–1794), was a polymath chemist chiefly interested in industrial and practical applications of chemistry. In 1809, he published an out-of-character piece in the *Annales de Chimie* on the analysis of seven samples of pigments taken from a Pompeii color shop.[12] The opening sentence of his paper explains this anomaly: he received these samples from her royal majesty, the Empress and Queen Joséphine de Beauharnais (1763–1814), first wife of Napoléon I. Hence, by royal decree, Chaptal was thrust into the position of being almost the first person to examine and identify the constituent elements of artists' pigments.

Chaptal's "analysis" of the first four samples was a visual one: he described their colors but performed no chemical or physical tests. The first was a greenish and soap-like raw earth that he called "terre de Vérone" since it was first discovered on Monte Baldo, near Verona, and described as early as 1574 in the Vatican mineral collection catalogue.[13] Its modern name is "green earth," either of the two minerals celadonite and glauconite, essentially hydrous iron, magnesium and aluminum potassium silicates.[14] Chaptal's second sample was a yellow material which he assumed was yellow ochre, the hydrated form of iron oxide.

He observed that, since intense heat turns this compound red, it could not have experienced high temperatures during the volcano's eruption. Sample three was a reddish-brown ochre obtained by heating yellow ochre. Sample four was a very light and white pumice stone. Chaptal subjected sample five, which he described as a brighter and more vivacious blue than *cendres-bleu* (natural and synthetic copper carbonates)[15] to the "acid test; " it exhibited light effervescence when treated with hydrochloric, nitric, and sulfuric acids. He remarked that it seemed to resemble a copper compound found on Egyptian hieroglyphs, namely copper oxide mixed with lime and aluminum. Sample six was a paler version of number five since it seemed to contain more lime and aluminum. (The latter two colors were more than likely Egyptian blue [**Chapter 4.6.8**].) Sample seven was not a metal-based pigment. He described it as "lacquer de Garance," which is the red color madder, precipitated on an alum substrate. He remarked that it had lasted intact for nineteen centuries.

7.3.2.2 Humphry Davy's Report. Some five years later, Sir Humphry Davy received his first Pompeian sample from the hands of a queen as well. This time, it was Maria Carolina of Austria (1752–1814), wife of Ferdinand IV, King of Naples and Sicily, and sister of the unfortunate Marie Antoinette (1755–1793). During his projected three-year tour of Europe and Asia Minor, Davy, along with his wife, Jane Apreece (1780–1855), a young Michael Faraday (1791–1867), and an entourage of servants, reached Rome, Naples and Pompeii in April, 1814. While attending an excavation at Pompeii in May, he received a small pot of a pale blue color from the queen.[16] He later found that it consisted of calcium carbonate mixed with what he called the "Alexandrian frit," namely, Egyptian blue.

Addiction is hardly a recipe for upward mobility, but Humphry Davy's obsession with laughing gas eventually afforded him a route to a position at the Royal Institution in London. Davy arose from modest beginnings by circumstances of his birth in Cornwall and subsequent initial education which, according to his first teacher, "failed to uncover anything extraordinary" in this future prodigy of science. His first brush with chemistry was some brief tutoring by Gregory Watt (1777–1804), son of the inventor-engineer, James Watt (1736–1819), who was a sometime lodger at Davy's home. At the age of 20 in 1798, Davy took the post of assistant to a Dr Thomas Beddoes (1760–1808) at the latter's newly-founded Pneumatic Institution. There he became a captive of nitrous oxide (laughing gas) and interested, even obsessed, with galvanism and the power of electricity – the latter skill led him within two years to the Royal Institution in London. From there, Davy moved from strength to strength, discovering five chemical elements by electrolysis and establishing himself as a spectacular chemical demonstrator and lecturer.[17] A laboratory accident in 1812 forced him to hire an assistant, one Michael Faraday, who, some say Davy claimed, was really his greatest discovery.[18]

In the paper he wrote in 1815 following his return to England,[16] Davy described the various sources of the ancient pigment samples that he analyzed in his portable laboratory in Rome. Due to his friend, the sculptor Antonio Canova (1757–1822), Inspector-General of Antiquities and Fine Art of the Papal State, Davy had had almost "carte blanche" in terms of sample acquisition. These samples consisted of fragments, pots of paint, lumps of pigment, *etc.* from historic sites like the Baths of Titus, the Baths of Livia, and the ruins of Pompeii, as well as the famous ancient Roman Aldobrandini Wedding Fresco now on display in the Vatican Museums (Figure 7.2).[19] Much of the paper also concerns itself with the writings of Vitruvius and Pliny on pigments. Davy also made it a showpiece to display his own scholarship on Greek

Figure 7.2 Aldobrandini Wedding. Augustan Age (43 BCE – 18 CE) fresco sampled by Humphry Davy. It was discovered on the Esquiline Hill in 1601 and acquired for the Vatican Museums in 1818. Reproduced from ref. 19.

and Roman art. However, his experimental work falls short and could even be termed "sloppy" in some instances. Some of the experiments he described are hard to follow, and the conclusions he reaches do not match the progress of the work. Rees-Jones remarks:[11]

> [S]ome experiments had so many chemicals added together that they were closer to alchemy than chemistry. He [Davy] did not fully explain many of the experiments, being more interested that the results should agree with his scholarly researches.

7.3.2.3 Summary of Ancient Pigments: Literature and Analysis. In Table 7.1,[20] we summarize the ancient literary source of Pliny for each of the major colors (column 2). Column 3 contains actual analyses of color samples from Pompeii by J.-A. Chaptal. Column 4 Humphry Davy's results of analyses of color samples from wall paintings, paint pots, frescoes, *etc.* from some representative sites in the ancient Greco-Roman world. For comparative purposes, we also present the set of analyses carried out in 1967 at the dawn of instrumental analysis (column 5),[21] and a more recent set of analyses using the customary battery of analytical instruments (column 6).[22] In the case of Humphry Davy, the analyses were done by heating, blowpipe oxidation–reduction reactions, and observation of changes upon addition of acids and bases; The names of the pigments given in the table are those used by the authors. While organic colorants were used in the ancient world, their absence from the analytical list is an indication of their chemical instability over the course of centuries.

In the course of Davy's analyses, he made two very astute observations. The first, based on his chemical knowledge, was that all the pigments that were completely oxidized, like the ochres, stood the test of time better than those, like massicot (yellow lead oxide, PbO), that were not. His second remark, based on his knowledge of art history, is that "the Greek and Roman painters had almost all the same colours as those employed by the great Italian masters at the period of the revival of the arts in Italy." This poses a conundrum: has nothing changed with respect to the makeup of the artist's palette or have the changes been subtle? Let's take a look at some of these subtleties.

7.4   COLOR EXPLOITATION AND PALETTE EXPANSION IN THE ANCIENT WORLD AND BEYOND

The pigments listed in Table 7.1 bear witness to the fact that the colors in use through the first century CE were virtually identical to those in use from the Bronze Age (*ca.* 3000 BCE forward). Except for Egyptian blue, the pigments were all prepared from naturally occurring materials. The basic color gamut was not great, but artists could skillfully extend the chromatic range by mixing, diluting, concentrating and grinding the pigments at hand. Painters were adept at obtaining a wide range of colors from one pigment; this skill was well known at Lascaux 17 000 years earlier (**Chapter 2.5.1**). They were also knowledgeable about the beneficial and deleterious effects of impurities and knew how to handle both situations. Finally, green was not a very common color, pointing to the possibility of the lack of a ready and reliable source.

Unless a pigment is bound firmly to its surface, it will not last. The techniques developed by the ancients for achieving a degree of permanence are typically called *buon fresco*, *fresco secco* and tempera. *Buon fresco* is the technique of applying a pigment mixed with water to fresh plaster that has a lime (calcium hydroxide, $Ca(OH)_2$) base. As the water dries, the lime migrates to the surface and combines chemically with atmospheric carbon dioxide, CO_2, to form chalk, calcium carbonate, $CaCO_3$. The newly formed chalk locks the pigment into place. The painter has to be very skilled to use this technique because the surface dries very quickly and once the pigment is applied, there is no room to correct any mistakes. True fresco requires rapid, sound and perfect judgment and a dexterous, resolute hand.[6] *Fresco secco* is really not fresco at all. It involves premixing the pigment with a medium, often lime, and then applying the mix to a dry or dampened surface. In the tempera technique, the pigments are mixed with a binder such as egg, casein (from milk), vegetable gums or animal glue and then applied to a hard, dry surface.[23] Some of the pigments were applied by the *mezzo fresco* technique, *i.e.*, brushed onto a dry wall but mixed with limewash which then formed a calcium carbonate layer on the surface. As time rolled on, the blur between true fresco and *fresco secco* widened and many unconventional mixes and binders found their way into

Table 7.1 Pigments used in the Mediterranean (Greco-Roman) world from the fourth century BCE to the first century CE. Adapted from ref. 21 with permission from American Chemical Society, Copyright 2015.

Color	Pliny (77–79 CE)	J. Chaptal (1809)	H. Davy (1815)	S. Augusti (1967)	I. Kakoulli (2009)
Black/Brown	Atramentum Ivory black Fossil black Sepia (cuttlefish)	N/A	Ochres Manganese oxide Carbon black	Atramentum	Charcoal Carbon black
White	Alum White clay Lead white Magnesium hydrosilicate or calcium carbonate	Pumice	White clay Fine chalk (calcium carbonate)	Calcium carbonate Diatomite Silicates	Calcium carbonate
Red/Orange	Red ochre Burnt ceruse (minium) Red ochre of Sinope	Red ochre Lacque de Garance	Minium Red ochre Cinnabar	Red ochre Burnt ochre Cinnabar Realgar	Red ochre Realgar Cinnabar
Yellow	Yellow ochre Orpiment	Yellow ochre	Yellow ochre Massicot	Yellow ochre Massicot Orpiment	Yellow ochre Orpiment Sulfur
Green	Terre verte Malachite Verdigris	N/A	Copper carbonates Green earth	Terre verte Malachite Verdigris	Malachite Verdigris Terre verte
Blue	Azurite **Indigo**	Blue frit (Egyptian)	Blue frit (Egyptian) Cobalt (glass) Copper oxide	Azurite **Indigo** Lapis lazuli Egyptian blue	Egyptian blue
Violet	Burnt ochres Purpurissum	N/A	Mixtures of red and blue	Dark purple Light purple	Tyrian purple Mixtures of red and blue

wall paintings including flowers, spices, human urine and even earwax.[24]

7.4.1 Pompeian and Roman Pigments Beyond Davy

Modern analytical techniques applied to these ancient monuments can yield so much more information than the visual examination and wet chemistry techniques available to Chaptal and Davy. Some examples follow.

7.4.1.1 Temple of Venus, Pompeii. A recent study of the Temple of Venus at Pompeii picked up some refinements in technique and identified a new pigment by instrumental means including **X-Ray diffraction** (XRD) and **Fourier transform infrared spectrophotometry** (FTIR). The investigators[25,26] found that:

- Colors, especially the reds and yellows, were made darker by addition of carbon black;
- Some of the colors, mainly the blacks, reds, greens and yellows, were prepared using different pigments. Adopting a different recipe for the same color indicates the possibility of several different workshops employed at the same or different times in decorating the temple;
- The presence of tin in the form of $CaSnSiO_5$ in the pale Egyptian blue showed that it was manufactured from recycled bronze, a copper-tin alloy, not from copper scraps;
- A yellow-brown pigment composed of volcanic glass was characterized for the first time. Its presence suggests a local production works for this otherwise undocumented pigment. The glass is chemically compatible with Vesuvian volcanic products unknown elsewhere;¶
- There was a gray recipe that used seven different pigments; there are no analogies in the literature of such a complex admixture of pigments;
- There were lime lumps in a large number of the samples, indicating that some of the lime did not react completely

---

¶High temperature Vesuvian fumaroles have been known to emit minerals peculiar to the volcano or even to a single eruption; at least 150 minerals, many rare or very rare, have been collected from previous eruptions, notably that of 1906. Some of the materials are subject to rapid alteration, which may be the case for the mysterious yellow-brown pigment found only as volcanic glass.

with water during the slaking; this provides strong evidence of careless or inept mixing;

- Analysis revealed that 16 different recipes were used *in toto*.[27]

7.4.1.2 The Villa dei Misteri. One of the most famous almost intact survivors of the 79 CE eruption, the Villa of the Mysteries, with its dreamlike and decadent background, has excited storytellers and film directors alike. Its most famous room, number 5, contains a decorative frieze depicting initiation into Bacchic rites (Figure 7.3)[28] that inspired the setting in the final scene of the 1983 film, *Scarface*. Both scenes are drenched in red – death-dealing cinnabar in the case of the villa, blood red walls saturated with real blood by the end of the film. In his 1969 film, *Satyricon*, set in ancient Rome, Federico Fellini (1920–1993) found the Villa of the Mysteries an important source of inspiration for his sets. Mini scenes full of red (reminiscent of blood again) pepper Tim Burton's 2007 *Sweeney Todd: The Demon Barber of Fleet Street*, particularly Judge Turpin's study. The nude women of the Dionysiac frieze have made cameo appearances in such fantastic re-creations and parodies of classical antiquity as Monty Python's satire, *Life of Bryan* (1979), and the television miniseries *The Last Days of Pompeii* (1984).[29,30] The painting cycle

Figure 7.3   Fresco from the Villa dei Misteri, Pompeii. This scene depicting a Bacchic rite is steeped in cinnabar. Reproduced from ref. 28.

in this famous room is thought to be a copy of an earlier Greek original.

The Villa was discovered as the result of an excavation begun in April, 1902. The chief archaeologist, Giulio de Petra (1841–1925), asserted that all the paintings were carried out by a very generous application of the binding agent, and not by *buon fresco*. He was very critical of the design, noting some anatomical errors such as the enormous wrists on the forearms, and some fingers drawn too large whereas others were too thin. He also disparaged the artist's palette, claiming that his unfamiliarity with the use of colors led him to render all four walls in a monotone: on a background of cinnabar, the overlaid browns, yellows, pinks and greens all had the same intensities without any variation in tonality.[31] However, the Villa dei Misteri is one of the best-preserved structures in Pompeii and gives us valuable information that the basis of the colors was red cinnabar painted over by green earth, red ochre, yellow ochre, carbon black, and white calcite (calcium carbonate). The palette of this and of all the other "domus" in town was the same except there was no blue (which would have been Egyptian blue) in this particular venue. By all appearances, this palette stood not only the test of time but also of a 300-mile per hour, high temperature pyroclastic blast that probably changed its appearance drastically – we will never know how the artist viewed the work originally.[32]

7.4.1.3 Possible Cases of Cinnabar Degradation on Pompeian Paintings. Cinnabar samples taken from the Villa Sora, near Pompeii, which had been buried by a pumice deposit in 79 CE, were still red when excavated in 1988. By 1990, a mysterious black crust, plus some white and gray crusts, developed. Archaeologists know well that cinnabar (HgS)-based red pigment undergoes photodegradation to black metacinnabar when exposed to light. However, this process is not systematic nor is it completely understood and may, in fact, be erroneously attributed in some cases. Unexpectedly, no metacinnabar was found on the degraded pigments at Villa Sora. Instead, the investigators identified several compounds of mercury bound to chloride – light gray to black corderoite ($Hg_3S_2Cl_2$), white calomel (Hg_2Cl_2) and yellowish terlinguaite (Hg_2ClO), suggesting reaction of the cinnabar with ordinary salt (NaCl) present in the seaside environment. However, what caused the black? The research team could only postulate that HgS, catalyzed by chloride, can yield Hg and S as the products of photodegradation. The sulfur can further react to form white gypsum, which they detected, but also degrade the integrity of the pigment surface leaving nooks and crannies to be filled by airborne soil particles. Furthermore, the black crust was only about 5 μm thick – a surface phenomenon that left the underlying cinnabar pigment intact.[33] A suggestion that the black may be caused by the release of black metallic mercury needs yet to be investigated.[34] This type of research is throwing light on the crucial issues involved with preserving cultural heritage and how to avoid similar problems in the future.

7.4.1.4   Some Additional Words About Cinnabar.   Given the splashes of cinnabar that occur around Pompeii, in contrast to its parsimonious use in Classical Antiquity,[35] it is only right and proper to make some remarks about the human cost involved in its production and use. Cinnabar is democratic: it is just as toxic to the Senator as to the slave. The toxicity quotient can be measured in terms of exposure time, and here is where the inequality begins. Cinnabar miners, often slaves and little children, would descend into the tunnels for days, sometimes weeks, on end. There they would labor in semi-darkness without fresh air, covered with moisture, and breathing constantly the toxic powder and volatile elemental mercury that accompanied it, their only protection a sheep's bladder to breathe through, if you were lucky to have one. Consequently, a miner's shift might turn out to be his last. Realization of these hazards prompted Jim Powell to compose the following poem, *Cinnabar*, (slightly abridged) commemorating William Brewer's 19th century descent into the New Idria Mercury Mine in California:[36]

The matrix rock is metamorphosed slate, porous, fractured,
the ore distributed capriciously.
Tunnels diverge in all directions threading unmined seams,
the veins of cinnabar diffused in streaks across drift faces
brilliant blood red under our candlelight,

the miners naked above the waist, their shoulders bur-
nished copper,

a hard sort? Cornishmen, Chileans. The mines

are profitable to stockholders. Nine hundred flasks per
month

are shipped in pairs by mule for San Francisco. In brick
furnaces

the ore, reduced and roasted, distills quicksilver.

The atmosphere it vents of arsenics, sulfurous acids,
and vapors of mercury is ruinous.

The men who go inside to clean the condensation chambers

do not recover, yet the higher wage commands fresh victims
yearly,

and all are poisoned by the furnace work.

Horrendous as these conditions were, they were undoubtedly far better than what befell the forced laborers of millennia
before.

But the privileged few, users and admirers of the cinnabar,
were not much better off. Closed off rooms of villas awash with
cinnabar were slow-acting, sublethal gas chambers that unaware
victims ate, drank, and slept in. There are no statistics on who
or how many may have succumbed to heavy metal mercury poisoning, but even if death was not the end result, the neurological
effects may have been massive.

7.4.1.5   More Archaeological Data from Pompeii.   When a group
of Italian archaeologists scoured Pompeian houses for artifacts,
they found nine bronze vessels containing black powders that
simply begged for analysis. Five of the nine occupied cylindri-
cal vessels called "theca atramentaria," or in modern parlance,
inkwells. And indeed, analysis (by **X-ray diffraction** (XRD),
Fourier-transform infrared spectrophotometry (FTIR) and **gas
chromatography/mass spectrometry** (GC/MS) among others)
showed that all of them contained a carbon soot mixed with
large amounts of polycyclic aromatic hydrocarbons (PAHs),
a virtual fingerprint for wood burned in a limited supply of

air. This evidence of carbon-based ink indicates that iron gall inks had not yet been introduced although the reaction that produces them was known from as early as the third century BCE.[37,38] Philon of Byzantium described an ink made from oak galls and "Roman vitriol," aka iron(II) sulfate, a plant nutrient and a staple of your local garden shop. It was, in effect, invisible ink: you wrote with a solution of nut-gall or oak gall on a surface and then traced over it with the vitriol to render it visible.[39] This type of ink supplanted the carbon-based inks in later Roman times and remained the ink of choice well into the 20th century (**Chapter 14.3**).

More data from the Villa of the Papyri, one of the most impressive architectural examples in Herculaneum, tell us that the typical Roman palette was used in the paintings, almost all of them completed using the *buon fresco* technique. Some of the paintings were done by tempera with an egg binder. The absence of beeswax indicated that no paintings were executed with a wax medium (encaustic), nor has any wax been found in any Pompeian or Roman paintings.[6]

One interesting observation was that the yellow ochre morphed into red due to the high temperatures during the eruption.[40] Chaptal had observed yellow ochre (goethite) in a paint pot from Pompeii. Another analysis of some paint pots found near the Temple of Venus support this find with three more intact goethite samples.[41] This might be valuable information in terms of the temperature differential during the eruption: Herculaneum suffered more from heat than Pompeii, where the yellow ochre remained intact. In fact, the minimum temperature necessary to transform yellow ochre to red ochre is 250 °C (400 °F), the temperature of a very hot oven. This may have been the minimum temperature of the pyroclastic cloud that reached Herculaneum. And indeed, measurements of the residual magnetization at many sites affected by the eruption showed that the temperatures in Pompeii ranged from about 180 °C to 300 °C, while those at Herculaneum were, on average, from 360 °C to 380 °C.[42] From these observations, it is quite possible that many of the red wall paintings in Pompeii and surroundings were once actually yellow. Archaeologists have subsequently confirmed that after the eruption, there were 246 red paintings and

57 yellow paintings, whereas before the blast, it was estimated that there were 165 reds and 138 yellows.[43] Subsequent **X-ray fluorescence** work using arsenic as a marker for the presence of the red form of ochre has enabled workers to identify original and transformed yellow ochre in wall paintings at several Pompeian sites.[44]

7.4.1.6 Beyond Pompeii: Rome and the World. Looking for Roman artistry as we move beyond Pompeii to Rome itself, the most spectacular decorations of the era occupied the Gilded Vault of the *Domus Aurea* (Golden House) at Rome. Built by the Emperor Nero (37–68 CE) following the Great Fire of 64 CE, it occupied over 80 hectares (200 acres) of central Rome covering an extent now encompassing the area around the Colosseum and far beyond. Following Nero's suicide, this great monument to unbridled opulence was an embarrassment to subsequent rulers, so much so that it was dramatically obliterated in a very short period of time. Because it was built over by newer construction, it was, ironically preserved. Accidentally discovered toward the end of the 15th century, it became a wonder of the art world as, one by one, great artists descended into the dark grotto‖ to study the world of antiquity first-hand. A 2011 study of the now-famous Gilded Vault showed that the major pigments were Egyptian blue, green earth, red ochre, cinnabar, a lead compound that could be either an oxide or lead carbonate, and madder supported on aluminum silicate. The colors were seldom used alone but were expertly mixed in complex blends that provided deep and brilliant hues. Instrumental analysis showed that a peculiar green-pink decorative effect was fashioned by superimposing four alternating layers of Egyptian blue-green earth and madder lake-cinnabar.[45]

Additional pigment anomalies have been found in some other investigations that may or may not be artifacts of the sensitivity of the methods used. For example, the discovery of the highly-prized purple pigment, sometimes called English red or *caput mortuum*** (death's head), in samples from a Romano-British

---

‖The word "grotesque" originated here but has now taken on many other meanings.
**Certainly not by the Romans themselves.

villa was a great surprise.[46] Although the pigment identity of the *caput mortuum* is spectroscopically hematite (Fe_2O_3), its darker hue remained a puzzle until further work demonstrated that it was a particle size effect: the particles of *caput mortuum* are about twenty times larger, on average, than so-called normal hematite.[47] This conclusion is not surprising since it has long been known that excessive grinding of a pigment to reduce particle size leads to a much paler shade.

Analysis of late-empire Roman pigments from a 1974 excavation at the Forum Boarium in Rome reported pots full of workhorse pigments and a complete absence of the precious types used in fine art. This may reflect the area in which they were found – a workshop in Rome's largest and busiest commercial center – and possibly the only surviving pigment shop in ancient Rome. Just the ticket for the home decorator.[48,49]

Scientists and archaeologists have come a long way from the pioneering analytical work of Chaptal and Davy, who were chiefly interested in the identities of the materials that the ancient artists used. Now that the "what" is clearly in hand, continuing work can deal with the "when," the "how," and the "why." Major advances in the dating of artifacts using radiocarbon dating techniques, when organic materials are present, and other specialized techniques when they are not, have complemented art historical stylistic comparisons. How the paintings were made, particularly by examining the binding materials, the underlying plaster, and the overlaid varnishes now comprises a large body of data. One troublesome problem that still needs a lot of research is the "why" of degradation: the capricious darkening of cinnabar is just one example. As progress is made in each of these areas, the body of know-how for restoration, conservation and preservation is continually being built up. Museums around the world are making a tremendous contribution to our cultural heritage by transmitting this information to the public in the form of well-documented permanent exhibits, special shows, lectures, and catalogues.

While the excavators at Pompeii took colored wall paintings for granted, it was not so for the many marble statues unearthed at the same time. The 1760 discovery of remains of color on a statue of Artemis (now at the Archaeological Museum, Naples) should have marked a turning point leading to a surge of research on this topic but, instead, a pall of silence descended for the next 200 years.[50]

7.5   ARCHITECTURE AND SCULPTURE

Departing from the *Domus Aurea*, a 20 minute sprint along the Via dei Fori Imperiali takes you to Trajan's Column, the imposing monument built to commemorate the Emperor's victories in the Dacian wars (101–106 CE). Acid rain and pollution over the course of two millennia have worn down the relief sculptures into indistinct and unrecognizable forms. However, in the 1500s, a set of plaster casts of the column were made before the damage was done. They are now conserved in the Museum of Roman Culture. Despite the toll of centuries, the quality of the fine Carrara marble, pristine, pure, unsullied white, renders the column one of the most distinctive landmarks in Rome. Not far from Trajan's Column, on top of the Capitoline Hill, the Palazzo Nuovo houses one of the world's largest and most breathtaking collections of Greek and Roman sculpture; the quality of the form and of the marble are exquisite.

Trajan's Column and the Palazzo Nuovo sculpture collection appear quite differently now than they did 20 centuries ago. This realization was one of the great surprises of modern art historical research, as we shall see in the next section.

7.5.1   Canons of Beauty

One of the key events of the Renaissance was the 1506 discovery of the Laocoön statuary group, now residing in the Vatican Museums. Sixty years afterward, Francesco da Sangallo (1494–1576), an eyewitness to the event, described its discovery in a vineyard on the Esquiline hill.[51]

The statue (Figure 7.4),[52] the work of three noted sculptors of Rhodes, made its way to the Palace of the Emperor Titus in the first century CE; how it got from there to be buried in a vineyard for 1400 years is anybody's guess. Although the statue may not have been completely devoid of its original polychromy, it was

Figure 7.4 Laocoön and his Sons. First Century CE Greek. Vatican Museums. Reproduced from ref. 52 under the terms of the CC BY-SA 4.0 license, https://creativecommons.org/licenses/by-sa/4.0/deed.en.

apparently overlooked. Subsequent rigorous cleanings and polishing would have obliterated it almost completely.

Art historian Mark Abbe was literally "ambushed by color" when working at the Aphrodisias archaeological dig, a Greek site in Turkey, in 2000 where he found unmistakable traces of polychromy on ancient marbles. He immediately realized that the idea that the ancients disdained bright color is the most common misconception about Western aesthetics in the history of Western art, "a lie we all hold dear."[53] He later elaborated: "The monochromatic appearance of these works gave rise to new, modern canons of sculpture characterized by an emphasis on form with little consideration of color. In antiquity, however, Greek and Roman sculpture was originally richly embellished... Such polychromy...was integral to the meaning and immediacy of such works..."[54]

For well into the 20th century, Johann Joachim Winckelmann (1717–1768) was the household word on classical beauty. In his 1764 definitive work on the subject, *Geschichte der Kunst des*

Altertums, he asserted the Greeks' preference for pure white untreated marble statuary as opposed to polychrome statuary.[55] Put succinctly, he said that "the ancients knew the secret of finishing statues, merely with the chisel."[56] Even though Winckelmann himself later changed his mind, this opinion held sway in art history classrooms and museum exhibits around the world. Some complain that the continued practice of ignoring polychromy through lack of information on object labels, in audioguides, or in gallery tours distorts the historical character of the ancient world by displaying works in ways that never existed in ancient Greece and Rome. Though Winckelmann's opinion bears the brunt of 21st century criticism nowadays, one must recall that Michelangelo (1475–1564) was well aware of polychromy's existence on ancient statuary, yet chose to create his own in monochrome.[57]

7.5.2   Setting the Record Straight: Polychromy Rediscovered

For many, the rediscovery of polychromy was a very disruptive event. It overturned everything we had ever learned in school about Greek and Roman sculpture and, by extension, architecture. Shifting a viewpoint from the unadorned marble of sculpted portraits and carved architraves to one of lavish polychromatic decoration was a new and unexplored topic. Polychromy was an essential feature of the ancient sculptor's art: added color extended, completed and transcended the sculptural form. Taking issue with Winckelmann, we now know that the chisel does not have the last word, *i.e.*, the final surface texturing and detailing performed by the sculptor is not the visual end goal.[58]

This fact hit me in the face when Marco Leona, David H. Koch Scientist in Charge of the Department of Scientific Research at The *Metropolitan Museum* of Art ("the Met"), guided a group of "us chemists" from the American Chemical Society around the Greek and Roman gallery a few years ago. He took great pains to make sure that we saw the polychrome remains on some of the statues, though without raking light and a good magnifying glass, some of us remained non-believers.

Fortunately, detecting and reconstructing ancient polychromy is not a task left to the naked eye which, as we realized from our experience with Leona, is clearly inadequate for this task. In order to clean off bothersome crusts of "dirt," too many works of art have been scrupulously dispossessed of their original bright coverings down to the minutest detail. But the scientist now has a battery of multispectral photographic techniques that can snoop out the remains of color down to the microgram level. **Induced luminescence**, whether by ultraviolet (UVL) or by visible light (VIL), are two of the workhorses of the conservation laboratory, aided by optical microscopy, **fiber optics reflectance spectroscopy** (FORS) and portable **X-ray fluorescence spectroscopy** (p-XRF). Many ancient pigments, such as madder and Egyptian blue, strongly fluoresce when irradiated by one or more of these sources.[59,60] This allows for positive detection of their presence even if only a tiny bit of material remains hidden in a marble crack on the sculpture.[61]

The scientific principle behind fluorescence methods goes back to an Irish physicist, George Gabriel Stokes (1819–1903), who first described the phenomenon in 1852.[62] It is easily observed in the laundry if you add a "fluorescent brightener" to your wash. When the clothing absorbs ultraviolet radiation from sunlight, it re-emits the energy in the form of visible light in the blue region of the spectrum. The emitted blue light, as the spectral complement to yellow, masks any yellowing that the clothes might undergo with aging (**Chapter 1.4**). Ancient pigments tend to absorb radiation in the red region of the spectrum and re-emit the energy in the infrared region, which is not visible to the human eye. However, infrared-sensitive cameras can detect the radiation and visually pinpoint the fluorescing pigment.

Another powerful tool based on the fluorescence principle is **fluorescence spectroscopy**. It can identify both organic and inorganic pigments by producing a full spectrum readout characteristic of each species. It has been used in a variety of contexts, including the Domus Aurea investigations.[63]

Every visitor to Rome's "Forum Romanum" has either entered or exited that site within just a few yards of the Spolia Panel on the interior of Arch of Titus on the Via Sacra. Colorless as it is today,[64] modern detective work shows that in 81 CE when it was erected, it was beautifully polychromed. A team from Yeshiva University made a series of spectroscopic measurements using the methods named above. They analyzed and compared the results to known color databases to determine the original colors. The reconstruction, compared with the colorless "original," is shown in Figure 7.5.[65]

Using similar methodology, investigators in the 1980s found that even pristine "white" Trajan's Column bore a fair amount of color.[66] They thus laid to rest over a century of controversy by demonstrating that deliberately added orange minium, Pb_3O_4, once graced the column.

Figure 7.5   Reconstruction of the Arch of Titus. Top: Spoils of the Temple: After a Relief from the Arch of Titus, Rome by Jean-Guillaume Moitte, *ca.* 1791; Los Angeles County Museum of Art. Reproduced from ref. 64 with permission from Museum Associates/LACMA. Bottom: A reconstruction of the polychromy of the Arch of Titus Spolia panel; VIZIN: The Institute for the Visualization of History and the Yeshiva University Arch of Titus Project. Reproduced from ref. 65, courtesy of Steven Fine.

And many a visitor to London has caught at least a fleeting glimpse of the famous pure white Parthenon sculptures, resident aliens in the British Museum. Victorian painter, Lawrence Alma-Tadema (1836–1912), seemed to be one of the few 19th century artists who recognized the truth of their original polychromy through his own work: we need only take a glance at his 1868 oil, "Phidias Showing the Frieze of the Parthenon to his Friends" (Collection of the Birmingham Museums Trust) to see the riot of colors that issued from his imagination. Taking a lesson from Alma-Tadema, there is a collaborative ongoing project to create a digital "Virtual Parthenon" that will incorporate all the pieces and colors of the original structure to generate a multi-level, interactive educational tool.[67]

The fruits of heritage research will find an eager audience worldwide. Ancient polychromy can be communicated to many different audiences through guided tours, videos, conferences, podcasts, workshops, and online digital reproductions.[55] Hypothetical reconstructions of polychromy are powerful images when compared to the present state of the artifact. Many recent exhibitions in Europe and the USA have played a tutorial role in this regard. At the forefront we can name the J. Paul Getty Museum, Malibu, California, the Ny Carlsberg Glyptotek, Copenhagen,[68,69] and the Met's major exhibit on ancient polychromy. The Boston Museum of Fine Arts' conservation of its two prize Etruscan sarcophagi, which included a public viewing room where they were scanned for color, has also gone far to set the record straight.[70]

Even closer looks at church façades followed by scientific examination have revealed lost color. This realization has prompted experiments with digital color projections on the façades of the cathedrals at Amiens and Chartres as a way of visualizing color that once was.[71]

7.6   CONCLUSION

It is clear that this type of research is intrinsically multidisciplinary, international and unbound by constraints of time or space. As new methods of investigation and communication develop, they will surely be put to work to enhance the unexpected flowering of a new phase in art history. When the gods that govern Vesuvius's moods decided to go on a destructive rampage, they ironically created a gift package that will keep on giving to civilization well into the future. The analytical impetus at the beginning of the 19th century continues to enhance our understanding of this most important piece of history.

However, just as Pompeii crumbled under the impact of the Vesuvian nightmare, a slower but more inevitable trauma destroyed Rome's hegemony and led to a power vacuum that brought on her downfall. It is the story of how civilization responded to and rebounded from this breakdown that is the subject of the next chapter.

REFERENCES

1. A. Rios, We Are of a Tribe, in *Poetry of Presence: An Anthology of Mindfulness Poems*, Grayson Books, West Hartford, 2017, p. 51.
2. A. Angela, *I tre giorni di Pompei*, Rizzoli, Milano, 2014.
3. E. M. Moormann. Literary evocations of ancient Pompeii, in *Pompeii: Stories of an Eruption*, ed. P. G. Guzzo, Mondadori Electa S.p.A., Milano, 2005, pp. 15–33; 18.
4. R. Scandone, L. Giacomelli and M. Rosi, *J. Res. Didact. Geogr.*, 2019, **2**(8), 5–30.
5. H. Sigurdsson, S. Cashdollar and S. R. J. Sparks, *Am. J. Archaeol.*, 1982, **86**(1), 39–51.
6. A. P. Laurie, *The Materials of the Painter's Craft*, T.N. Foulis, London, 1910.
7. Karl Brullov – The Last Day of Pompeii, https://commons.wikimedia.org/w/index.php?title=File:Karl_Brullov_-_The_Last_Day_of_Pompeii_-_Google_Art_Project.jpg&oldid=438242075, accessed June 2021.
8. Pliny the Elder, *The Natural History*, ed. J. Bostock and H. T. Riley, http://www.perseus.tufts.edu/hopper/text?doc=Perseus:-text:1999.02.0137, accessed September 2020.
9. Letters of Pliny the Younger to Tacitus. The A.D. 79 Eruption at Mt. Vesuvius, https://igppweb.ucsd.edu/~gabi/sio15/lectures/volcanoes/pliny.html, accessed March 2021.
10. J. J. Nadolny, *Stud. Conserv.*, 2003, **48**(suppl. 1), *Rev. Conserv.* 4, 39–51.
11. S. G. Rees-Jones, *Stud. Conserv.*, 1990, **35**(2), 93–101.
12. J.-A. Chaptal, *Annu. Chim.*, 1809, **70**(2), 22–31.

13. N. Eastaugh, V. Walsh, T. Chaplin and R. Siddall, *The Pigment Compendium*, Elsevier, New York, 2004, p. 387.

14. R. J. Gettens and G. L. Stout, *Painting Materials: A Short Encyclopedia*, Dover Publications, New York, 1966, p. 117.

15. R. Harley, *Artists' Pigments C. 1600–1835*, Butterworths, London, 1970, p. 48.

16. H. Davy, *Philos. Trans. R. Soc. London*, 1815, **105**, 97–124.

17. J. A. Paris, *The Life of Sir Humphry Davy, Bart.*, Henry Colburn and Richard Bentley, London, 1831.

18. Humphry Davy, https://en.wikipedia.org/wiki/Humphry_Davy, accessed March 2021.

19. Aldobrandini wedding, https://commons.wikimedia.org/w/index.php?title=File:Aldobrandini_wedding.JPG&oldid= 428236502, Accessed June 2021.

20. M. V. Orna, Historic Mineral Pigments: Colorful Benchmarks of Ancient Civilizations, in *Chemical Technology in Antiquity*, ed. S. C. Rasmussen, American Chemical Society, Washington, DC, 2015, pp. 17–69.

21. S. Augusti, *I Colori Pompeiani*, De Luca, Rome, 1967.

22. I. Kakoulli, *Greek Painting Techniques and Materials from the Fourth to the First Century BC*, Archetype Publications, London, 2009.

23. R. E. Jones and E. Photos-Jones, *Technical Studies of Aegean Bronze Age Wall Painting: Methods, Results and Future Prospects*, British School of Athens Studies 2005, pp. 199–228.

24. F. Casadio, I. Giangualano and F. Piqué, *Rev. Conserv.*, 2004, **5**, 63–80.

25. R. Piovesan, R. Siddall, C. Mazzoli and L. Nodari, *J. Archaeol. Sci.*, 2011, **38**, 2633–2643.

26. R. Piovesan, *Plinius*, 2009, **35**, 157–163.

27. See A. Pelloux, *Am. Mineral.*, 1927, **12**(1), 14–21.

28. Roman fresco Villa dei Misteri Pompeii, https://commons.wikimedia.org/w/index.php?title=File:Roman_fresco_Villa_dei_Misteri_Pompeii_008.jpg&oldid=335982442, accessed June 2021.

29. B. Bergmann, Seeing women in the Villa of the Mysteries: A modern excavation of the Dionysian murals, in *Antiquity Recovered: The Legacy of Pompeii and Herculaneum*, ed. V. C. Gardner Coates and J. L. Seydl, J. Paul Getty Museum, Los Angeles, 2007, pp. 231–269.

30. F. Pesando, Shadows of light: Cinema, peplum and Pompeii, in *Pompeii: Stories of an Eruption*, ed. P. G. Guzzo, Mondadori Electa S.p.A., Milano, 2005, pp. 34–45.

31. F. Esposito, C. Falcucci and D. Ferrara, *Rivista di Studi Pompeiani*, 2011, **22**, 149–158.

32. S. De Caro, Excavation and conservation at Pompeii: A Conflicted History, *The Journal of Fasti Online, Archaeological Conservation Series*, 2015, www.fastionline.org/docs/FOLDER-con-2015-3.pdf, accessed September 2020.

33. M. Cotte, J. Susini, N. Metrich, A. Moscato, C. Gratziu, A. Bertagnini and M. Pagano, *Anal. Chem.*, 2006, **78**, 7484–7492.

34. R. Nöller, *Stud. Conserv.*, 2015, **60**(2), 79–87.

35. H. Brecoulaki, *Revue Archéologique*, 2014, **1**, 1–35.

36. J. Powell, *Cinnabar, Substrate*, Pantheon, New York, 2009, pp. 112–113. Reprinted with permission.

37. C. Canevali, P. Gentile, M. Orlandi, F. Modugno, J. J. Lucejko and M. P. Colombini, *et al.*, *Anal. Bioanal. Chem.*, 2011, **401**, 1801–1814.

38. T. Christiansen, *Bull. Am. Soc. Papyrol.*, 2017, **54**, 167–195.

39. F. Flieder and M. Duchein, *Livres et documents d'archives: sauvegarde et conservation*, UNESDOC, UNESCO Digital Library, 1986, p. 21, https://unesdoc.unesco.org/ark:/48223/pf00 00056224?posInSet=1&queryId=992578b5-58c6-42a9-bdcb-bb30d2c21e91, accessed September 2020.

40. M. L. Amadori, S. Barcelli, G. Poldi, F. Ferrucci, A. Andreotti, P. Baraldi and M. P. Colombini, *Microchem. J.*, 2015, **118**, 183–192.

41. D. Cottica and G. A. Mazzochin, Pots with coloured powders from the forum of Pompeii, in *Vessels inside and outside*, ed. K. T. Biró, V. Szilyág and A. Kreiter, EMAC07, 24–27 October 2007, Hungarian National Museum, Budapest 2008, pp. 151–158.

42. R. Cioni, L. Gurioli, R. Lanza and E. Zanella, *J. Geophys. Res.*, 2004, **109**(B2), B02207.

43. C. D. Harris, Cinnabar: the symbolic, seductive, sublethal shade of Pompeii, Master's thesis, Brandeis University, Waltham, MA, 2015, p. 36.

44. I. Marcaida, M. Maguregui, S. Fdez-Ortiz de Vallejuelo, H. Morillas, N. Prieto-Taboada, M. Veneranda, K. Castro and J. M. Madariaga, *Anal. Bioanal. Chem.*, 2017, **409**, 3853–3860.

45. C. Clementi, V. Ciocan, M. Vagnini, B. Doherty, M. L. Tabasso and C. Conti, *et al.*, *Anal. Bioanal. Chem.*, 2011, **401**, 1815–1826.
46. S. E. J. Villar and H. G. M. Edwards, *Anal. Bioanal. Chem.*, 2005, **382**, 283–289.
47. L. F. C. Oliveira, H. G. M. Edwards, R. L. Frost, J. T. Kloprogge and P. S. Middleton, *Analyst*, 2002, **127**, 536–541.
48. R. F. Beeston and H. Becker, Investigation of ancient Roman pigments by portable X-ray fluorescence spectroscopy and polarized light microscopy, in *Archaeological Chemistry VIII: American Chemical Society Symposium Series 1147*, ed. R. A. Armitage and J. H. Burton, American Chemical Society, Washington, DC, 2013, pp. 19–41.
49. H. Becker, Color Technology and Trade, in *Cultural History of Color*, ed. D. Wharton; series eds. C. Bigham and K. Wolf, Bloomsbury Publishing, New York, 2021, vol. 1, pp. 35–48.
50. J. S. Østergaard, *Polychromy, Sculptural, Greek and Roman, Oxford Classical Dictionary*, https://oxfordre.com/classics/view/10.1093/acrefore/9780199381135.001.0001/acrefore-9780199381135-e-8118, accessed September 2020.
51. Letter of Francesco da Sangallo, quoted in L. Barkan, *Unearthing the Past: Archaeology and Aesthetics in the Making of Renaissance Culture*, Yale University Press, New Haven, 1999, p. 3.
52. Laocoon and His Sons, https://commons.wikimedia.org/w/index.php?title=File:Laocoon_and_His_Sons.jpg&oldid=445114392, accessed June 2021.
53. M. Talbot, *The New Yorker*, 29 October 2018, 44–51.
54. M. B. Abbe, *Polychromy of Roman Marble Sculpture, Heilbrunn Timeline of Art History*, 2000. http://www.metmuseum.org/toah/hd/prms/hd_prms.htm, accessed September 2020.
55. C. Brøns and M. Papadopoulou, True colors: polychromy in ancient Greek Art and its dissemination in museum collections, in *Color Culture Science*, ed. M. Godýn, B. Groborz and A. Kwiatkowska-Lubańska, Akademia Sztuk Pięknych im. Jana Matejki, Kraków, 2018.
56. J. J. Winckelmann, *Reflections of the Painting and Sculpture of the Greeks: With Instructions for the Connoisseur, and an Essay on Grace in Works of Art*, trans. H. Fusseli, A. Millar and T. Cadell, London, 1767, p. 269.

57. M. Combs, The polychromy of Greek and Roman art; an integration of museum practices, MA Dissertation, City University of New York, 2012.

58. A. Skovmøller, Portraits and colour-codes in ancient Rome: the polychromy of white marble portraits, PhD Dissertation, Faculty of Humanities, University of Copenhagen, 2015, p. 243, 54.

59. D. Magrini, S. Bracci, G. Bartolozzi, R. Iannaccone, S. Lenzi and P. Liverani, *Heritage*, 2019, **2**, 2160–2170.

60. J. Dyer and S. Sotiropoulou, *Heritage Sci.*, 2017, **5**, 24–45.

61. I. Kakoulli, R. Radpour, Y. Lin, M. Svoboda and C. Fischer, *Dyes Pigm.*, 2017, **136**, 104–115.

62. B. Valeur and M. N. Berberan-Santos, *J. Chem. Educ.*, 2011, **88**, 731–738.

63. A. Romani, C. Clementi, C. Miliani and G. Favaro, *Acc. Chem. Res.*, 2010, **43**(6), 837–846.

64. Jean-Guillaume Moitte – Spoils of the Temple – After a Relief from the Arch of Titus, Rome, https://commons.wikimedia.org/w/index.php?title=File:Jean-Guillaume_Moitte_-_Spoils_of_the_Temple-_After_a_Relief_from_the_Arch_of_Titus,_Rome.jpg&oldid=118091120, accessed June 2021.

65. S. Fine, P. J. Schertz and D. H. Sanders, *Biblical Archaeology Review*, 2017, **43**(3), 28–3561.

66. M. Del Monte, P. Ausset and R. A. Lefevre, *Archaeometry*, 1998, **40**(2), 403–412.

67. D. Williams, P. Higgs, T. Opper and M. Timson, A virtual Parthenon metope: restoration and colour, in *Gods in Colour: Painted Sculpture of Classical Antiquity*, ed. V. Brinkmann and R. Wunsche, Stiftung Archeölogie, Glyptothek München, 2007, pp. 112–117.

68. M. Bradley, *Art History*, 2009, **32**(3), 427–457.

69. E. M. Moormann, *La pittura parietale Romana come fonte di conoscenza per la cultura antica, Scrinium: Monographs on history, archaeology and art history*, no. 2, Van Gorcum, Assen, 1988.

70. Conservation in Action, Boston Museum of Fine Arts, https://www.mfa.org/collections/conservation/conservationinaction_etruscansarcophagi/june2013, accessed June 2021.

71. B. Kiilerich, *J. Art Historiography*, 2016, **15**(December), 1–18.

CHAPTER 8

Monastery Mysteries: Illuminating the Dark Ages[†]

Ignorance is the curse of God; knowledge is the wing wherewith we fly to heaven.[1]

William Shakespeare

8.1 INTRODUCTION

From feeble beginnings that counterbalanced the disintegration of the Roman Empire, a great flowering of culture, literature, scholarship and art brought stability to the early medieval world in the form of monasteries. Some claim that this development actually saved Western civilization from irreversible destruction.

8.2 SAVING CIVILIZATION: 5TH – 6TH CENTURIES CE

While the ancient Roman Empire was splintering under the onslaught of Lombards, Huns and Ostrogoths pouring across its exposed borders, a fleeting glimmer of light was flickering on the

---

[†]Glossary entries and chapter cross-references are in boldface.

March of the Pigments: Color History, Science and Impact
By Mary Virginia Orna
© Mary Virginia Orna 2022
Published by the Royal Society of Chemistry, www.rsc.org

far-off isle of Eire. It was here on the far western edge of Europe
that inspired saints like Brigid (*ca.* 451–525) and Colmcille
(Columba, *ca.* 520–597) were founding those exemplars of stability and peace, monasteries. While the Empire was imploding
under its own shortages of manpower, finances and supplies,
the Irish were welcoming refugees fleeing the chaos that reigned
elsewhere. Among them may have been monks from Gaul who
learned the art of manuscript illumination from monks from
Egypt.[2] While the newcomers displayed the wealth of books and
manuscripts they had carried with them, the Irish were exchanging their Ogham markings[‡] for Hebrew, Greek and Latin texts.
Within a few decades, the monasteries grew into not only economic and agricultural centers, but also establishments of literacy and learning. Not only did the monks copy and preserve
the precious heritage in foreign books, but they also began to
craft their own with unmatched creativity. The Irish monastic
movement was instrumental in preserving much of the written
heritage of Western civilization and then reintroducing it to continental Europe during the centuries that followed. The illuminated manuscript was the torch that set the gloom of the Dark
Ages ablaze with light again.[3]

In addition to the manuscripts' precious textual content and
their legacy of literacy, the images they contain present us with
a most fruitful resource for the identification of the medieval
and early Renaissance artists' palette. For the most part, they
have not suffered from environmental, physical or restorative
damage, so "what is" in a manuscript now is "what was" when
it was created.[4] In this chapter, we will follow the building up of
this reliable body of work, early attempts to examine it, and the
development of non-invasive techniques for its definitive study.
We will follow the marvelous adventures of a single pigment,
ultramarine, as its presence in manuscripts has signposted wom-
en's role in manuscript illumination, cultural dissemination,
forgery, and the roots of its own disease. Finally, we will revel
in the gigantic structural outcome of illumination through a
look at the stained glass palette of the great medieval cathedrals.

[‡]Ogham was an early form Irish writing that consisted of marks scratched on sticks to
indicate phonetic sounds that could be interpreted for words.

This journey will drop us off at the early Renaissance where we will take up wall painting and easel painting once more.

8.3 A TUMULTUOUS TIMELINE

Though the dates for the divisions of the Middle Ages are elastic, key events are handy for recall's sake. For our purposes, we can date the beginning of the Early Middle Ages (the so-called Dark Ages) from the fall of the Roman Empire, in 476 CE, until the accession of Charlemagne as Holy Roman Emperor in 800 CE. For the next three centuries of the so-called Carolingian Renaissance, until about the end of the eleventh century, commerce, urbanization, and scholarship gradually continued to rebound. Former slaves, barbarians, migrants and refugees morphed into the landed gentry of the countryside and the artisans and craftsmen of commercial centers. Most historians date the High Middle Ages from 1100 CE until the onslaught of the Black Death in 1347. The Late Middle Ages range from 1350 until Columbus's first voyage to the New World in 1492. During this very colorful and tumultuous period, the world's greatest volume of manuscript production took place, from Ireland in the West to Armenia and Persia in the East. Two events in the fifteenth century, just prior to Columbus, marked the illuminated manuscript's dénouement: the invention of the printing press in 1440 and the Ottoman conquest of Constantinople in 1453. Printing effectively put an end to the slow, labor-intensive, error-prone exercise of manual copying; Islam's hegemony in the east ended the activity, and often the existence, of monasteries within its boundaries.

8.4 TWO IRISH MANUSCRIPTS

Colmcille was a man of great energy but also from a well-positioned family that could support his efforts with generosity. Within 41 years he had founded 41 monasteries, including two of the most famous, Durrow and Kells. Both became great seats of artistry and learning, the sources of two of the world's most famous books or, strictly speaking, codices.[§]

[§]A codex, the successor of the scroll, was a stack of individual sheets of parchment or vellum, loosely bound.

The Book of Durrow, dating from the late 7th century, is a gospel book on vellum decorated with beautiful calligraphy and iconography. Presently lodged at the Trinity College, Dublin library, legend has it that while it was out in the "public domain," its spiritual power was invoked by a farmer who dipped it into a vat of water in order to make holy water to feed to his sick cows.[5] One can daydream about the spirit of Colmcille browsing through its pages as he sat embowered under the spreading branches of his beloved cathedralesque oak trees, the non-negotiables around which he built his monastery (Box 8.1).[6]

Durrow was the model for the Book of Kells (Figure 8.1)[7] that was crafted at the island monastery of Iona in the late 8th century, but the scribe was no copyist. Rather, he took the opportunity to create what has been called "the most beautiful book in the world." A masterpiece of Celtic art, complete with intricate knot designs, interlocking patterns of foliage, animals real and imaginary, and a depth of color that is still wondered at today, it is also a wonder that it has survived fires, Viking pillage, and the passage of time. What it did not survive was a conscientious librarian who, in the early 1800s, cut the pages down "to size," all neat and tidy, trimming irreplaceable embellishments out of existence.[6] Today, Kells also resides at Trinity College Dublin, not quite enjoying a quiet monastery existence. Second only to the Guinness Warehouse as the most popular tourist attraction in Dublin, visitors, by reservation, wait for hours in snaking lines to enter a crammed display center filled with outsized reproductions of some of the most beautiful pages. In the next room, to

---

Box 8.1   Parchment and Vellum

Parchment is a general term for any animal skin, *e.g.*, sheep, goat or calf, that is prepared for writing. Vellum, derived from the Old French for calf, "vélin," is specifically calfskin, considered much finer than other types of skin. Skin preparation involves soaking in water for one day followed by several days longer in a solution of lime to eliminate the hairs and residual flesh. The skins are then dried and stretched on a frame for further manual de-hairing. Of the skins' two sides, the inner side is finer and often used for the "recto" side in a manuscript; the outer side, which may bear vein marks or scars, is often used as the "verso" side of a manuscript page. Between two and three hundred skins were needed to produce enough material for a large Bible.

Figure 8.1   Book of Kells, Folio 32v, Christ Enthroned. Scanned from Treasures of Irish Art, 1500 BC to 1500 CE. From the Collections of the National Museum of Ireland, Royal Irish Academy, & Trinity College, Dublin. Reproduced from ref. 7.

view the actual manuscript, you just might succeed if you first honed your elbows in a New York City subway car at rush hour – I, personally, found the visit more difficult than those to the Mona Lisa and the Rosetta Stone. The legend, or at least the venue, of the Book of Kells found new life in an animated 2009 film, "The Secret of Kells," produced by Cartoon Saloon, a studio that specializes in hand-drawn rather than computer-generated graphics. The film was a runner-up for Best Animated Feature at the Oscars.[8]

8.5   MANUSCRIPTS: THE MONASTIC LEGACY

In addition to the scriptorial activities of the Irish monasteries, two other movements, both in Italy, served to provide the crumbling social order with a sort of life raft to weather the storm of the Empire's collapse. Benedict of Nursia (*ca.* 480–550), to escape Rome's licentious society, retreated to Subiaco, and later to Monte Cassino, to lead an eremitic life. Joined by likeminded confrères, he wrote a rule toward the end of his life that included in the daily order prayer, manual labor, and reading. Thus, literacy became a given if you wanted to be a monk. Farther south in Italy, Cassiodorus (*ca.* 485 – *ca.* 585) added writing to his rule, rendering the copying of texts an important part of monastic life. Gradually, with the deterioration of infrastructure and especially the famous Roman roads (there was no army to use them anymore), small localized self-sufficient centers developed, many of them ordered on the monastic model. In time, both in these monastic enclaves and those in Ireland, texts of both pagan and Christian works filled their own bookshelves or were exported to new foundations. For all the centuries that the lamps of learning remained dimmed, this important service continued.

8.5.1   Painting Manuscripts: What the Experts Say

Some of the references that may have been consulted by the late-Roman and early medieval manuscript compilers are given here. We must keep in mind that the resources used by the Irish artists are largely unknown because they arrived randomly, brought by happenstance refugees for the most part. On the other hand, the prolific manuscript painters in the Byzantine, Armenian, Turkish, Persian, Iranian and Indian worlds far to the east, held fast to their own traditions, some of which will be differentiated in what follows.

8.5.1.1   Pliny the Elder.   Pliny the Elder's massive 59 chapters of Book XXXV in his "Natural History,"[9] describe all that was known in the ancient world about the art of painting. This work, practically the only one that describes the methods and materials of artists of the time, is used as a reference for the history of art even today. Among his comments about the actual materials, he also tells a few tales about some famous artists. In Chapter 36, there is one about Alexander the Great's favorite painter, Apelles of Cos, a rather competitive chap who felt he was the most realistic painter in the world. To prove his point, he once took a group of horse paintings done by other artists as well as one that he himself had painted. One by one, he held each painting up to be viewed by a group of horses. It was only when they saw his own painting that they began to whinny (one assumes with desire, not with horror). In Chapter 40, Pliny mentions, among other women painters, Iaia of Cyzicus, who did her art at Rome, painting with the brush, and also engraving upon ivory. A specialist in portrait painting, she even captured her own image with the aid of a mirror. Pliny expatiates: "[H]er artistic skill was such that her works sold at much higher prices than those of the most celebrated portrait-painters of her day." His accompanying book on art history contains many other delightful tales and interesting studio gossip.[10]

8.5.1.2 Other Key Resources. Two third century CE Alexandrian Greek papyri, now held at Leyden and Stockholm, are the last surviving technical manuscripts written in a European language until the reappearance of the technical tradition around the year 800.[11,12] These manuscripts describe, among other things, metals, alloys, and artists' colors. The basic material they contain somehow survived over five centuries only to re-emerge in two Latin manuscripts in the late eighth or early ninth century. One of these, *Mappae Clavicula*, or *A Little Key to the World of Medieval Techniques*, contains numerous recipes for making dyes and pigments.[13] The other, contemporaneous with the Mappae, the Lucca Manuscript or *Codex Lucensis 490*,[14] contains much of the same material. Together with these two manuscripts, two others, the tenth century work of Heraclius, *De diversibus artibus Romanorum*,[15] and the twelfth century treatise of Theophilus (Roger of Helmarshausen),[16] comprise what would have been available to the late medieval manuscript painter.

8.6 WHY STUDY MANUSCRIPTS? WHAT CAN WE LEARN?

The first and most compelling reason for manuscript study is that of heritage: custodians of a piece of priceless patrimony need to gain as much knowledge about it as possible. There are so many unknowns with respect to a medieval manuscript, beginning with its origin, production method, date executed, history, techniques and materials used, the painter or painters' identities, or number of painters, iconography, lines of influence between and among painting schools and cultures, conservation and handling history, restoration history, and present and former environments. Among the major materials used are pigments and their binders. Their identification can often differentiate what was shared and what was distinctive to a given time and place, can ferret out anachronistic materials indicating either restoration or forgery, and can enable comparison with modern reference standards.[17] Pigment identification is also key to shedding light on artists' concerns with texture, color permanence, compatibility when placed near other pigments, fastness in various media, transparency or opacity, wettability, oil absorption, drying effects and toxicity.[18] Finally, there is the time-bomb of degradation: pigment identity is essential for determining pigment vulnerability under a variety of circumstances. For example, a major degradation product of red lead, Pb_3O_4, is black galena, PbS, and/or dark brown plattnerite, PbO_2.[19] This author has witnessed holes eaten through as many as 20 parchment pages of a 10th century manuscript because of the application of synthetic verdigris: the copper decomposed into black copper oxide, and the acidic medium caused the physical damage. Unraveling these questions and intervening when necessary can provide us with new insights into the expansion of culture and technology through artistic media.

8.7   SCIENTIFIC EXAMINATION OF MANUSCRIPTS

Prior to the early 1960s, manuscript examination was the purview of the art historian who relied primarily on formal stylistic analysis for describing artistic works and the interrelationships among artists' schools.[20] Things changed rather quickly in the following decades.

8.7.1   Beginnings. Visual Comparison with Standard Samples

Actual pigment analysis by a form of material examination began with the work of Heinz Roosen-Runge and Tony Werner in the early 1960s. Their approach to the Lindisfarne Gospels, a precious "untouchable" Gospel book, was visual comparison of the various pigment surfaces in the manuscript with standard samples manufactured according to instructions in medieval painters' manuals.[21] They reported the use of verdigris, minium, yellow ochre, ultramarine, orpiment, **indigo** and several mixtures (**Table 4.1**). Such visual assessment, as opposed to a quantitative spectroscopic method, is notoriously error-prone. A 2002 study showed definitively that ultramarine and yellow ochre were not present in the manuscript. In fact, despite many assertions to the contrary, ultramarine did not appear in European manuscripts until the early 11th century.[22] Any ultramarine pigment used in Europe prior to this date was more than likely purchased as a pre-prepared pigment[23] and, incidentally, around that time, vermilion gradually replaced minium.[24,25] However, to be fair, Roosen-Runge and Werner did a pretty good job with the means at their disposal.

The analytical history of the Book of Kells parallels that of the Lindisfarne Gospels. In 1960, Roosen-Runge and Werner used the same methodology and came up with comparable (sometimes erroneous) results.[21] Two additional examinations by Cains in 1989[26] and Fuchs and Oltrogge in 1991–1992[27] enlarged upon the original report, bringing the number of observed pigments to 28. Finally, in 2008, a team from Trinity College Dublin issued their results based upon micro-**Raman spectroscopy** and came up with a surprisingly very simple list of pigments: red lead, orpiment, **indigo**, gypsum, carbon black, and iron gall ink (**Chapter 14.3**). However, this simple palette mirrors the circumstances under which the manuscript was made: local sources provided the materials. The scribes used whatever was to hand where the manuscript was made, presumably on the Isle of Iona.[28] What is amazing is the magnificent effect that has been accomplished with so little.

8.7.2 Approach by Sampling Manuscripts

Although sample extraction from manuscripts is on the proscribed list today, the rationale in the 1980s made it the method of choice: it can be accomplished in the manuscript's home base as a one-off operation, particles are not discernible by the naked eye, and definite identification by means of **X-Ray Diffraction** and **Fourier Transform Infrared Spectroscopy** is possible. For the first time, the palette of an early 14th century Armenian manuscript, the Gladzor Gospel Book, was done in this way. Fortunately, much of the pigment had transferred to parts of the opposite pages as offsets, so these could be sampled a bit more generously. Stylistically, five different hands were identified and the differences in their pigment usage revealed the presence of two different workshops. The master painter and his assistant worked together; the three other painters, apparently apprentices, worked in a separate place, and possibly at a different time. Except for madder **lake** and gamboge, all pigments were **inorganic** (mineral) in origin. For the blue pigment, the master painter used azurite, his assistant used high quality ultramarine, and the apprentices all used a lower quality ultramarine. All used orpiment and vermilion. Only the apprentices (Figure 8.2)[30] used white lead. All of them used finely beaten thin gold leaf to lay down a background upon which all the other pigments were painted.[30] The most rewarding result from this analysis is that stylistic distinctions between workshops coincided with differences in the chemistry of the palette.

Twenty-three additional Armenian manuscripts ranging in date from 908 to 1575 CE of different provenances[†] were analyzed and, with the exception of two **organic** pigments, **indigo** and madder **lake**, all the pigments were mineral in origin. This fact was evident by the general appearance of the manuscripts. Invariably, every page was brightly illuminated, looking like it may have been painted yesterday.

By way of contrast, the colors in the Byzantine manuscripts analyzed appeared more restrained. Eight manuscripts from the University of Chicago Special Collections ranging in date from the 10th century to the late 13th century made up this cohort. While the page borders glowed with enamel-like reds and blues and the skies shone with gold, the figures were painted in pastel shades with muted backgrounds. Our analysis found that the palette was the reason for the difference. Byzantine artists relied heavily on **organic** pigments which generally yield weaker and

---

[†]The manuscripts were from the Freer Gallery of Art (3), University of Chicago (1), the Morgan Library (2), the Metropolitan Museum of Art (1), the Monastery of San Lazzaro, Venice (4), the Monastery of Saint James, Jerusalem (8), and the Walters Art Gallery (4).

Figure 8.2   Gladzor Gospel Book of UCLA, p. 312: Visitation of Mary to Elizabeth, attributed to one of the "apprentice" painters, T'oros of Taron, one of the few medieval artists who actually signed his name to his work. The blue is ultramarine, the red is vermilion, the yellowish is orpiment and massicot, the green is mixed orpiment and ultramarine. Reproduced from ref. 29 with permission from the University of California at Los Angeles Special Collections.

more transient colors. The mineral pigments were the bright red vermilion and the brilliant blue ultramarine that lit up the borders; all the rest of the colors were organic in origin, which can be quite vivid when first applied, but molecular degradation over a period of many centuries has led to the subdued color effects we observe today.[31]

These observations are an object lesson for conservators. Knowing the identities of the pigments can give clues to how the manuscripts should be displayed. Armenian manuscripts do not need subdued lighting because their mineral colors will not fade. On the other hand, Byzantine manuscripts must be displayed in subdued light and each page should have a limited time of exposure.

8.7.3 Subsequent Work

Over the years, from these first forays into manuscript analysis, the analytical data have mushroomed, largely due to new and improved instrumental methods of analysis and people wanting to take a fresh look at the enormous volume of manuscript holdings in the major libraries of Europe. The acknowledged pioneer in this effort is Robin Clark of University College London whose groundbreaking work with **Raman spectroscopy** on a 13th century Paris Bible sowed the seeds for all future work.[32,33]

Some selected interesting discoveries are described here. The use of ultramarine in a 6th century Byzantine manuscript has been proven[34] although it took five more centuries for this pigment to find its way into northern Europe. A surprise discovery was the presence of Tyrian purple in the 6th century *Vienna Genesis* (Ms. Codex Vindobonensis Theol.gr.31). Another 6th century Greek manuscript provides the first evidence of the use of elderberry **lake** as a pigment in a manuscript.[35] Direct results on analysis of Anglo-Saxon manuscripts (600–1066 CE) revealed, with the exception of **indigo**, the exclusive use of **inorganic** pigments.[36] Almost 200 samples of Italian manuscripts, dated from the 12th to 17th centuries, showed consistency of a fixed palette regardless of region or date of composition, as well as regular addition of hematite to azurite paints, and the use of a crystalline form of iron gall ink as a pigment.[37] Analysis of some Persian manuscripts revealed use of realgar and pararealgar, pigments very seldom found in Western manuscripts.[38] Working from a 2003 proposal that brazilwood pigments were characteristic of 15th century French manuscript illumination workshops,[39] they were successfully identified in three such manuscripts for the first time. Because of its low concentration in the paint, brazilwood is invisible in infrared spectra; it was successfully identified by a combination of **fiberoptic reflectance spectroscopy** and micro-**fluorescence spectrometry**, the only two techniques capable of this analysis.[40] And finally, the purported revival of dyeing manuscript pages with Tyrian purple during the Carolingian Renaissance revealed instead the use of a more practical substitute, folium, extracted from the plant *Chrozophora tinctoria*.[41] Figure 8.3 [42] is a colorful graphic of the delights to be found in so many of these illuminated masterpieces.

Figure 8.3 Major colorants characterized in medieval manuscripts (eleventh to the fifteenth centuries). Reproduced from ref. 42 with permission from Walter de Gruyter & Co., Copyright, 2019.

In the Late Middle Ages, red lead or vermilion often marked the end or beginning of a section in the manuscript; red was also used in books of the Mass or liturgical texts to denote section headings or to give directions: sit, stand, kneel, *etc.* From the Latin word for red, "rubrica," these directions were called rubrics (and they still are). The meaning of the word has expanded into other contexts as well.

Analysis has revealed that the major binding media for pigments in manuscripts were egg white (glair), egg yolk, and gums, chiefly gum Arabic. The latter was the exclusive binder used for inks. A major additive to all media was honey.[43]

We now single out one major medieval pigment for a closer look. Its origin in the virtually inaccessible valleys of the Hindu Kush was extraordinary enough (as told in **Chapter 5.5.2**). Its presence in European painting is no less so.

8.8 NATURAL ULTRAMARINE

Ultramarine has a long history in European painting both in its Nature-given form and in a commercially-instigated modification. In the literature, it has many names that speak of its origin, processing and use: lapis lazuli, lapis, ultramarine, azure, azure blue, ultramarine blue, lazurite, *etc.*[44] We single it out here for extended discussion due to the role it has played in providing some very interesting data both by its presence and its absence in certain contexts.

8.8.1 Finger Lickin' Good

In 1327, reading Aristotle's *Poetics* could have been risky business at Abo of Fossanova's northern Italian monastery, especially if you licked your fingers in order to turn the pages. The latter were all laced with some quick-acting and lethal toxin that left the victim's tongue and fingers black.[45] Two centuries later, just handling a manuscript by its binding could put you in touch with arsenic in the form of orpiment, As_2O_3, or worse yet, its volatile derivative, arsine, AsH_3.[46] And in the 20th century, we could tune into the "Lip, dip, paint. Lip, dip, paint," mantra sung by the women dial painters at their new jobs at the US Radium Corporation in Orange, New Jersey. As they painted, their camel-hair brushes would lose their shape, so they naturally pointed their brushes with their lips. They were enjoying an economic boom brought on by the Defense Department's contract for luminous watch dials and instruments for the military from 1917 to 1925. However, many of these workers began to notice constant bleeding of their gums, gradual anemia, and necrosis of the jaw. An estimated 4000 women were recruited into this work in the US and Canada; many of them fell sick; an unknown number died of radiation sickness.[47] Any of these toxins could enter the body by ingestion, inhalation or absorption through the skin.

8.8.2 Women's Work?

Zoom back to 1160 CE to a nun's burial in a monastery cemetery in Dalheim, Germany. In 2014, following excavation of the remains, identified as Individual B78, by the Westphalian Museum of Archaeology, the archaeologists began looking for dietary clues *via* starch granules that may have been entrapped in her teeth. To their astonishment, they found numerous blue particles dispersed throughout her dental calculus. When they sought help from physicists, the response was "We don't know what you're talking about." Art restorers asked, "Why are you working with plaque?" Finally, with some help from scientists at the University of York, they learned that the particles were lazurite, the mineral derived by grinding down the semiprecious stone, lapis lazuli. "How can this be?" they wondered. This woman was in very close and personal contact with one of the most precious materials in the medieval world.

When the researchers Christina Warinner and Anita Radini suggested that this individual painted manuscripts and shaped her paintbrush in the time-honored manner described for the Radium Girls, they were dismissed by the art history community. In fact, one person remarked that the nun was probably just the cleaning lady.[48] Several other scenarios were possible: (1) she was grinding lazurite and inhaled some particles (close to the cleaning lady hypothesis); (2) she was using the lazurite as a medicament; (3) the particles entered her mouth through devotional osculation, *i.e.*, frequent kissing of the sacred text. Given the incredible expense involved in importing this virtually inaccessible material, scenario 3 is highly unlikely, as is scenario 2 since there is very little evidence for either practice in medieval Germany. If we accept the facts at their face value, they mean that this woman represents the earliest direct evidence of lazurite pigment usage by a religious woman in Germany.[49] When Warinner dug further, she learned that historians have been cataloging hitherto overlooked women's contributions to medieval book production, even finding some surviving manuscripts signed by women.[48,50] One example, MS GER 4, written in the Alsatian dialect, was created and signed by a woman religious.[51] Figure 8.4[52] shows the nun's dental calculi under discussion.

8.8.3 Ultramarine: Nature's Gift to the Art World

Though Nature saw fit to put one of her most beautiful and unique creations in a remote and inaccessible valley in Afghanistan, the very challenge in obtaining it has made it all the more appreciated (**Chapter 5, BOX 5.1** documents the journey). Some notable occurrences of natural ultramarine's presence attest to its use from ancient times and in many places, originally as a pharmaceutical. Artistically, archaeologists have unearthed thousands of cylinder seals, necklaces, and statuettes from the great cemetery at Ur in

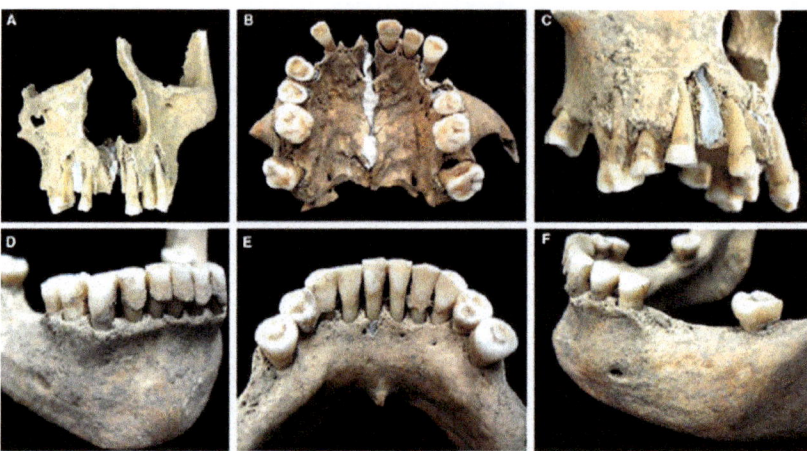

Figure 8.4   Distribution of dental calculus deposits on the dentition of individual B78. (A) Anterior, (B) inferior, and (C) lateral views of the
maxillary dentition prior to sampling. (D) Anterior, (E) partial pos-
terior, and (F) lateral views of the mandibular dentition prior to
sampling. Credit: Christina Warinner, Institute for Evolutionary
Medicine, University of Zürich, Zürich, Switzerland. Reproduced
from ref. 52 with permission from the American Association for
the Advancement of Science, Copyright 2019.

Mesopotamia – created 4000 years ago from material transported
more than 2500 km from its original source.[53] Although the current literature places its first recorded use in the wall paintings
of Mansur Depe, Turkmenistan, Central Asia in the second century BCE,[54] this record has been superseded by use at the island
citadel of Gla (Greece) in the Late Bronze Age. Not only is this use
a first for wall paintings, but also for its use as *buon fresco*. This
method put the pigment in intimate contact with lime, a practice
that later practitioners said would cause fading – however, evidence collected from many sites show that this is not so. Lapis
lazuli was also mixed with red iron(III) oxide (red ochre) to form
a purple pigment as well.[55] This discovery moves that date for
earliest occurrence on plaster back by about 2000 years.[56]

Not surprisingly, considering its proximity to the source, it was
also identified on the famous 6th century CE Bāmiyān Buddhas
of Afghanistan that were destroyed in 2001.[57] It was found, by
Raman spectroscopy, on a mummy coffin lid from the Egyptian
17th Dynasty (1580–1550 BCE), although the scientists involved
were surprised by its presence and speculate that this unusual

pigment could have its origin in a 20th century restoration.[58] Natural ultramarine was also a constant feature of high-end jewelry in the Middle East. It was used almost universally in Byzantine and Armenian medieval manuscripts.[31]

In the West, ultramarine was first identified on a 2nd century CE Gallo-Roman statuette found in an excavation at Argenton-sur-Creuse, France,[59] and in blue enamels of Roman glass.[60] As already noted in this chapter, its unusual presence in the dental calculi of a known German manuscript painter of the late 11th – early 12th century points to its use in manuscripts of that time as well.[47] Because of its expense, it was not lavishly used in Western art although many instances of its use among Renaissance painters have been found in the National Gallery, London.[61] Michelangelo (1475–1564) used it sparingly when painting the ceiling of the Sistine Chapel in 1511 (Figure 8.5),[62]

Figure 8.5 Michelangelo's Zacharias the Prophet on the Sistine Chapel Ceiling (1511). Note the blue collar rendered in natural ultramarine pigment (arrow). Adapted from ref. 62.

but quite lavishly in the Last Judgment (1534–1541) when the Pope agreed to underwrite his efforts.[63] A famous attempt at forgery on the part of the 20th century sometime artist, Han van Meegeren (1889–1947), backfired when he purchased very expensive natural ultramarine from an art dealer, unaware of the fact that it had been adulterated with a modern cobalt-containing pigment.[64]

8.8.4   Is There a Doctor in the House? Ultramarine Disease

Artists have noted over the years that ultramarine pigment suffers from a peculiar "disease" whereby the bright blue color turns a spotty gray. Chemists have puzzled over this phenomenon as well, at first thinking that the degradation was due to loss of ultramarine's sulfur content. In 2006, a research team at New York University tested the possibility of degradation of the pigment's silicate-aluminate cage. This would release the enclosed **chromophore**, the color-bearing portion, the trisulfide radical anion, S_3^-, which is unstable outside its cage.[65] Another approach was recognition that the pigment shares structural characteristics with a class of chemical compounds called zeolites that exhibit catalytic activity. The catalytic function of ultramarine was examined in a 2020 study of the dehydration of methanol to dimethyl ether which found a twelve-fold higher product yield in the presence of the pigment. These results suggest that ultramarine, acting as a catalyst, is capable of hastening the degradation of its organic binding medium, usually an oil, which opens up micro-cracks in the pigment surface, thus enhancing light scattering. Microscopic examination of *in situ* situations have shown that the pigment is still there, as blue and bright as ever, but shadowed by the surface phenomenon that promotes light scattering rather than coherent reflection.[66,67] The occurrence of the 'disease' is unpredictable and why some samples suffer and others do not is not clearly understood.

8.8.5   Ultramarine's Absence Fingers a Forger

Since our team's analyses of a dozen Byzantine manuscripts from the University of Chicago Special Collections found a 66.7%

likelihood of ultramarine's appearance, we were curious when one of the manuscripts being analyzed contained no ultramarine at all. However, Ms. 972, identified as a 12th century Byzantine manuscript, was awash with a different blue pigment as illustrated in Figure 8.6, folio 40r.[68]

Fourier transform infrared spectra of several blue pigment samples from this folio (and many others) compared exactly with a reference spectrum of the blue pigment, Prussian Blue, sometimes known as iron blue. Its presence everywhere in this manuscript raised doubts about its authenticity.

The iron blues are among the earliest of the artificial pigments with a known history and an established date of first preparation. The Berlin color makers Johann Jacob Diesbach and Johann Konrad Dippel (1673–1734) manufactured it in or around 1706[69] (**Chapter 11.3**). Furthermore, there is little likelihood that it would have been discovered independently elsewhere due to its

Figure 8.6   Ms. 972, the Archaic Mark, folio 40r. Special Collections Research Center, University of Chicago Library. Reproduced from ref. 68 with permission from the Special Collections Research Center, University of Chicago Library.

complexity of composition and mode of synthesis.[70,71] There is additional evidence that links Ms. 972 with a gospel fragment from the Hermitage Museum, St. Petersburg[72] suggesting that they were copies of a late 12th century gospel book, Codex 93, held in the National Library of Greece.[73] Furthermore, neither of these manuscripts has a genealogy prior to 1930. This leaves little doubt that they belong to a group of Athenian forgeries dating from about 1920 to take advantage of gullible tourists.[74]

The University decided to lay to rest once and for all the questions that arose from our study by initiating a comprehensive chemical examination of all components of the codex (parchment, ink, paints and coatings), utilizing the most current technologies available. The final report unambiguously reconfirmed our initial finding of Prussian Blue, an invention of the early 18th century, as well as additional incongruent materials, namely, zinc white (ZnO), not used as a pigment until about 1780, and blanc fixe, a synthetic form of barium sulfate ($BaSO_4$), introduced in the 1820s. The report also said that the combination of zinc white and blanc fixe suggested the presence of lithopone, or Orr's white, which was not invented until 1874. These findings alone thrust the manuscript, at its earliest, into the latter part of the 19th century. Furthermore, infrared spectroscopy also indicated the presence of cellulose nitrate most likely used as a binding medium; it was only in 1920 that this material became widely available. Finally, the report on the radiocarbon dating of the manuscript parchment gave a date range of 1461–1640, indicating that the forgers used old parchment, but apparently, not old enough. De-authentication of the Archaic Mark was important from a scholarly point of view as well: it had the highest degree of correspondence with the oldest complete text of the Gospel of Mark, Codex Vaticanus (4th century). Apparently, the forgers learned from the Codex and not the other way around.[75,76]

This discovery was one of the most satisfying justifications of pigment analysis in our own experience. The contribution to New Testament research by a group of chemists was a wonderful vindication of the utility of our science. Many more sophisticated methods of manuscript analysis have been developed in the intervening years. A helpful review of eight major instrumental approaches specifically aimed at illuminated manuscript characterization along with their advantages and disadvantages is cited here.[77]

8.9   WAYWARD MANUSCRIPTS

Sometimes manuscripts can wander out of their happy homes and never come back. However, some do. A rare 15th century Dutch-commissioned prayer book written in low German "went missing" for decades from the Peabody Library at Johns Hopkins University. In 2012, it mysteriously turned up in a plain brown wrapper and it fell to a graduate student and a professor emeritus of art history to piece together its provenance and its weird journey home.[78] Then again, some don't. Take the case of the silk purse made out of illuminated bifolia recently acquired by the Boston Public Library. The gracious uncial lettering is still visible, though carelessly chopped up to form backing for a modern silk covering.[79] Apparently this was the fate of many manuscripts consigned to "book waste;" the parchment backing was considered far more valuable than the manuscript's contents. *Sic transit gloria mundi*!

8.10   THE HIGH MIDDLE AGES REACH HIGHER

The great feast of light and nature – stained glass – became the hallmark of a Europe emerging from the Dark Ages. In the tenth century watershed years, people looked back on five centuries of Muslims, Vikings and Slavs attacking Europe from the west, north and east. They looked forward to a more peaceful time with an improving climate, a revival of agriculture, and a concomitant population explosion. While these were powerful factors that gave rise to the flowering of medieval culture, the driving force was existential relief. Conventional wisdom had predicted, with its usual certainty, the end of the world as the bells rang in the year 1000.[80] (We may recall a similar fear at the turn of the 21st century when the same "wisdom" predicted total internet failure.) When people woke up to the fact that this would not happen, within about two or three years, according to the monk Ralph Glaber (985–1047) "every nation of Christendom rivaled with the other, which should worship in the seemliest buildings. So it was as though the very world had shaken herself and cast off her old age, and were clothing herself everywhere in a white garment of churches. Then indeed the faithful rebuilt and bettered almost all the cathedral churches."[81]

8.10.1 Gothic Architecture

Much of this building spate followed the tried and true Romanesque style, but in 1137, when Suger (*ca.* 1081–1151), the innovative abbot of the Abbey of Saint-Denis, near Paris, decided to rebuild the abbey church, he incorporated the major features of a new Gothic (a pejorative term meaning "barbarian") architecture. Thus the first building with soaring pointed arches, a ribbed vault, clustered columns and flying buttresses was dedicated only seven years later – and became the prototype for all future major churches in northern France. Over the next 500 years, more stone was quarried in France to build cathedrals than was used to build the Pyramids.

8.10.2 Stained Glass Window Art

The newly-opened high arches cried out to be filled with an equally light and airy substance – and thus arose the greatest flowering of stained glass window art ever seen. Stained glass has been an architectural element since the ancient Roman era, but never has the medium been so effectively exploited as both an appealing work of art and an instructive medium as in the medieval cathedral. These creations, increasingly important as symbols of prestige and culture, became one of the most widespread forms of painting[82] and remain a lively industry even today.

Medieval glaziers took over the compounds of copper and iron that were ideal colorants for ceramics materials. Copper compounds produced turquoise blue and green colors; iron compounds produced yellow, green, and brown colors. Progress in finding additional colorants was slow. Copper and iron compounds remained the mainstay of stained glass colorants until around the year 1200 CE. Around this time, additional colors like red copper(I) oxide, Cu_2O, and later, yellow lead antimonate, $Pb_3(SbO_4)_2$, blue cobalt silicate, Co_2SiO_4, and purple-brown manganese silicate, Mn_2SiO_4, were added to the mix – literally the mix of the glazier's crucible.[‖] The Moors introduced tin(IV) oxide as a white colorant, but colloidal gold was unknown in Europe as

---

[‖]Because glass was formed at very high temperatures, only inorganic (mineral) pigments could be used. Organic compounds from plants or animals would immediately decompose in the processing, as would some of the more fragile mineral colors like vermilion.

a pink colorant until the latter part of the sixteenth century.[70] An interesting colorant known from ancient times and still in use today as a high-grade glaze colorant of unsurpassed clarity and brightness is "Purple of Cassius." It is produced by adding a solution of tin(II) chloride to a very dilute solution of gold chloride, $AuCl_3$, producing a precipitate of hydrated tin(IV) oxide interspersed with **nanoparticles** of elemental gold, the reduction product. The gold colors the tin oxide precipitate brown, purple, or red, depending upon the original concentration of the solution[83] (**Chapter 12, Box 12.2**). Given the great cost of this color, you will never see much purple in truly medieval stained glass. These magnificent masterpieces (Figure 8.7),[84] the products of ingenious chemistry, doubled as works of art and religious

Figure 8.7   Canterbury Cathedral, Canterbury, England. 12th century stained glass, among the oldest stained glass windows in England. Seth and Adam from a series of small windows illustrating the "Ancestors of Christ" and now incorporated into a large perpendicular window. Reproduced from ref. 84 under the terms of the CC BY-SA 3.0 license, https://creativecommons.org/licenses/by-sa/3.0/deed.en.

instruction "manuals" for a largely illiterate society for centuries. Having survived iconoclastic attacks, fires, and wars, they remain a part of our precious cultural heritage and major tourist attractions to this day.

A characteristic feature of typical stained glass windows are grooved strips of strong, yet malleable lead (lead cames) which hold the glass pieces in place. Over time, the weight of glass supported by the cames causes metal fatigue. The lead may buckle to the point that the glass actually falls out of the frame and has to be replaced. An example of this unfortunate occurrence and an even more unfortunate restoration (when you can't find the right pieces to put back in) is shown in Figure 8.8. But modern chemistry has fixed this problem: a lead-antimony alloy, and other lead-metalloid alloys, greatly strengthen the lead's crystal structure and extend its life. So, instead of a stained glass window having to be re-camed every 100 years, every 150–200 years will do.

Figure 8.8   Stained and Painted Glass (*ca.* 1200 CE), Cartmel Priory, Cumbria, England. Reproduced from ref. 70 with permission by Springer Nature, Copyright 2013.

Unlike diamonds, stained glass is not forever. Constant vigilance to monitor potential deterioration and ongoing examination to document previous history of the windows is vital. One great obstacle to achieving this *in situ* is the working distance between the window and the measuring instrument. A new technical adaptation has overcome this difficulty at Canterbury Cathedral, revealing a great deal of information about the windows such as the degree of previous glass recycling and discovery of previously unidentified repairs. This methodology promises to enable differentiation of groups of glass of medieval origin and opens the way for future studies on the movement of glass in the medieval period.[85]

The glazier's palette expanded considerably after a late 14th century innovation: silver stain became commonplace. A dilute silver salt such as silver nitrate is painted onto the reverse (exterior) glass surface with a carrier such as Venice turpentine and is fired in a kiln at a temperature range of 500 to 650 °C. The process promotes silver ion penetration into the glass by ion exchange with potassium or sodium and subsequent reduction to metallic silver which forms submicroscopic clusters or **nanoparticles**. These particles selectively absorb light to produce a stain that can range in intensity from pale yellow to dark orange. The final color depends on such variables as firing time, kiln temperature, type of paint and glass composition.[86] This technique gave the term "stained glass" to the entire process of glass coloration.[87] The invention, termed "jaune d'argent" initiated a rapid and fundamental aesthetic and stylistic evolution in glass painting.[88] It is quite prominent in Figure 8.8. A new range of vitreous enamel pigments came into the play with a red called "Cousin's Red," followed by the introduction of brilliant blues, greens and purples from newly available sources, transition elements like chromium, manganese and nickel, and rare earth elements like neodymium and erbium. Two major American glassmakers, John LaFarge (1835–1910) and Louis Comfort Tiffany (1848–1933), are practically household names as a result of their innovative designs and introduction of new colors.

If you ever plan to visit Chartres Cathedral, the iconic magnet for stained glass lovers worldwide, don't overlook the International Stained Glass Centre,[89] only a few steps away (Figure 8.9). The only museum of its kind in France, it features collections of

Figure 8.9   Left: International Stained-Glass Centre, Chartres. Reproduced from ref. 90 with permission of American Chemical Society, Copyright 2014. Right: Example of Modern Stained-Glass on Exhibit at the Centre.

ancient and contemporary stained glass windows, a permanent exhibit explaining the windows in the cathedral, and master glaziers' studios. It also houses a stained glass professional training center and cultural heritage school.[90]

8.11   CONCLUSION

Conservation and examination of precious documents such as illuminated manuscripts affords a window onto the medieval world that can be gradually expanded as data builds. The scientific community is well aware of the need to assemble databases not only of contemporary commercially available artists' pigments but also of historic ones. Groups of researchers are now conducting "experimental archaeology" by reconstructing the conditions in which medieval artists worked with respect to synthesizing the pigments they used from recipes taken from medieval technical treatises, manufacturing the parchment and vellum they used as their support, and reformulating the binding media from recipes and from instrumental analysis of extant materials. When the pigments are painted out on parchment their reflectance spectra can be acquired for comparison with real art works.[91] This method is particularly useful for organic pigments because different preparations of the colorants can produce very different hues depending on the production circumstances. With such a systematic approach, we can expect to see a great deal more information about this resource in the future. Stained glass is also the subject of intense research with a view to understanding provenance, workshop technique, conservation, and perpetuating this art form into the future.

In Italy, where church architects of the 14th century seldom used the French Gothic style, fresco and panel painting rather than glass painting experienced a creative spurt from the talented hands of Giotto, Cimabue and Duccio. They and many others laid the groundwork for the incomparable art of the Renaissance in the following century.

REFERENCES

1. W. Shakespeare, *Henry VI, Part 2*, IV:vi.
2. G. Bazin, *A History of Art*, Bonanza Books, New York, 1959, p. 127.
3. T. Cahill, *How the Irish Saved Civilization*, Doubleday Anchor Books, New York, 1995.
4. S. Panayotova and P. Ricciardi, *Masters' Secrets in Colour – the Art and Science of Illuminated Manuscripts*, Harvey Miller/ Brepols, London and Turnhout, 2016, pp. 119–161.
5. B. Meehan, *The Book of Durrow: A Medieval Masterpiece at Trinity College Dublin*, Town House, Dublin, 1996, pp. 13–16.
6. E. P. O'Connor, *Ireland – Herself*, Dodd, Mead & Co., New York, 1918.
7. Book of Kells, Folio 32v, Christ Enthroned, https://commons. wikimedia.org/w/index.php?title=File:KellsFol032vChris- tEnthroned.jpg&%20oldid=428227879, accessed June 2021.
8. M. O'Connell, *The New Yorker*, 21 December 2020, pp. 26–30.
9. Pliny the Elder, *The Natural History*, trans. John Bostock, http:// www.perseus.tufts.edu/hopper/text?doc=Plin.+Nat.+toc, accessed September 2020.
10. Pliny the Elder, *The Elder Pliny's Chapters on the History of Art*, trans. K. Jex-Blake and ed. E. Sellers, Macmillan, London, 1896.
11. E. R. Caley, *J. Chem. Educ.*, 1926, **3**, 1149–1156.
12. E. R. Caley, *J. Chem. Educ.*, 1927, **4**, 979–1002.
13. C. S. Smith and J. G. Hawthorne, *Mappae Clavicula, A Little Key to the World of Medieval Techniques*, Trans. Am. Phil. Soc., New Series, 1974, Part 4, p. 64.
14. H. Hedfors, *Compositiones ad Tingenda musiva (Codex Lucensis 490)*, Almqvist & Wiksells, Uppsala, 1932.

15. J. C. Richards, *Speculum*, 1940, **15**, 2550271.
16. Theophilus, *De Diversis Artibus*, Manuscript of *ca.* 1123, trans. C. R. Dodwell, Thomas Nelson and Son, London, 1961.
17. E. Savage-Smith, S. Neate and R. Ovenden, Curatorial Issues and Research Questions: Current Research on Western Medieval Manuscripts and Oriental Manuscripts in the Bodleian Library, in *The Technological Study of Books and Manuscripts as Artefacts: Research Questions and Analytical Solutions*, ed. S. Neate, D. Howell, R. Ovenden and A. M. Pollard, Oxford, British Archaeological Reports, 2011, pp. 3–7.
18. R. J. H. Clark, *Chem. Soc. Rev.*, 1995, **24**, 187–196.
19. C. Miguel, A. Claro, A. P. Gonçalves, V. S. F. Muralha and M. J. Melo, *J. Raman Spectrosc.*, 2009, **40**, 1966–1973.
20. M. V. Orna and T. F. Mathews, *Anal. Chem.*, 1988, **60**, 47A–52A.
21. H. Roosen-Runge and A. E. A. Werner, The Pigments and Medium of the Lindisfarne Gospels, *Evangeliorum Quattuor Codex Lindisfarnensis*, ed. T. D. Kendrick, Urs Graf Verlag, Lausanne, 1960, vol. II, pp. 261–295.
22. The earliest European *recipe* for the preparation of ultramarine blue dates from the 13th century. D. W. Singer, *Isis*, 1929, **13**(1), 5–15.
23. S. Bucklow, *Z. Kunsttechnol. Konserv.*, 2000, **No. 14**, 5–14.
24. M. Clarke, *Gazette du livre médiéval*, 2004, **44**, 50–53.
25. M. Clarke, *Gazette du livre médiéval*, 2002, **40**, 36–44.
26. A. Cains, *The Book of Kells, Fine Art Facsimile*, Fine Art Facsimile Publishers of Switzerland, Faksimile Verlag, Luzern, 1990, pp. 211–227.
27. R. Fuchs and D. Oltrogge, in *The Book of Kells. Proceedings of a Conference at Trinity College Dublin, 6–9 September 1992*, ed. F. O'Mahony, Scolar Press, Aldershot, UK, 1994, pp. 133–171.
28. S. Bioletti, R. Leahy, J. Fields and B. Meehan, *J. Raman Spectrosc.*, 2009, **40**, 1043–1049.
29. UCLA Library, *Gladzor Gospel Book*, p. 312, http://digital2.library.ucla.edu/viewItem.do?ark=21198/zz0009hj74, accessed June 2021.
30. M. V. Orna and T. F. Mathews, *Stud. Conserv.*, 1981, **26**, 57–72.
31. S. L. Merian, T. F. Mathews and M. V. Orna, The Making of an Armenian Manuscript, in *Treasures in Heaven: Armenian Illuminated Manuscripts*, ed. T. F. Mathews and R. S. Wieck, Princeton University Press, Princeton, 1994.
32. S. Best, R. Clark, M. Daniels and R. Withnall, *Chem. Br.*, February 1993, 118–122.

33. S. Everts, *Chem. Eng. News*, 2012, (December 17), 36.
34. M. Aceto, A. Agostino, G. Fenoglio, P. Baraldi, P. Zannini, C. Hofmann and E. Gamillscheg, *Spectrochim. Acta, Part A*, 2012, **95**, 235–245.
35. M. Bicchieri, *Environ. Sci. Pollut. Res.*, 2014, **21**, 14146–14157.
36. M. Clarke, *Stud. Conserv.*, 2004, **49**, 231–244.
37. L. Burgio, R. J. H. Clark and R. R. Hark, *Proc. Natl. Acad. Sci. U. S. A.*, 2010, **107**(13), 5726–5731.
38. V. S. F. Muralha, L. Burgio and R. J. H. Clark, *Spectrochim. Acta, Part A*, 2012, **92**, 21–28.
39. P. Roger, I. Villela-Petit and S. Vandroy, *Stud. Conserv.*, 2003, **48**(3), 155–170.
40. M. J. Melo, V. Otero, T. Vitorino, R. Araújo, V. S. F. Muralha, A. Lemos and M. Picollo, *Appl. Spectrosc.*, 2014, **68**, 434–444.
41. C. Denoël, P. R. Puyo, A.-M. Brunet and N. P. Siloe, *Heritage Sci.*, 2018, **6**, 28.
42. M. J. Melo, P. Nabais, R. Araújo and T. Vitorino, *Phys. Sci. Rev.*, 2019, **4**(8), 20180017.
43. S. Kroustallis, *Revista de História da Arte N.º Especial*, 2011, 105–117.
44. G. Frison and G. Brun, Ground lapis lazuli. A new approach to the colour term 'azure' and the pigment ultramarine blue up to the 13th century, in *Colour and Colorimetry. Multidisciplinary Contributions*, ed. M. Rossi and D. Casciani, *Proceedings of the 11th Conferenza del Colore, 10–11 September 2015*, Politecnico di Milano, Milano, 2015, vol. XIB, pp. 265–276.
45. U. Eco, *The Name of the Rose*, trans. W. Weaver, Harcourt, New York, 1983.
46. T. Delbey, J. Povl Holck, B. Jørgensen, A. Alvis, V. Haight Smith and G. M. Kavich, *et al.*, *Heritage Sci.*, 2019, **7**, 91–109.
47. K. Moore, *The Radium Girls*, Sourcebooks Publishers, New York, 2017.
48. S. Zhang, Why a medieval woman had lapis lazuli hidden in her teeth, *The Atlantic*, 9 January 2019, https://www.theatlantic.com/science/archive/2019/01/the-woman-with-lapis-lazuli-in-her-teeth/579760/, accessed February 2021.
49. A. Radini, M. Tromp, A. Beach, E. Tong, C. F. Speller and M. McCormick, *et al.*, *Sci. Adv.*, 2019, **5**(1), eaau7126.
50. D. K. Coveney, *A Descriptive Catalogue of Manuscripts in the Library of University College*, University College, London, 1935, Introduction, pp. 29–37.

51. L. Burgio, D. A. Ciomartan and R. J. H. Clark, *J. Mol. Struct.*, 1997, **405**, 1–11.
52. A. Radini, M. Tromp, A. Beach, E. Tong, C. Speller, M. McCormick, J. V. Dudgeon, M. J. Collins, F. Rühli, R. Kröger and C. Warinner, *Sci. Adv.*, 2019, **5**(1), eaau7126.
53. J. Wyart, P. Bariand and J. Filippi, Lapis-lazuli from Sar-e-Sang, Badkhshan, Afghanistan, *Revue de Géographie physique et de géologie dynamique, 2 serie*, 1972, **14**(4), 443–448Republished in Winter, *Gems Gemol.*, 1981, 184–190.
54. N. Lapierre, *Arts Asiatiques*, 1990, **45**(1), 28–40.
55. A. Brysbaert, *Stud. Conserv.*, 2006, **51**, 252–266.
56. M. C. Gaetani, U. Santamaria and O. Seccaroni, *Stud. Conserv.*, 2004, **49**(1), 13–22.
57. C. Blänsdorf, S. Pfeffer and E. Melzl, The Polychromy of the Giant Buddha Statues of Bāmiyān, in *The Giant Buddhas of Bāmiyān: Safeguarding the Remains. Monuments and Sites XIX*, ed. M. Petzet, International Council on Monuments and Sites, Paris, 2009. pp. 237–264; 254.
58. H. G. M. Edwards, S. E. Jorge Villar and K. A. Eremin, *J. Raman Spectrosc.*, 2004, **35**(89), 786–795.
59. F. Delamare, *Blue Pigments: 5000 Years of Art and Industry*, Archetype Publications, London, 2013. p. 105.
60. S. Grieff and J. Schuster, *J. Cult. Herit.*, 2008, **9**, 27–32.
61. J. Plesters, Ultramarine blue, natural and artificial, in *Artists' Pigments: A Handbook of Their History and Characteristics*, ed. A. Roy, National Gallery of Art, Washington, DC, 1993, vol. 2, pp. 37–65, at 54.
62. Zacharias (Michelangelo), https://commons.wikimedia.org/w/index.php?title=File:Zacharias_(Michelangelo).jpg&oldid=519902776, accessed June 2021.
63. G. Chiari, Measurement Techniques to Uncover Mysteries in Art, Plenary Lecture delivered on Monday, 10 December 2018, *3rd International Conference on Techniques, Measurements and Materials in Art and Archaeology, Jerusalem, Israel*, 9–13 December 2018, Book of Abstracts, p. 21.
64. F. Wynne, *I Was Vermeer: The Rise and Fall of the Twentieth Century's Greatest Forger*, Bloomsbury, New York, 2006, p. 118.
65. New York University, Researchers Find Why Ultramarine Blue Fades, *ScienceDaily*, 10 October 2006, www.sciencedaily.com/releases/2006/10/061002214727.htm, accessed September 2020.

66. K. Schnetz, A. A. Gambardella, R. Elsas, J. Rosier, E. E. Steenwinkel, A. Wallert, P. D. Iedema and K. Keune, *J. Cult. Herit.*, 2020, 25–32.

67. J. Urquhart, Luxury blue paint pigment catalyses its own 'disease'. Chemistry World, Royal Society of Chemistry, https://www.chemistryworld.com/news/luxury-blue-paint-pigment-catalyses-its-own-disease/4011853.article, accessed September 2020.

68. Goodspeed Collection Ms. 972 (Archaic Mark), University of Chicago Special Collections.

69. A. Kraft, *Bull. Hist. Chem.*, 2008, **33**(2), 61–67.

70. M. V. Orna, *The Chemical History of Color*, Springer, Heidelberg, 2013.

71. R. J. Gettens and G. L. Stout, *Painting Materials: a Short Encyclopedia*, Dover, New York, 1966, pp. 20–21.

72. K. Treu, *Die grieschischen Handschriften des neuen Testaments in der USSR*, Akademie-Verlag, Berlin, 1966, pp. 229–230.

73. A. Marava-Chatzinicolaou and C. Toufexi-Paschou, *Catalogue of the Illuminated Byzantine Manuscripts of the National Library of Greece*, Publications Bureau of the Academy of Athens, Athens, 1978, vol. 1, pp. 224–243.

74. M. V. Orna, P. L. Lang, J. E. Katon, T. F. Mathews and R. S. Nelson, Applications of infrared microspectroscopy to art historical questions about medieval manuscripts, in *Archaeological Chemistry IV*, ed. R. O. Allen, American Chemical Society, Washington, DC, 1989, pp. 265–288.

75. M. M. Mitchell, J. G. Barabe and A. B. Quandt, *Novum Testam.*, 2010, **52**, 101–133.

76. J. G. Barabe, A. B. Quandt and M. M. Mitchell, Analysis of the "Archaic Mark" Codex: a collaborative study in authentication, in *Collaborative Endeavors in the Chemical Analysis of Art and Cultural Heritage Materials*, ed. P. Lang and R. A. Armitage, American Chemical Society, Washington, DC, 2012, pp. 197–217.

77. S. Pessanha, M. Manso and M. L. Carvalho, *Spectrochim. Acta, Part B*, 2012, **71–72**, 54–61.

78. B. McCabe, Return to Lender, *Johns Hopkins Magazine*, Spring, 2012, https://hub.jhu.edu/magazine/2012/spring/return-to-lender/, accessed September 2020.

79. J. L. Hester, How a Trashed Italian Manuscript Got Sewn into a Sweet Silk Purse, 14 February 2020, https://www.atlasobscura.com/articles/manuscript-fragments-purse-lining, accessed September 2020.

80. L. Lee, G. Seddon and F. Stephens, *Stained Glass*, Crown Publishers, New York, 1976, p. 64.
81. Fordham University, *Ancient History Sourcebook*, https://sourcebooks.fordham.edu/source/glaber-1000.asp, accessed September 2020.
82. Khan Academy, Stained Glass: History and Technique, https://www.khanacademy.org/humanities/medieval-world/gothic-art/beginners-guide-gothic-art/a/stained-glass-history-and-technique, accessed September 2020.
83. L. B. Hunt, *Gold Bull.*, 1976, **9**, 134–139.
84. Canterbury cathedral-stained glass 03 Seth and Adam, https://commons.wikimedia.org/w/index.php?title=File:Canterbury,_Canterbury_cathedral-stained_glass_03_Seth_and_Adam.JPG&oldid=386875614, accessed June 2021.
85. L. W. Adlington, I. C. Freestone, L. Seliger, M. Martinón-Torres, F. Brock and A. Shortland, *In situ* methodology for compositional grouping of medieval stained glass windows: introducing the "WindoLyzer" for pXRF, in *Archaeological Chemistry: a Multidisciplinary Analysis of the Past*, ed. M. V. Orna and S. C. Rasmussen, Cambridge Scholars Publishing, Newcastle upon Tyne, 2020, pp. 176–201.
86. J. Delgado, M. Vilarigues, A. Ruivo, V. Corregidor, R. C. da Silva and L. C. Alves, *Nucl. Instrum. Methods Phys. Res., Sect. B*, 2011, **269**, 2383–2388.
87. Boppard Conservation Project – Glasgow Museums, https://boppardconservationproject.wordpress.com/2013/07/28/facts-about-glass-silver-stain/, accessed December 2020.
88. C. Lautier, *Bull. Monum.*, 2000, **158**(2), 89–107.
89. Chartres, International Stained-Glass Center, https://www.centre-vitrail.org/en/, accessed January 2021.
90. M. V. Orna, Paris: a scientific theme park, in *Science History: a Traveler's Guide*, ed. M. V. Orna, American Chemical Society, Washington, DC, 2014, pp. 135–158.
91. M. Aceto, A. Agostino, G. Fenoglio, A. Idone, M. Gulmini, M. Picollo, P. Ricciardi and J. K. Delaney, *Anal. Methods*, 2014, **6**, 1488–1500.

CHAPTER 9

Botticell's Bottega: The Glory of the Renaissance†

Art establishes the basic human truths which must serve as the touchstone of our judgment.[1]

John F. Kennedy

9.1 INTRODUCTION

The very word "Renaissance" conjures up ideas of art, beauty and a universal humanistic outlook. The forces that gave rise to this movement had their roots in the realism and monumental works of Giotto di Bondone. These totally new approaches to the entire body of human endeavor signaled the death knell of the Middle Ages and the emergence of the modern world.

9.2 ACHIEVING REALISM AND ITS CONSEQUENCES

Not many art teachers would recommend that their students throw objects at their paintings if they didn't like them. However, sometimes it works. The ancient Greek painter, Protogenes, tried

---

†Glossary entries and chapter cross-references are in boldface.

March of the Pigments: Color History, Science and Impact
By Mary Virginia Orna
© Mary Virginia Orna 2022
Published by the Royal Society of Chemistry, www.rsc.org

every trick in his paint box to portray a dog realistically foaming at the mouth. In a rage, he threw a sponge at the "hateful spot," and lo! Upon impact, the sponge rearranged the picture, leaving behind the exact effect that Protogenes had intended. Thus chance became the mirror of nature.[2]

This passionate quest for realism faded from artistic intent over the centuries culminating in the High Middle Ages with a stylized, almost frozen, approach to painting in the manner of Greek icons. Toward the end of the 13th century, this style began to thaw with the advent of Giotto, whose work presaged the great flowering of humanism in art that we call the Renaissance. Whole books have been written on individuals who populated this period – spanning the 14th through the 17th centuries – so we must pick and choose. The path we will trace through this chapter will be to offer a glimpse of key individual painters and their colors or techniques: Giotto (yellow), Botticelli (copper-green and blue pigments), Leonardo da Vinci (unusual use of oils) and Titian (a rich Venetian palette).

9.3   GIOTTO DI BONDONE

The course of painting in Italy toward the end of the 13th century changed when, quite by chance, the Florentine artist Cimabue (1240–1302), took a detour through a field in the late 1270s. There he marveled to see a little ragged shepherd boy incising the image of a sheep on a stone slab in a completely ingenuous and naturalistic way. He immediately offered an apprenticeship to the lad and, with his father's consent, little Giotto di Bondone (1266–1337) entered Cimabue's workshop. There, according to his biographer, Giorgio Vasari, "in a short time, assisted by nature and taught by Cimabue, the child not only equaled the manner of his master, but became so good an imitator of nature that he...revived the modern art of painting."[3] And thus, another chance happening became the mirror of nature.

9.3.1   Talent is as Talent Does

Relatively unschooled, following his natural talent, Giotto quickly changed the "language" of painting from Greek (Byzantine, iconic) to Latin (classical, naturalistic) and took Cimabue along with him on his coattails. His talent for realism can be encapsulated in this little story: Giotto once painted a fly on the nose of one of Cimabue's figures so true to nature that when Cimabue came back to work on the painting, he tried more than once to shoo the fly away before he realized his mistake. Today, we recognize Giotto as the key figure in Western painting, his hallmark being the bulk and monumentality of his figures. As his style developed, he set his figures in credible pictorial space, framed by architecturally convincing structures in a coherent composition that draws the viewer's eye to a focal point even before the laws of perspective were formalized.[4,5] His innovations set the course of the Western practice of art for the next five hundred years.

9.3.2 Giotto's Yellow

As he matured, Giotto received his own commissions for major works, chief among them are the frescoes in the Scrovegni (Arena) Chapel, Padua, acknowledged as his masterpiece. The first of another series of seven panels painted for the Villa Poniatowski, Rome, completed in 1320, is shown in Figure 9.1.[6]

In this painting, and in so many others by Giotto or his workshops, the outstanding color is yellow in its varying shades. We have to rely on analysis of what lies on the painted surface for our pigment identification since there is very little knowledge about where 14th century artists obtained their colors.[7] The second of the three kings in this painting is arrayed in a bright gold color that, on analysis, is what chemists call lead-tin yellow, Type II. It was identified instrumentally by **X-ray diffraction** (XRD).[8]

Lead-tin yellow, regardless of type, was an unknown to Giotto and his contemporaries. Other words like *zallolino* or *giallorino* would have fit his vocabulary better. Giallo is the word for yellow in Italian. Today, as a noun, a giallo is a detective mystery, so designated because paperbacks of this genre are printed in Italy with yellow covers. If the diminutive – *lino* or *rino* – is attached to *giallo*, it expresses "a little yellow," or a pale yellow color.

An early 15th century recipe stipulates melting one pound of lead and two pounds of tin together in a furnace. Reheating

Figure 9.1   Giotto di Bondone. Adoration of the Magi. Metropolitan Museum of Art. John Stewart Kennedy Fund, 1911. Reproduced from ref. 6 under the terms of the CC0 1.0 license, https://creativecommons.org/publicdomain/zero/1.0/deed.en.

two pounds of this mix with 2.5 pounds of minium (red lead, Pb_3O_4), and a half-pound of sand, produces lead-tin yellow, Type I (Pb_2SnO_4).[9] Since no consistent system of chemical nomenclature yet existed, there was massive confusion. Artists rarely knew what they were using and had to trust the supplier or they had to manufacture their own pigments. For example, in northern Europe, they called lead-tin yellow, Type I, massicot, whereas yellow lead monoxide, PbO, goes by the same name. Since the latter has never been identified in any paintings between the 13th and 20th centuries, we can conclude that the "massicot" of northern Europe is lead-tin yellow, Type I. Its lifetime on the artists' palette dates from about 1450 to 1750.

But we still haven't zeroed in on Giotto's yellow, which is Type II (sounds like a recipe for diabetes). This particular pigment has been very little studied because it's non-stoichiometric. What does that mean? Up till now, all of our chemical formulas have had definite whole number ratios associated with them. For example, the formula for calcium sulfate, $CaSO_4$ tells us that calcium and sulfur and oxygen are present in this compound in a ratio of $1:1:4$. Non-stoichiometric compounds can't be represented by these small whole numbers. In lead-tin yellow, Type I with the formula Pb_2SnO_4, lead, tin and oxygen are present in the ratio of $2:1:4$. But when this compound is heated with a fair amount of sand, SiO_2, it turns out that some of the silicon, Si, insinuates itself into the compound, taking the place of some oxygen and even some of the tin; the formula turns out to be indefinite and has to be represented algebraically with an x: $PbSn_{1-x}Si_xO_3$.[10] This is getting complicated. Essentially, this formula means that as a greater proportion of silicon (x) enters the compound, some tin $(1-x)$ has to make way for it. Another way of looking at it is that these compounds form an enormous three-dimensional lattice in which some of the atoms are missing or that too many atoms are packed in and distort the lattice. More simply, Giotto's yellow pigment was a little bit more sandy or glassy than the original lead-tin yellow. He didn't know that, and we wouldn't know it either if it were not for modern sophisticated instrumentation.

Lead-tin yellow, Type II is a misnomer from another perspective: it predates lead-tin yellow, Type I as an artists' pigment, by about 150 years. However, it was certainly known to the glass industry well before that. It was even identified in a 4th – 5th century CE glass shard by **X-Ray fluorescence** (**XRF**) and **XRD**.[11] Although prior to 1300, the major yellow artists' pigments were orpiment, As_2S_3, and massicot, PbO, lead-tin yellow, Type II was on a parallel track as a favorite glaze in the ceramics industry. It could be applied to a piece of pottery without running and could be fired at standard kiln temperatures without changing color. Probably due to the proximity of the glass industry in Florence, it "sidled" over to the artists' palette with ease and soon became a staple about 1300 and remained firmly entrenched there for another 350 years.

A very interesting find regarding Giotto's lead-tin yellow is that it had never been made from a glass – there is no evidence of pulverized glass fragments. This could only mean that his pigment was purposely synthesized for the palette.[12] A comprehensive study by Martin and Duval reveals that Type I had almost completely replaced Type II by about 1450.[13]

Just to add to the confusion, another leaded yellow, called Naples yellow, arrived on the palette in the early 18th century, replacing tin with antimony. Lead antimonate yellow, $Pb_2Sb_2O_7$, is its common chemical name. Well known to the ancient Egyptian glass makers, like Egyptian blue, it is one of the oldest synthetic pigments known.[11] It was the only yellow pigment in general use during the first 1700 years of glassmaking.[14] The pigment's "glassiness" put off its use by artists since it was hard to grind and stubbornly retained its grittiness. But sometime in the early 1600s, the nature of Naples yellow changed; the glassy component disappeared. This was a game-changing moment in the development of color.[15] From then on, its place on the palette was assured, from Luca Giordano's (1634–1705) *Esther and Ahasuerus* to Pierre-Auguste Renoir's (1841–1919) *Chrysanthemums*.[16]

As if this were not enough, another yellow pigment containing the three metals, lead, tin, and antimony, with the formula $Pb_2SnSbO_{6.5}$, *zallo de depenzer*, appeared in the glass workshops of Murano (Venice) in the sixteenth century.[17,18] Too late for Giotto, it occurs in many 17th century Renaissance paintings.

By varying the reagents and the temperature, all of these yellow pigments could be prepared in different shades. When using the ancient recipes, the results seldom produced a pure material: they would often contain lead oxide or tin oxide, giving the mix a paler hue.[19]

Giotto's school was creative in another interesting way. For the stars, halos and drapery forms, they used gilded tin, but this method fails when the tin tarnishes. In Figure 9.1, there are two stars over the manger structure, one of which is blackish, which shows the tarnish, and one slightly to the right of it that is still golden – the latter was painted later using a method called mordant gilding (gold applied over an adhesive); the star's tail is hiding cracks in the panel.[20] As the Middle Ages drew to a close, Giotto furnished the innovative gateway that would lead to a whole new world – both literally and figuratively.

9.4   DAWN OF THE RENAISSANCE

It is always difficult to pin down exactly when something new is beginning. Giotto gave us more than a slight hint that something old was disappearing by his different approach to nature and his realistic depiction of people and settings. He may have changed visual representation, but something much deeper and universal was churning away in the background. People's whole outlook on life was gradually unfolding into exploratory curiosity, a renewed optimism, openness to new ideas as well as new materials, and a grasp of the whole picture, not merely the details. Masaccio (1401–1428) was at the forefront of this movement, painting figures that looked like they had real flesh and blood, unlike Giotto's carved monumental forms. Had he lived beyond his brief 27 years, he may have been the sustaining leader of a renascent art with his bold new style and perfect command of scientific perspective. As it is, he only had a chance to get his foot in the door of this new age.

In 1436, on the cusp of the transition from the medieval period to the Renaissance, Leon Battista Alberti's (1404–1472) volume *Della Pittura*[21] appeared – just ten years before our protagonist, Sandro Botticelli, was born. The first treatise on art theory since antiquity, Alberti resurrected the classical writers' canons of art and dressed them in a whole new raiment. Not simply a "how to" work, though it contains plenty of that, it also stresses the importance of attention to beauty in nature, harmony in all parts of a work, and the insistence that each work contains within itself the entire universe of beauty, truth and goodness – warts and all. It was this philosophy of beauty, but particularly Alberti's connection of perspective art as a secular physical reality to the whole human moral order that created the paradigm shift we call the Renaissance.[22] A fine example of this new philosophy is Domenico Ghirlandaio's (1448–1494) warm and loving portrait of an old man and his grandson (Figure 9.2).[23]

Figure 9.2   Domenico Ghirlandaio. An Old Man and his Grandson, 1490. Musée du Louvre. Reproduced from ref. 23.

9.5   SANDRO BOTTICELLI

Botticelli is the liminal figure, the hinge that snapped shut the door to the medieval world. Cennino Cennini's (b. 1370) contemporaneous *Libro dell'arte*[24] looked back to the past. Alberti, future-oriented, sought to universalize and popularize the new humanist practice of art. All this was taking place in Florence, the crucible in which the Renaissance took shape. When Botticelli arrived, there was no going back.

Alessandro di Mariano di Vanni Filipepi (1445–1510) acquired his soubriquet from an older brother, it seems, because his anatomy precisely matched a little barrel or wine cask, a *botticella*. His artistic style derived from the humanistic approach of his teacher, Filippo Lippi (1406–1469), just at the time that the formalized style of the Middle Ages was beginning to wane. What took its place was precisely the warmth, tenderness, ambiguity and sense of mystery that Botticelli bestowed on his masterpieces – qualities that would be the model for all that followed.

As Botticelli matured and his reputation grew, he became so well known that in 1481, Pope Sixtus IV invited him to Rome to take charge of decorating his newly constructed eponymous chapel. Botticelli contributed three major paintings to this project based on episodes in the Hebrew Scriptures. If they were anywhere else, they would be so acclaimed that they could populate a stand-alone museum. As it is, every eye of every visitor to that chapel lasers onto Michelangelo's ceiling, and *basta*! Unfortunately, Botticelli was no manager – although the Pope lavished gifts and money galore on him, by the time he returned to Florence, it was all gone – and Botticelli had no idea of where it all went.

9.5.1 Botticelli's Two Most Famous Masterpieces

Obviously in need of ready cash, soon after his return to Florence, Botticelli secured a commission from the Medici for his two most famous paintings, *Primavera* (*ca.* 1482) and *The Birth of Venus* (*ca.* 1486). Completed in reverse order according to their themes, *Venus* spoke to the universality of love, and *Primavera* to the fruits of love through joy and fecundity. These themes and the rendition of them hit Florence like a bombshell. No one, much less a practicing Catholic, had ever dared to celebrate sexually charged pagan ideas with such abandon. *Primavera*, Italian for "spring," (Figure 9.3)[25] and *Venus* are very large; although the latter is one

Figure 9.3 Sandro Botticelli, Primavera, *ca.* 1482. Uffizi Gallery, Florence. Tempera on panel; 6.6 × 10.3 ft. Reproduced from ref. 25.

of the largest paintings ever done on canvas, it is dwarfed by its sister painting on wood panel. Both hang in the same room in the Uffizi Gallery, Florence, usually the most crowded room in the museum, high season or not.

9.5.2 Botticelli's Copper Greens and Blues

Primavera is a very colorful work with outstanding flesh tones and reds and oranges for the 100 or so species of flowers that you can pick out in the painting. However, about 40% of the surface area is taken up by dark, almost black, tree foliage and forest floor. Unfortunately, the colors you see now are not those that Botticelli painted. These, and many of the colors in his other paintings have lost the richness of their beauty due, at times, to over-eager restoration and, at others, to the pigments' degradation. As one art historian laments:[26]

> As to the Primavera, if you look at it on a misty afternoon...you will feel as if the silvery mist coming from the nymph's wood makes the whole room hazy. Though this grey is preciously Botticelli's own, yet it is to a great extent the work of time, and we cannot conclude from it anything very definite as to the painter's sense of colour.

Copper blues and greens are especially susceptible to the changes referred to, and in particular, copper resinate, a Botticelli favorite, tends to turn irreversibly from green to brown. This reaction results in not only a color change, but also upsets the contrast and subtle gradations in dark and light originally conveyed by the artist.[27]

Copper resinate is not a common pigment by any means. It is never listed among those for sale by color merchants, indicating that artists had to manufacture it themselves. The earliest known recipe, from 1601, specifies boiling pine **resin**, mastic and a bit of "new wax" over a charcoal fire "until it no longer squeaks." After cooling, straining, and adding verdigris a little at a time, the mix is reheated and used while still hot.[28] Pine **resin** is chiefly abietic acid, $C_{19}H_{29}COOH$, a diterpenoid **resin** acid capable of forming copper salts with acceptable pigment properties,[29] *e.g.*, with verdigris, basic copper acetate, $Cu(CH_3COO)_2 \cdot 2Cu(OH)_2$, the oldest of the manufactured copper-green pigments.[30]

Perhaps Botticelli was captivated by the novelty of this approach or by the transparency and shade of this pigment. Maybe he wanted to express his spirit of adventure in experimenting with new techniques and pigments. But perhaps it was made by accident, by simply adding some pine **resin** or turpentine to his tempera mix of malachite or verdigris to produce a brighter green for his forest scene. We can lament the folly of this approach in the obvious darkening of the scene, brought on, we surmise, by the copper playing its usual trick of slowly exchanging its carbonate, resinate, acetate or hydroxide partners with ever-present oxygen to form black copper(II) oxide, CuO. Another example of Botticelli's use of copper resinate is the San Marco Altarpiece, the *Coronation of the Virgin* (1490–1492), also in the Uffizi Gallery. Saint Augustine, second figure from the left, is wearing a once-green vestment on which the copper resinate glazed over malachite has darkened, flattening out the folds in the garment into an undistinguished mass.[27,31] Works of other users of this unfortunate pigment, such as those of Cima (1459–1517), Raphael (1483–1520) and Paolo Veronese (1528–1588), have suffered similar consequences. By the end of the sixteenth century, artists were well aware of copper resinate's tendency to discolor and its use slowly declined.

Some late works of Botticelli held by the National Gallery, London, contain natural azurite, a copper-blue pigment, $2CuCO_3 \cdot Cu(OH)_2$, and artificial malachite, a copper-green pigment, $CuCO_3 \cdot Cu(OH)_2$.[27] Since some of his earlier works, notably his beautiful "Trionfo d'Amore" and the *tondo*, "Madonna col Bambino e S. Giovannino" of *ca.* 1480, also contain these pigments, we can presume that he used them throughout his career.[32,33] Both of the naturally occurring versions of these compounds occur together in ores and are widely available. Azurite was the "workhorse" blue pigment of choice through the centuries due to its brightness, stability, and above all, its low cost. Its stability vis-à-vis a commonly used cobalt glass has been attested to by comparison of two Veronese blue sky paintings; the one done with the cobalt has degraded to a grayish tone; the one done with azurite is still bright blue.[34] In addition, azurite's versatility is stunningly illustrated by the shimmering "Voronet Blue" that decorates the exteriors of eight Moldavian 15th –16th century churches. Only recently it has been shown that the artists cleverly coated the church walls with a thick charcoal covering and then painted over them with azurite, a trick that enhanced its color intensity enormously.[35]

The synthetic versions are called respectively, blue verditer and green verditer (from the French "*vert de terre*"). In Botticelli's day, on average, the price of ultramarine was about 27 times greater than an equivalent amount of azurite, and absent a generous patron or commissioner of the work, it is no surprise that he chose the less expensive route. But if his *Primavera* had, indeed, been commissioned by the Medici, why did Botticelli seem to cut corners by using such an inexpensive and fugitive pigment as copper resinate?

Botticelli's use of synthetic malachite, *i.e.*, green verditer, gives us pause as well. According to the accepted recipes for its preparation, copper nitrate (made from elemental copper and dilute nitric acid) was poured on whiting (calcium carbonate, $CaCO_3$) to form copper carbonate, $CuCO_3$. However, subsequent addition of lime was necessary to form the basic copper carbonate, $CuCO_3 \cdot Cu(OH)_2$. This recipe is similar to that recommended by Bertrand Pelletier[36] for the production of blue verditer. In any case, both were manufactured under very harsh conditions, *i.e.*, use of a strong acid and then of a strong base. The likelihood of achieving neutrality for the finished product was remote given the rudimentary chemistry of the day, so application to any surface could, in time, cause it to deteriorate.[37]

In a sense, we can't lay too much blame at Botticelli's doorstep for selecting fugitive greens. Both blue and green have been scarce colors since the world began – they do not belong to the so-called earth colors that can be collected on the Earth's surface. To get them, you must dig deep, mix them with yellow, or be a clever chemist. The following box is a digression on green choices, or the lack thereof (Box 9.1).

Box 9.1 It's Not Easy Being Green

Throughout history, artists have tried using a variety of tricks to obtain their favorite shade of green. Organic, inorganic and synthetic mixes have come and gone, and almost all of them were unsatisfactory, be it due to changing shade, substrate damage, and plain old dullness.[38] It would be two and a half centuries after Botticelli was laid to rest that another satisfactory green would hit the market in 1778. Known from the name of its German-Swedish discoverer, Carl Wilhelm Scheele (1742–1786), as Scheele's Green or copper arsenite, $CuHAsO_3$, its bright shimmering color endeared itself to artists and wallpaper manufacturers alike.[39] However, covering one's bedroom four walls with an arsenic compound turned out to be so detrimental to one's health that there was an upsurge in the mortuary business.[40,41] There was speculation that the pigment may have played a role in Napoleon Bonaparte's (1769–1821) death on Saint Helena where, you guessed it, the wallpaper was green. Napoleon died of stomach cancer; gastric carcinoma is one of the hazards of arsenic poisoning. Quantitative analysis of the actual wallpaper showed that it may have caused illness, but that there was not enough to cause death. If Napoleon died of arsenic poison, it would have had to have been deliberately administered.[42] Arsenic pigments were also used extensively in clothing dyes, soap, food, dried flowers, paper and cardboard, with a concomitant toll of chronic illness and numerous deaths.[43] A 20th century example of arsenic poisoning hit the *New York Times* in 1956. Over a period of about a year, US Ambassador to Italy, Clare Boothe Luce (1903–1987), suffered from a mysterious illness eventually attributed to arsenic poisoning. Apparently, small flakes of an arsenical paint fell from painted roses on the ceiling of her Rome apartment. In her first public statement about the issue made on 21 July, she said that the paint was probably shaken by people walking overhead in a laundry and that flakes probably fell into her food, onto her table, *etc.*44

A Swedish chemist, Sven Rinman (1720–1792), published the details of a more benign green pigment, cobalt green, $CoO·nZnO$, in 1780. It turned out to be well-accepted by artists who described it as a pure, but not very powerful green, durable in water and oil and a good drier. In 1814, another brilliant green pigment, copper aceto-arsenite, Emerald Green, sometimes also known as Paris Green or Schweinfurt Green, $Cu(CH_3COO)_2·3Cu(AsO_2)_2$, was developed as an attempt to improve upon Scheele's Green. It turned out to be more durable and had a more intense color, although of a shade seldom encountered in nature.[30] Scheele's Green enjoyed limited use by artists, most notably by J. M. W. Turner (1775–1851) and Édouard Manet (1832–1883). Emerald Green, on the other hand, found its way onto the palette of numerous artists from its introduction until well into the 20th century. Both compounds were banned from use on objects that would come into personal contact such as toys and clothing. They have both been used with success as insecticides and fungicides.

9.5.3   Botticelli's Later Career

The Medici in Botticelli's mature period promoted a liberal and permissive culture in Florence to which he, while not a member of the Academy that Cosimo de' Medici had set up, surely subscribed. The growth of humanism and the revival of Neoplatonic philosophy signaled a shift across all disciplines: literature, science, theology, ethics and the arts. The pervading icon of the period, representing this new vision of *humanitas*, was Venus, a subject that Botticelli frequently took up. Thus, his paintings gave concrete form to a rising wave of secular thought that would soon put an end to the dying medieval world and usher in the High Renaissance. This new world view, emphasizing classical form and rational science, would develop into what we now call the Enlightenment a few centuries later on.

In 1492, Lorenzo de' Medici (the Magnificent), virtual dictator of Florence and patron of the Academy, died unexpectedly of a gangrene infection, changing almost instantaneously the whole political climate. Resistance to the Medici-sponsored progressive and liberal atmosphere that had given rise to a certain moral breakdown had been growing. The family's influence waned considerably after Lorenzo's death, while the movement that was the antithesis of the Medici was getting stronger and shriller. This clash culminated a few years later.

Surefire power in Renaissance Florence at this point lay in the mightiest brow, the harshest voice and the most apocalyptic cause. In 1497, that power was wielded by Fra Girolamo Savonarola, a Dominican friar who shouted, screamed, inveighed and bullied a cowed populace, fearful of eternal damnation, into surrendering their most precious goods to his bonfire of vanities. Even Sandro Botticelli fell under his thrall, and who knows how many of his warm and tender masterpieces shriveled to ashes, consigned there by his own hand. We can only thank the Medici for keeping custody of *"Primavera"* and *"The Birth of Venus"* in their country villa – far beyond the reach of the hordes of misguided fanatics who went in search of them.

It is not certain if Botticelli continued his adherence to Savonarola's ideology, especially since the latter was executed the year after his great bonfire. However, during the last decade of his life, although his paintings still exuded warmth and imagination, they were also more religious in content. Since he was a known squanderer of his resources, it is likely that he died in poverty; the day was 17 May 1510.

9.5.4 Botticelli's Legacy

As the High Renaissance got into full gear, Botticelli's reputation went into total eclipse for three centuries. He was alternately dismissed as a poor colorist or denigrated for painting very bad landscapes until he was rediscovered by the Pre-Raphaelites in the 19th century. From then on, Botticelli's star has risen to the heights of the pop culture world. The Victoria and Albert Museum, South Kensington, London, headlined its 2016 exhibition, "Botticelli Reimagined" with the words:[45] "Sandro Botticelli is recognised as one of the greatest artists of all time. His celebrated images are firmly embedded in public consciousness and his influence permeates art, design, fashion and film." The show went on to document how his masterpieces have been reinterpreted by artists and designers many times over and how he became an icon of pulp fiction. In film, the moment in which Ursula Andress breaches the sea in that famous white bikini has been cited as an iconic moment in cinematic history – 500 years after Botticelli's *Venus* caused a similar stir in Renaissance Italy[46] and lent her name to a 2003 best-selling novel. The name? You guessed it: *The Birth of Venus*. Botticelli, little barrel, is now a household word.

9.6 LEONARDO DA VINCI

Few personalities hailing from the sixteenth century have more visibility and clout in the modern world than Leonardo (1452–1519). In the popular imagination, Mona Lisa and *"The Da Vinci Code"* seem to overshadow everything else we know about him: his left-handedness, his mirror-writing, his engineering skills, his almost workable flying machines and war machines, his various other contraptions for hoisting weights and drawing water, his theories on proportion, and even his other two most famous works, *"Il Cenacolo"* (*The Last Supper*) and his drawing, *"Vitruvian Man"* (Figure 9.4). Often cited as one of the greatest influences ever on art, he was a polymath who excelled not only in sculpting,

Figure 9.4   Model of the *"Vitruvian Man,"* Vinci, Italy. The original is in pen and
ink on paper showing a male figure with arms and legs extended
inscribed in both a circle and a square, a Canon of Proportions
based on the work of the Roman architect, Vitruvius Pollio (81–15
BCE). It is held at the Galleria dell'Accademia, Venice.

painting and drawing, but also in mechanics, architecture, metallurgy, carpentry and chemistry. His "technical reach" in terms
of his all-pervasive distinctive compositional and visual inventions has been well documented.[47]

Leonardo was born in the hamlet of Anchiano on the outskirts
of Vinci, a Tuscan hill town, to Caterina di Meo Lippi, a young
peasant girl who had been, as we would say, compromised, by
the rich and powerful lawyer, ser Piero da Vinci. Leonardo spent
his first five years with his mother, from whence he moved to
live in his father's household in Vinci where he had an informal
education in Latin, geometry and mathematics. At 14, he became
a "gofer" for a few years in Andrea del Verrocchio's (1435–1488)
Florence workshop, graduating into an apprenticeship that lasted
another seven years. It is certain that during that time he rubbed
elbows with Botticelli, Ghirlandaio, Perugino (1446–1523) and
Lorenzo di Credi (1459–1537). Besides his many talents, Vasari
mentions in his biography of Leonardo that he was a compas-
sionate young man who, whenever he passed a market where
birds were being sold, bought them and set them free. He kept
a number of notebooks in which he drew, wrote observations

and took general notes on virtually everything that caught his attention. He would often follow a person whose face caught his fancy for a whole day, and then go home and draw that face from memory. Vasari speaks most of Leonardo's great talent – virtually everything he put his hand to, he accomplished with such grace and beauty that people could only sit back and marvel. He tempers this hagiographic account by also observing that Leonardo hardly ever finished his commissions before he moved on to a better offer. Once, when he was being pushed to complete a job, he said to his patron, "Men of lofty genius sometimes accomplish the most when they work the least."[48]

There is a certain irony in the fact that Leonardo had no compunctions about dissecting human cadavers to study anatomy first-hand in order to make his own paintings true-to-life. He was able to commandeer unclaimed bodies, usually those of vagrants or drunkards. He even befriended a centenarian who, on his death, bequeathed his body to Leonardo for further study. However, since almost all the bodies available to him were male, he was stumped regarding the female reproductive apparatus, which caused some of his drawings to be inaccurate. He was so observant and painstaking in his studies that his findings are still up-to-date. His passion for knowledge overcame the obvious drawback for this type of investigation: fifteenth century Florence had no refrigeration.[49,50] In 1494, when Leonardo was in mid-career, Duke Ludovico Sforza (1452–1508) of Milan, commissioned him to paint a *Cenacolo*, or "Last Supper," for the refectory of the Dominican monastery of Santa Maria delle Grazie. This subject was *de rigueur* for the dining areas of religious communities of the time; it commemorated for Catholics Jesus's institution of the Eucharist as well as his farewell to his Disciples, two themes to meditate on during the enforced silence of the meal. Leonardo's design projected the dramatic moment when Jesus announced that one of his own would betray him. His final study shows 12 male figures expressing their shock in a maelstrom of taut, twisted, tension-filled poses (Figure 9.5).[51]

The wall surface of the refectory was typical for the time. Finished off with plaster that could be coated with several layers of marble dust, sand and slaked lime ($Ca(OH)_2$), it would be ready to receive the artist's *buon fresco* mix of lime and pigments that would permanently lock them to the wall by reacting with atmospheric

Figure 9.5 Leonardo da Vinci, Il Cenacolo (1494–1498), Oil and tempera on plaster. Monastery of Santa Maria delle Grazie, Milan. Reproduced from ref. 51.

CO_2 to form calcium carbonate ($CaCO_3$). However, this medium did not suit Leonardo's style. He wanted to work at his usual leisurely pace; *buon fresco* required quick decisive action that allowed for very little change of mind. He also especially desired to endow his work with a mysterious sfumato, or smoky, appearance where "shadows and highlights fuse without hatching or strokes, as does smoke."[52] To achieve this, Leonardo reverted to a mixed medium of egg tempera and oil, a method typical of painting on wood panel. To understand the subsequent fate of "Il Cenacolo," we have to first learn what these two media are all about.

9.6.1   Leonardo's Choice of Medium: Egg Tempera and Oil

Though Jan van Eyck (1390–1441) is often credited with introducing the practice of oil painting, it was actually well known among northern European painters of the twelfth century, long before his time.[53] Another myth is that Antonello da Messina (1430–1479) brought the technique to Italy. But in all likelihood, Antonello himself was instructed in oils by the Neapolitan, Niccolò Colantonio (b. 1420) who had learned it from a Netherlandish visitor to Naples in 1440, Bartélemy d'Eyck (1420–1470).[54] Whatever the history of its transmission, these painters became virtuosos in its use, displaying its versatility in both heavy impasto and thin glazes. One of oil's great advantages was that it was slow drying, thus allowing for a degree of improvisation and change not possible with tempera painting and fresco. But, by acclamation, oil's ability to confer shimmering, jewel-like quality to every facet of a pigment particle catapulted it to the top of every artist's shopping list.

Renaissance artists' favorite drying oils were linseed (flax seed), poppy seed and walnut oils, although linseed's tendency to yellowing was a drawback. Leonardo liked to mix pressed mustard seed with his linseed oil and often added distilled cypress oil (a turpentine) mixed with juniper gum.[55] All oils and fats contain long-chain fatty acids, that is, carbon atoms linked together into chains about 18 carbons long bonded to glycerol in what chemists call an ester linkage. Generally, fats are solids and oils are liquids at room temperature. The common fatty acids in drying oils contain double bonds between some of the carbons in the chains, such as $-C=C-C=C-$; these are called unsaturated, or polyunsaturated if they have more than one double bond. The fatty acids in fats have all single bonds, such as $-C-C-C-C-$ and are called saturated. Oils containing more double bonds dry more quickly because the mechanism involves breaking these bonds, and subsequent re-bonding with an adjacent molecule – a form of **polymerization** called cross-linking. As these bonds break and reform with neighboring acids, they gradually form a web or film that coats the painting and encloses the pigments – drying "to touch" in a few days,[‡] but the entire process would take decades or possibly centuries.[56] The Renaissance artists knew how to shorten the drying time of these oils by boiling them with metallic compounds such as litharge, lead oxide, PbO. The heat would accelerate the **polymerization** process and the lead would react with the fatty acids to form lead soaps, giving the oil medium greater smoothness and easier workability. As the transition from tempera to oil took place, so also did the move from using wood panels for supports to canvas.[57]

Egg tempera is one of the oldest continuously employed methods known to fix pigments onto a surface, with roots in ancient Greece, and possibly even earlier. Tempera comes from the Latin "temperare," to moderate. That which is moderated is a mix of egg yolk as

---

[‡]Laurie tells of a method that buyers of Old Dutch Masters used to detect forgeries: test the picture with the point of a pin. If the surface is hard right through, it is genuine; if it is only hard on the surface, it is more than likely a modern forgery.

a binder combined with a thick paste of pigment ground in water to make a workable paint. Egg yolk is about half water, 33% lipids (such as fats and oils) and 17% protein. After application, the water dries and the protein loses its normal structure that maintains it as a semi-liquid. This process, called denaturation, causes a hardening of the yolk into a tough, durable film that holds the pigment in place. (If any of you has ever had your car windshield "egged" on Hallowe'en, you know how hard it is to remove the film next morning.) The lipids, consisting mainly of unsaturated fatty acids, and the pigments as well participate in further cross-linking to form a strong **polymeric** network similar those formed by oils (Box 9.2).[58]

Box 9.2   Modern Polymeric Media

Obviously, the use of egg tempera and oil was empirical and serendipitous. That they stood the test of time over many centuries is witness to the expertise of the practicing artists who perfected them. They only gave way in the 20th century to polymeric materials after a serendipitous discovery of polyethylene in 1898 by the German chemist Hans von Pechmann (1850–1902). Revealing the nature of this high polymer and many others was due to the groundbreaking work of Hermann Staudinger (1881–1965) and Wallace Carothers (1896–1937). These two chemists tackled the polymer problem from opposite ends of the spectrum: Staudinger began by focusing his experimental work on naturally-occurring polymers, whereas Carothers began with smaller molecules to build up entirely synthetic long chains. Together they laid the groundwork for one of the most important developments of the 20th century, both economically and theoretically.

Building on the experimental and theoretical work of these two giants, polymeric media such as acrylics and alkyds opened up new vistas for artists, both professional and amateur. Chemically, the **acrylic** polymer is polymethylmethacrylate, usually marketed in the form of emulsions. They are water-based, quick drying, flexible, impervious to cracking, mix well with other media, and are inexpensive. With an almost limitless spectrum of vivid colors embedded into an acrylic binder, many artists looked no further for their perfect medium. **Alkyd** resin formation involves reaction of an anhydride such as phthalic anhydride with a monoglyceride ester formed from glycerol and a fatty acid.

These two binders have changed the way art is done. Professionals like Josef Albers (1888–1976), David Hockney (b. 1937), Loris Cecchini (b. 1969), José Gutiérrez (1900–1968) and Gunther Gerzso (1915–2001) have taken advantage of the many qualities of these and other polymeric materials to give free rein to their creativity. An iconic user of alkyd resins was Jackson Pollock (1912–1956): these resins allowed for his famous drip paintings.[59] These media have also opened the door to thousands of amateurs for whom oil painting would be most daunting. We can only imagine the direction the fine arts would have taken if these media had been available to our ultra-creative Renaissance artists.

9.6.2 "The Last Supper:" A Failed Experiment and a Lifetime of Vicissitudes

Oil turned out to be a near-perfect medium for canvas, but sadly, not for plaster. And certainly not for the exterior plaster wall of the delle Grazie refectory that was saturated with moisture due to Milan's typical humidity and exposed to steam and heat from the nearby kitchen. Even during his own lifetime, Leonardo's decision to use a mix of egg tempera and oil for a wall painting of his greatest masterpiece was cause for regret.

True to his desultory wont, Leonardo did not complete the work until 1498, but compared with many of his other commissions, this was lightning speed. Apparently, he spent a lot of time wandering around Milan in search of just the right faces to use for the Apostles, and even threatened to paint the Prior's face for Judas if he pushed him too far to finish. Although it turned out to be his greatest masterpiece, it was little appreciated in his time. The monks cut a door through it, removing Christ's legs and feet along with part of the wall in the process. By 1517, long after its creator had gone off to France to work for King Francis I (1494–1547), paint began flaking off parts of the painting. The painting's Achilles heel was the ground layers, in touch with a plaster wall instead of with wood or canvas. The fundamental physico-chemical incompatibility between the plaster and ground caused the latter to split longitudinally in many places. This initial deterioration led to a series of attempts at restoration, one more erroneous or incompetent than the next. Meanwhile, the room flooded several times and was used on and off as a stable, a prison, and for target practice. To cap off all of these insults, on 16 August 1943, the building was destroyed in an Allied bombing raid; the Last Supper survived, due to the careful placement of some sandbags, and languished for some months practically out in the open air. In 1979, a team working on it said "what we have before our eyes today is now only the ghost of the original work of Leonardo; a ghost nonetheless significant enough for our culture to be worthy of every effort to conserve it."[60]

That last attempt at restoration, begun in 1978 and finished in 1999, suffered criticism for having removed some of the earlier restoration work; the critics claimed that the earlier work was integral to the painting's legacy. About 20% of Leonardo's original paint remains, prompting some to say that we can no longer say that this is his work.[61] Nonetheless, over 3 30 000 visitors come each year for their 15-minute harangue before this legend; they are the lucky ones because each year, an equal number of requests are denied. This visitor was able to obtain a ticket for October 2000 by booking as late as August; that timeline is no longer possible. There was still some scaffolding up at the time and the venue was sterile, very much like the barn that it once doubled as.

In an effort at ongoing conservation, Eataly, the Italian market chain, has contributed to the installation of a filtration system that ports 10 000 cubic meters of air per day into the site. One of the great concerns is the high CO_2 level caused by visitor pressure. There is a proposal afoot to try to reduce these levels by phytoremediation,[§] but it is still on the drawing board.[62]

Prior to the 1978–1999 intervention, some samples were removed by a team from the Opificio delle Pietre Dure restoration laboratory, in Florence. Their analysis of Christ's blue mantle was ultramarine mixed with some azurite and lead white; the bright red of Christ's sleeve was pure vermilion.[63] Later work identified the pink areas as madder **lake** (*Lac de Garance*) on the basis of the particles' fluorescence.[64] Still later, **Raman spectroscopy** applied to these same samples identified the presence of an aluminum complex of kermesic acid, or kermes.[65] Both madder, of vegetable origin, and kermes, of insect origin, are derivatives of **anthraquinone**. Their chemistry will be highlighted in the next chapter.

Although *The Last Supper* is a cultural casualty of war, as well as of ineptness and the initial failed experiment of its creator, we are very fortunate that it survived. There are many other great works that have not (Box 9.3).

No foray into the art of the High Renaissance would be complete without a nod to the most famous colorist of them all, Titian. As we shall see in the analysis that follows, Titian's mastery of color

---

[§]Phytoremediation is a process that uses living green plants to remove contaminants and pollution *in situ*.

Box 9.3   Lead White

A staple of Renaissance artists, lead white (sometimes called white lead) is the most important of the lead-based pigments and one of the first to be synthesized artificially. In fact, its entire history is one of intentional manufacture. It is a basic lead carbonate with the formula $2PbCO_3 \cdot Pb(OH)_2$. Its naturally occurring form is lead carbonate, $PbCO_3$, corresponding to the mineral cerussite, which has rarely been used as a pigment. Lead white has been prepared for centuries according to methods described in Pliny, Vitruvius, and Theophrastus and many other manuscripts. Essentially, lead sheets or plates are exposed to vinegar in a moist environment of fermenting manure or tanbark (which provides heat and ammonia to maintain alkalinity) and atmospheric CO_2 for a period of about three months to yield the formulaic pigment almost always mixed with some side products like lead acetate. (This very dangerous process was traditionally "women's work.") Historically, it is the most important of all the white pigments and has been mentioned in pigment catalogs from ancient times up to the present day. Though lead white reigned supreme through the ages up to the 19th century, it came into its own as the darling of Renaissance painters not only for its luscious, creamy color, but also for its ease of handling, its opacity and its bulk drying properties. Once married to an oil, lead can chemically bond to long chain fatty acids to form lead soaps, usually lead linoleate, one of the smoothest substances ever to be handled by a paintbrush. As more of the linoleate forms over the centuries, the pigment becomes more transparent. Titian used it habitually in almost all of his paintings, combining it with blues in his skies and seas to make his clouds skitter and his waves frolic. Johannes Vermeer's signature expansive lead white areas glow with a light that has become iconic. Lead white has two serious drawbacks: it is highly toxic, and it darkens to form black lead sulfide, PbS, upon exposure to sulfur-containing compounds. Since, according to Laurie,[66] it was an almost universal practice among artists of the Renaissance to rub the ground of a painting thoroughly with garlic to ensure successful binding of the size with the oil, and since garlic contains over 30 sulfur compounds, one wonders if discoloration of lead white might be attributable to the garlic content, and not necessarily to incompatible pigments. Laurie also says that in preparing grounds on canvas, flour, walnut oil and lead white mixed together and applied by hand is most effective,[66] but it would also be effective in cutting one's life short as well.

---

lay in his ability to astutely mix his pigments, a practice frowned upon in the early Renaissance. Painters knew that they could achieve almost any hue by mixing only three suitably selected pigments. However, they also knew that they paid a price – the more pigments mixed, the duller the result. The technical term is loss of saturation.[67]

9.7   TITIAN

Tiziano Vecellio, or Titian (*ca.* 1488–1576) was for 60 years the undisputed superstar of Venetian painting. It is an open question if this would have been so if his rival, Giorgione (1478–1510), had not been cut off at the age of 32. One of his major commissions included a cycle of paintings for the Doge of Venice's palace, that delicate, lacelike pink wonder snuggled beside the San Marco basilica. He also painted mythological scenes for the private enjoyment of Alfonso D'Este, Duke of Ferrara.

9.7.1   Titian's Masterpiece

From the Ferrara collection comes Titian's great masterpiece, *Bacchus and Ariadne*, painted between 1520 and 1523 (Figure 9.6).[68]

The scene involves Ariadne, princess of Crete, who, after helping the Athenian, Theseus, slay the Minotaur on her home island, ran off with him (or perhaps was kidnapped? Ovid (b. 43 BCE) in his *Metamorphoses* is vague about this.) but was abandoned on the Greek island of Naxos. Bacchus, the god of wine

Figure 9.6   Titian, Bacchus and Ariadne (1520–1523). National Gallery, London. Oil on canvas. Reproduced from ref. 68.

and revelry, arrives with his outrageous entourage, immediately falls in love with her, and promotes her eventually to become the constellation, *Corona borealis*, which you can see above her head in the painting. The story is gleaned from Ovid and the poetry of Catullus (84–54 BCE).

Our involvement with this painting has nothing to do with the mythic, but with the clear, hard facts of chemical analysis. The work is now in the hands of the National Gallery, London. It was badly treated and underwent several restorations after traveling, rolled up, halfway around the world. Though it did not suffer as much as the *Last Supper*, it was never treated with the reverence it deserves. Now it is pampered daily, given its delicate condition. But the great bonus is the Gallery's careful analysis of the pigments that Titian used and publication of these results in a visual rendition where each of the analyzed sites is numbered and described.[69] This is the only example of this kind of visual detail freely available that I am aware of. Would that every museum follow this lead – for every work of art, if possible!

The identities of the pigments are clearly laid out: the sky, the distant landscape and the blue draperies of the figures are painted with very high quality ultramarine; the seascape is azurite; the green is complex, but what you see on the surface is malachite which has been covered with an overglaze of copper resinate. Can you detect any discoloration of the copper resinate? Do you think it would have been exempted from the same deterioration as suffered by *Primavera*? The brown of the tree foliage is yellow and brown ochre? You remember that yellow ochre is goethite. What do you think that brown ochre is? We have not encountered it before.

For Ariadne's dramatic red sash, Titian used fine vermilion; for the rose pink color of Bacchus's cloak, the chemists discerned a red **lake**, but could not tell which kind. In the *Technical Bulletin* paper that details the analysis, they say, "It is a matter for regret that at the time the picture was under treatment and it was possible to take samples, the methods which have since been developed in the Scientific Department for the identification of the dyestuffs of red lake pigments were not available."[70] These **lakes** and how they can be differentiated will be dealt with in the next chapter.

Where did Titian get all these fine colors? The answer? He was Venetian. Location, location, location.

9.7.2   Venice – the Colorist's Entrepôt

During the Renaissance, Venice was the place to come for the widest range of not only purpose-made pigments but just about any other material, refined or raw. This availability built upon Venice's growing reputation as a center where celebrated artists congregated. But which came first? Venice was also at the nexus of east and west with a tradition of seamanship that morphed into a seafaring trade that was second to none by the late 15th century.[71] There is even archaeological evidence of this Venetian trade in coloring materials, the Gnalić shipwreck of 1583, of which the largest proportion of its cargo was art supplies.[72]

   Not only was there trade. Venice gave rise to a whole new profession, the so-called *vendecolori*, or colormen, who specialized in manufacturing and selling every form of art supply from their shops. Since these *vendecolori* were clustered around the Rialto Bridge area, they became "destination shops" for every type of artisan across Europe.[73,74] They were unique to Venice and also to the business: most other art suppliers were also apothecaries who dispensed many other types of goods, especially medications. In fact, if you weren't careful, you could get the two mixed up as this little anecdote relates: Once upon a time, Paduan artist, Dario Varatori (1539–1596), while under the care of a doctor, was also working on a painting. Bringing his medicine to work, Varatori sniffed the contents of his medicine bottle, immediately dipped in his brush and began to spread the solution on the wall apparently without compromising either the painting or his own health.[75] Naturally, the Venetian authorities had to set laws governing the color trade in order to protect their own reputation. All of Europe subsequently followed suit regarding standards of workmanship and regulation of the quality of art materials.[76] Safeguards against fraud, which was rife with respect to the most valuable pigments, was a priority.[77]

9.8   CONCLUSION

Our journey through the Renaissance was necessarily brief; otherwise, we could linger there a lifetime. The Renaissance – the rebirth – though initially centered in Italy, was taking place throughout Europe. Eventually, its philosophy would affect the evolution of ideas globally. Household words like Michelangelo and Tintoretto extend equally to Rembrandt, Vermeer and many others. A visit to virtually any art museum in the world reveals the enduring presence of this momentous movement that shaped the imagination of millions.

We have seen how the Renaissance artists depended heavily on materials that were in no way innovative but of very fine quality. Their greatest contribution is how they used these pigments to express a whole new way of thinking. But just as the High Renaissance was gearing up, an event that would refocus the vision, transform the economy, and compromise the moral compass of the entire European world took place. The year was 1492. The protagonist was Christopher Columbus. And the world of art was no less caught up in its wake, as we shall learn in the next chapter.

REFERENCES

1. John Fitzgerald Kennedy, Address, Amherst College, 26 October 1963.
2. Pliny the Elder, *The Elder Pliny's Chapters on the History of Art*, trans. K. Jex-Blake, ed. E. Sellers, Macmillan, London, 1896, pp. 137–139.
3. G. Vasari, *The Project Gutenberg eBook of Lives of the Most Eminent Painters*, vol. 1, p. 74, https://www.gutenberg.org/files/25326/25326-h/25326-h.htm, accessed October 2020.
4. Metropolitan Museum of Art, https://www.metmuseum.org/art/collection/search/436504, accessed October 2020.
5. D. Bomford, J. Dunkerton, D. Gordon and A. Roy, *Art in the Making: Italian Painting Before 1400*, National Gallery Publications, London, 1989, pp. 64–71.
6. The Adoration of the Magi, https://commons.wikimedia.org/w/index.php?title=File:The_Adoration_of_the_Magi_MET_DT3.jpg&oldid=465012477, accessed June 2021.
7. B. H. Berrie, M. Leona and R. McLaughlin, *Heritage Sci.*, 2016, **4**, 1–9.
8. H. Kühn, Lead-Tin Yellow, in *Artists' Pigments: a Handbook of Their History and Characteristics*, ed. A. Roy, National Gallery of Art, Washington, DC, 1993, vol. 2, pp. 83–112.
9. M. P. Merrifield, *Original Treatises on the Arts of Painting*, John Murray, London, 1849, vol. 2, p. 529.

10. R. J. H. Clark, L. Cridland, B. M. Kariuki, K. D. M. Harris and R. Withnall, *J. Chem. Soc., Dalton Trans.*, 1995, 2577–2582.

11. I. N. M. Wainwright, J. M. Taylor and R. D. Harley, Lead antimonate yellow, in *Artists' Pigments: a Handbook of Their History and Characteristics*, ed. R. L. Feller, National Gallery of Art, Washington, DC, 1986, vol. 1, pp. 219–254.

12. *Painting techniques: history, materials and studio practice. Contributions to the Dublin Congress 7–11 September 1998*, ed. A. Roy and P. Smith, IIC, London, 1998, pp. 160–165.

13. E. Martin and A. R. Duval, *Stud. Conserv.*, 1990, **35**(3), 117–136.

14. R. H. Brill, The Scientific Investigation of Ancient Glasses, in *Proceedings, Eighth International Congress on Glass, London, 1–6 July 1968*, Society of Glass Technology, Sheffield, 1969, pp. 47–68.

15. B. Berrie, *Early Sci. Med.*, 2015, **20**, 308–334.

16. R. J. Gettens and E. W. FitzHugh, *Stud. Conserv.*, 1974, **19**(1), 2–23.

17. C. Sandalinas and S. Ruiz-Moreno, *Stud. Conserv.*, 2004, **49**(1), 41–52.

18. H. G. M. Edwards, *Spectrochim. Acta, Part A*, 2011, **80**, 14–20.

19. G. Agresti, P. Baraldi, C. Pelosi and U. Santamaria, *Color Res. Appl.*, 2016, **41**(3), 226–231.

20. R. Billinge and D. Gordon, *Natl. Gallery Tech. Bull.*, 2008, **29**, 76–80.

21. L. B. Alberti, *On Painting (1435–1436)*, trans. J. R. Spencer, Yale University Press, New Haven, 1970.

22. S. Y. Edgerton, *Hist. Cienc. Saude-Manguinhos*, 2006, **13**, suppl. 0, 151–179.

23. Domenico Ghirlandaio: An Old Man and his Grandson, https://commons.wikimedia.org/w/index.php?title=File:Ghirlandaio,_Domenico_-%20_An_Old_Man_and_His_Grandson_-_Louvre_-_Google_Art_Project.jpg&oldid=484060236, accessed June 2021.

24. *Cennino Cennini's Il Libro dell'arte, a new English Translation and Commentary with Italian Transcription*, ed. L. Broecke, Archetype Publications, London, 2015.

25. Sandro Botticelli: Primavera, https://commons.wikimedia.org/w/index.php?title=File:Botticelli-primavera.jpg&oldid=434862134, accessed June 2021.

26. Y. Yashiro, *Sandro Botticelli and the Florentine Renaissance*, Hale, Cushman and Flint, Boston, 1929, vol. 1, p. 205.
27. J. Dunkerton and A. Roy, *Natl. Gallery Tech. Bull.*, 1996, **17**, 20–31.
28. Paraphrased from P. Ball, *Bright Earth*, Farrar, Straus and Giroux, New York, 2001, p. 114.
29. R. E. LaFever, B. S. Vogel and R. Croteau, *Arch. Biochem. Biophys.*, 1994, **313**(1), 139–149.
30. R. Harley, *Artists' Pigments: 1600–1835*, Butterworths, London, 1970.
31. Web Gallery of Art, https://www.wga.hu/frames-e.html?/html/b/botticel/8smarco/11smarco.html, accessed May 2021.
32. C. Andalò, M. Bicchieri, P. Bocchini, G. Casu, G. C. Galletti and P. A. Mandò, *et al.*, *Anal. Chim. Acta*, 2001, **429**, 279–286.
33. D. Bersani, P. P. Lottici, A. Casoli and D. Cauzzi, *J. Cult. Herit.*, 2008, **9**, 97–102.
34. E. S. Uffelman, E. Court, J. Marciari, A. Miller and L. Cox, Handheld XRF Analyses of Two Veronese Paintings, in *Collaborative Endeavors in the Chemical Analysis of Art and Cultural Heritage Materials*, ed. P. Lang and R. A. Armitage, American Chemical Society, Washington, DC, 2012, pp. 51–73.
35. M. D. Leonida, *The Materials and Craft of Early Iconographers*. Springer, Cham, 2014, p. 95.
36. B. Pelletier, Examen chimique des cendres bleues, et procédé pour les préparer, in *Mémoires et observations de chimie de Bertrand Pelletier recueillis et mis en ordre par C. Pelletier et Sédillot jeune*, Imprimerie de la République, an VI, Paris, 1798, vol. 2, pp. 1–21.
37. M. V. Orna, Archaeological Blue Pigments: Problem Children from the Get-Go, in *Archaeological Chemistry: a Multidisciplinary Analysis of the Past*, ed. M. V. Orna and S. C. Rasmussen, Cambridge Scholars Publishers, Newcastle upon Tyne, 2020, pp. 339–402; 355.
38. P. Ricciardi, A. Pallipurath and K. Rose, *Anal. Methods*, 2013, **5**, 3819–3824.
39. K. Eschner, Arsenic and Old Tastes Made Victorian Wallpaper Deadly, *Smithsonian Magazine*, https://www.smithsonian-mag.com/smart-news/victorian-wallpaper-got-its-gaudy-colors-poison-180962709/, accessed October 2020.

40. I. Fiedler and M. A. Bayard, Emerald Green and Scheele's Green, in *Artists' Pigments: a Handbook of Their History and Characteristics*, ed. E. W. FitzHugh, National Gallery of Art, Washington, DC, 1997, vol. 3, pp. 219–272.

41. M. Pastoureau, *Green: The History of a Color.* Princeton University Press, Princeton, 2014, p. 183.

42. D. E. H. Jones and K. W. D. Ledingham, *Nature*, 1982, **299**, 626–627.

43. A. Meharg, *Nature*, 2003, **423**(12 June), 688.

44. *New York Times*, Sunday 22 July, 1956, p. 22.

45. Victoria and Albert Museum, https://www.vam.ac.uk/exhibitions/botticelli-reimagined, accessed October 2020.

46. 1962 film Dr. No. https://en.wikipedia.org/wiki/White_bikini_of_Ursula_Andress , accessed October 2020. *Venus* even graced two *New Yorker* covers, 4 August 2014 and 25 May 1992.

47. M. Spring, A. Mazzotta, A. Roy, R. Billinge and D. Peggie, *Natl. Gallery Tech. Bull.*, 2011, **32**, 78–112.

48. G. Vasari, *Lives of the Most Eminent Painters, Sculptors and Architects*, vol. 4, *Leonardo da Vinci*, pp. 50–57; 53. Project Gutenberg, http://www.gutenberg.org/ebooks/28420, accessed October 2020.

49. S. Pappas, *Human Body Part that Stumped Leonardo da Vinci Revealed*, LiveScience, 2012, https://www.livescience.com/20157-anatomy-drawings-leonardo-da-vinci.html, accessed October 2020.

50. S. K. Ghosh, *Anat. Cell Biol.*, 2015, **48**(3), 153–169.

51. Leonardo, ultima cena (restored), https://commons.wikimedia.org/wiki/File:Leonardo,_ultima_cena_(restored)_01.jpg, accessed June 2021.

52. C. Nicholl, *Leonardo da Vinci: Flights of the Mind*, Viking, New York, 2004, p. 266.

53. Theophilus, *On Divers Arts*, trans. J. G. Hawthorne and C. S. Smith, Dover, New York, 1979, pp. 27–28.

54. S. Jones, Painting in Oil in the Low Countries and Its Spread to Southern Europe, in *Heilbrunn Timeline of Art History*, New York, The Metropolitan Museum of Art, 2000, http://www.metmuseum.org/toah/hd/optg/hd_optg.htm, accessed October 2002.

55. L. da Vinci, *The Notebooks*, trans. E. MacCurdy, Jonathan Cape, London, 1938, vol. II, pp. 355–356.

56. A. P. Laurie, *The Pigments and Mediums of the Old Masters*, Macmillan, London, 1914, p. 172.
57. C. S. Tumosa and M. F. Mecklenburg, *Stud. Conserv.*, 2005, **50**, suppl. 1, *Rev. Conserv.*, **6**, 39–47.
58. D. A. Peggie, The Chemistry and Chemical Investigation of the Transition from Egg Tempera Painting to Oil in Italy in the 15th Century, in *Science and Art: The Painted Surface*, ed. A. Sgamellotti, B. G. Brunetti and C. Miliani, Royal Society of Chemistry, Cambridge, 2014, pp. 209–229.
59. *Science and Art: The Contemporary Painted Surface*, ed. A. Sgamellotti, B. G. Brunetti and C. Miliani, Royal Society of Chemistry, Cambridge, 2020. pp. 1–17, 67–94, 316–337, 374–389, 404–430.
60. M. Matteini and A. Moles, *Stud. Conserv.*, 1979, **24**(3), 125–133.
61. P. B. Barcilon and P. C. Marani, *Leonardo, The Last Supper*, trans. H. Tighe, University of Chicago Press, Chicago, 2001.
62. E. Salvatori, C. Gentile, A. Altieri, F. Aramini and F. Manes, *Sustainability*, 2020, **12**, 565–581.
63. A. Dal Fovo, M. Oujja, M. Sanz, A. Martínez-Hernández, M. Vega Cañamares, M. Castillejo and R. Fontana, *Spectrochim. Acta, Part A*, 2019, **208**, 262–270.
64. A. Gallone, Analisi Stratigrafica di campioni di colore dell'Ultima Cena, in *Leonardo Ultima Cena, Indagine, Ricerche, Restauro*, ed. G. Basile and M. Marabelli, Nardini Editore, Firenze, 2007, pp. 145–154.
65. I. Osticioli, M. Pagliai, D. Comelli, V. Schettino and A. Nevin, *Spectrochim. Acta, Part A*, 2019, **222**, 11723.
66. A. P. Laurie, *The Materials of the Painter's Craft*, T.N. Foulis, London, 1910.
67. R. Kuehni and A. Schwarz, *Color Ordered*, Oxford University Press, New York, 2008, p. 54.
68. Titian Bacchus and Ariadne, https://commons.wikimedia.org/w/index.php?title=File:Titian_Bacchus_and_Ariadne.jpg&oldid=424068338, accessed June 2021.
69. Colourlex, https://colourlex.com/project/titian-bacchus-ariadne/, accessed October 2020.
70. A. Lucas and J. Plesters, *Natl. Gallery Tech. Bull.*, 1978, **2**, 25–47.

71. L. Matthew and B. Berrie, "Memoria de colori che bisognino torre a vinetia:" Venice as a Centre for the Purchase of Painter's Colors, in *Trade in Artists' Materials: Markets and Commerce in Europe to 1700*, ed. J. Kirby, S. Nach and J. Cannon, Archetype Publications, London, 2010, pp. 245–252.

72. K. Batur and I. R. Rossi, Archaeological Evidence of Venetian Trade in Colouring Materials: The Case of the Gnalić Shipwreck, in *Trading Paintings and Painters' Materials 1550–1800*, ed. A. H. Christensen and A. Jager, Archetype Publications, London, 2019, pp. 111–120.

73. L. C. Matthew, *The Burlington Magazine*, 2002, vol. 144(1196), pp. 680–686.

74. J. DeLancey, *Wallraf-Richartz-Jahrbuch*, 2011, vol. 22, pp. 193–232.

75. R. King, *Il papa e il suo pittore: Michelangelo e la nascita avventurosa della Cappella Sistina*, Rizzoli, Milano, 2003, p. 124.

76. J. Kirby, The Price of quality: factors influencing the cost of pigments during the Renaissance, in *Revaluing Renaissance Art*, ed. G. Neher and R. Shepherd, Ashgate Publishing, London, 2000, pp. 19–42.

77. S. Barker, *"Che Altri Che Lui non lo Fa,": Making Ultramarine Blue in Grand Ducal Florence, Trading Paintings and Painters' Materials 1550–1800*, ed. A. H. Christensen and A. Jager, Archetype Publications, London, 2019, pp. 130–135.

CHAPTER 10

Aztec Red and Maya Blue: Secrets of the New World[†]

When I have no blue, I use red.[1]

Pablo Picasso

10.1 INTRODUCTION

The mysterious origins of two exotic pigments from the New World were the subject of piracy, espionage and stories of human sacrifice. One of them was destined to eclipse its Old World counterparts; the other propelled 21st century research into exploring the nature of **nanomaterials**.

[†]Glossary entries and chapter cross-references are in boldface.

March of the Pigments: Color History, Science and Impact
By Mary Virginia Orna
© Mary Virginia Orna 2022
Published by the Royal Society of Chemistry, www.rsc.org

10.2   RED AND BLUE JUXTAPOSED

Red seeps in
and slaps you
in the face
dances on your grave   Blue! 'Tis the life of heaven,–the domain
makes you go   Of Cynthia,–the wide palace of the sun,–
forward   The tent of Hesperus and all his train,–
frightens the wits   The bosomer of clouds, gold, grey
out of you   and dun.
makes you faint   Blue! 'Tis the life of waters–ocean
in the emergency   And all its vassal streams: pools
room   numberless
seduces you   May rage, and foam, and fret, but
into   never can
forbidden zones   Subside if not to dark-blue nativeness.
wide-eyed and wanton Blue! gentle cousin of the forest green,
for what secrets   Married to green in all the sweetest flowers,
it has to tell   Forget-me-not,–the blue-bell,–and, that queen
and rants wild   Of secrecy, the violet: what strange powers
speaking of love   Hast thou, as a mere shadow! But how great,
all the while.   When in an Eye thou art alive with fate!
Red by Ann Naito   *Sonnet on Blue* by John Keats.[70]
Haney.[2]
Reproduced with
permission from
Ann Haney.

If Picasso can conflate blue and red – mix the progress, enlightenment, spaciousness, spirituality and freedom of blue with the activity, rage, rebellion, power, authority, fire, blood, crime and sex of
red, then these two colors can encompass all the commotions and
emotions of humanity. In the Bible, blue and purple are treated as
identical, so why not blue and red? We shall see that in the New
World iconography of the Aztecs and the Maya, they did just that.

10.3   INTRODUCING AZTEC RED

A few years ago, a friend and I were enjoying the marvelous flora
displayed in Phoenix's Desert Botanical Garden. While observing the prickly pear cacti, I spied a whitish fluffy film on a cactus

labeled *Opuntia decumbens*. "Look!" I cried. "That looks like a cochineal bug on that plant." Sure enough, the characteristic cottony film that envelops the cochineal scale insect was sitting in plain sight (Figure 10.1).[3,4] The film plays a protective role in sheltering the insect from its many predators. I had no sooner pointed this out when a staff member immediately pounced on the unfortunate plant, did his best to remove the offending insects, but finally carted the plant off to an unknown fate. That was an occasion to reflect on the saying "One man's meat is another man's poison."[5]

From at least the 10th century CE, indigenous Mexicans had cultivated the *Dactylopius* scale insect for its brilliant red dye, even sharing their food source, the nopal cactus, with this voracious predator. The botanical garden had good reason to fear – once a female scale insect sinks its enormous proboscis into the nopal, it can desiccate the fluid-swollen succulent down to a mere fiber in days. And its reproductive prowess is such that it can spread its progeny around in about the same amount of time. Historically, the best nopaleries (cactus gardens to cultivate the scale insects) were in Oaxaca on Mexico's southwest coast, where centuries of domestication had produced plump rounded female insects with up to 25–30% of their body weight comprised of the coveted red color. In addition to textile dyeing, the Aztecs used it as food color, body paint, medicine, and for burials.

Called *nochezli* (meaning "blood of the prickly pear") by the Aztecs, the Spaniards named it *grana cochinilla*, after the Latin term *grana* for seeds and a diminutive form of *coccinus* for scarlet

Figure 10.1   Left: *Dactylopius coccus* growing in La Palma, Canary Islands. Right: *Dactylopius coccus* on *Opuntia ficus-indica*, Andalucia, Spain. Reproduced from ref. 3 and 4 under the terms of the CC BY-SA 3.0 license, https://creativecommons.org/licenses/by-sa/3.0/deed.en.

color: little scarlet seeds. The French word for the color, *coche-nille*, came into English as cochineal. The chemical name for the dye/pigment molecule is carminic acid, and the colorant, usually in the form of a **lake** precipitated with alum, is called carmine. Cochineal carmine is the colorant derived from the New World insect species; kermes cochineal, Polish cochineal and Armenian cochineal are the names of the colorants derived from scale insect species of the Old World. They are among the oldest organic colorants known.[6]

10.3.1 Aztec Red Goes Global

How the beautiful red pigment became one of the most coveted and ubiquitous colors in Europe is one of the saddest stories in human history. It begins with a brash and ambitious young Spanish thug, Hernán Cortés (1485–1547). In 1519, Cortés invaded the Mexican mainland and managed to overthrow the Aztec Empire by playing different factions of the indigenous people against one another. While he and his cohorts were after precious metals, his reigning sovereign, Charles V (1500–1558) urged Cortés to send him some of the red *"grana cochinilla"* to assess its possible economic value – no surprise because the Emperor was always short of cash. Forthwith, a cargo of the new colorant, exacted as tribute from the conquered Aztecs, reached Seville by the next argosy and within four years, it was a major item of global trade.[7] Access to the New World cochineal along with other foreign dyestuffs such as logwood, brazilwood, annatto and indigo transformed the European textile industry.[8]

The Spanish took great pains to conceal cochineal's true nature to thwart the bioespionage and biopiracy they knew would threaten their monopoly. Although the other cochineals had been known and used for thousands of years, they were no match for this luscious new product, highly prized for its greater durability, brilliance, fastness and sheer efficiency. One pound of *grana cochinilla* outdid the dyeing power of twelve pounds of Old World *kermes cochinilla*. These qualities were due to the centuries of care and cultivation that the Aztecs had lavished on the *Dactylopius* scale insects, so much so that they became a new species, *Dactylopius coccus* Costa, a far cry from uncultivated wild species in either the Old or New World.

Figure 10.2 Left: Manufacturing and painting with cochineal. Florentine Codex, Book 11, fol. 216v. Reproduced from ref. 10. Right: Producing and painting with blue from indigo. Florentine Codex, Book 11, fol. 221v. Reproduced from ref. 11.

Franciscan friar Bernardino de Sahagún documented this industry in a massive twelve-volume ethnographic work compiled over 30 years, *Historia general de las cosas de la Nueva España*,[9] or the *Florentine Codex*, for short. The only extant copy of the Codex presently resides at the Biblioteca Laurenziana in Florence. The Codex is lavishly illustrated with over 2600 images in color and black and white; Figure 10.2 [10,11] documents pigment preparation for the two major colors used by the indigenous peoples, Aztec Red, *i.e.*, cochineal, and indigo (to come later in the story).[12,13]

The insects in question were sessile; once hatched, they settled in for good. Only the female produced the rich red carmine, pampered until it comprised up to 25–30% of the insect's body weight concentrated in the egg sac, presumably as a deterrent to predators, put off by its taste. Dyers also loved it because it contained near to none of the fat that the Old World insects had in abundance that literally gummed up the dye bath.[14]

10.3.2   What is it?

Just as the Israelites in the desert asked about the manna, *Ma'n Hu?* What is it?, so the whole European world was asking this question about *grana cochinilla*. The very word "*grana*" (grain) was very effective at concealing its true nature: many people thought it was seeds or berries from an exotic plant or perhaps a little worm bred by the cactus plant itself. This was a matter of mystery for almost two centuries. Exasperated by ignorance, in 1685, Robert Boyle (1627–1691), the quintessential chemist of the 17th century, persuaded the greatest microscopist of all time, Antonie van Leeuwenhoek (1632–1723), to take a look. Boyle was bitterly disappointed when the Dutchman reported that what he saw looked like "a dried black currant."[14, p. 147] Finicky, meticulous, talented, and secretive as van Leeuwenhoek was in his methods, he could not discern the nature of this creature even with the best lenses in existence. Boyle urged him to keep on looking. And look he did, until with a new instrument he began to descry little wings and tiny legs, and eventually masses of insect larvae resting in their mothers' bellies – this mysterious *"grana"* was not by any means a berry, but a bug! Van Leeuwenhoek's drawings are a marvel even to this day.[15] However, if we dig a little deeper, we find that another Dutchman, Nicolaas Hartsoeker (1656–1725), working in Paris as a physicist and lensmaker, illustrated these enigmatic little creatures and clearly identified them as insects as early as 1694. It seems that van Leeuwenhoek merely confirmed these findings without acknowledgment, reigniting an old enmity over a priority dispute from thirty years earlier.[16]

In 1756, the great Swedish botanist, Carl Linnaeus (1707–1778), was keen to add cochineal to his growing list of formally named genera and species. Eagerly awaiting a sample of the nopal cactus supporting the insects, he was out of town when a former student, Daniel Rolander, arrived with it, having nursed it with care all the way from Surinam. Linnaeus's ever-assiduous assistant was horrified to find the precious plant covered with "vermin" and immediately purged them before his boss returned. Fortunately, he did not do a thorough job, and one little bug remained – giving Linnaeus the chance to enter its formal name into the roll call of the living. *Coccus cacti* Linnaeus appeared for the first time in the 10th edition of his *Systema Naturae* in 1758.[17] However, the incident drove Rolander mad.[18]

10.3.3   New World Cochineal Deployed and Processed

Once its superb qualities were recognized, Aztec Red quickly eclipsed the older insect-derived dyes and plant-based madder, from the root of the perennial *Rubia tinctorum*. Both State and Church quickly legislated protective measures against these "colors of the devil" by leveling high fines or even threatening closure of studios and workshops.[19] Nevertheless, demand for this new color continued to pour revenue unabated into insatiable Spanish coffers, while Spain's envious neighbors plotted by any means to gain the secret (and a viable supply) of its essence.

Frenchman Nicolas-Joseph Thiéry de Menonville (1739–1780) pulled off one of the most foolhardy, but successful, attempts to break Spain's cochineal monopoly. In order to penetrate as far as the city of Oaxaca where the best of this dye was being produced, he had to traverse the 500-mile round trip from the port of Veracruz on foot. Who else but he, the possessor of a suave, arrogant, persuasive personality, could accomplish this task? And accomplish it he did – through lying, bribery, stealth and all the common methods of modern commerce, he actually managed to "export" large boxes of living insects to French territory under the noses of the Spanish officials in Mexico.[20]

Unlike the Old World kermes insects that were dead on arrival when harvested, the New World species was very much alive and had to be killed by several methods that affected the colorant's quality. The most effective was steaming in a specially built container, a *temazcal*, designed by the Aztecs, the fruit of their centuries-long experience. The resultant raw material was the crushed and dried insects – 70 000 of them were needed per pound of dye. They contained the main colorant, carminic acid $(C_{22}H_{20}O_{13})$, mixed with a little kermesic acid $(C_{16}H_{10}O_8)$, both derivatives of the parent compound, **anthraquinone** $(C_{14}H_8O_2)$.

10.3.4 The Chemical Nature of the Anthraquinoid Dyes

The **anthraquinone** family of dyes is many and varied. It is a biogenetic feat of Mother Nature who, over the course of evolutionary ages and continents apart, seemed to enjoy attaching **functional groups** onto the eight available carbon atoms of the parent structure, which is colorless, in order to dress it up in reds and yellows. The biological sources of these colors have been examined with respect to their quantitative yield for commercial purposes. The five important insect species[21] and the one major plant species[22] contain about seven major constituents although as many as three dozen minor colorant constituents have been identified in some of the species[23] (Table 10.1).

Table 10.1   Dye content of major sources of the **anthraquinoid** colorants.

Dye	Source	Species	Major component	Minor components
American cochineal (Aztec Red)[20,22]	Insect	*Dactylopius coccus*	Carminic acid, 94–98%	Kermesic and Flavokermesic acids, 0.4–2.2%
Kermes cochineal[20,22]	Insect	*Kermes vermillio*	Kermesic acid, 75–100%	Flavokermesic acid, 0–25%
Polish cochineal[20,22]	Insect	*Porphyrophora polonica*	Carminic acid, 62–88%	Kermesic and Flavokermesic acids, 12–38%
Armenian cochineal[20,22]	Insect	*Porphyrophora hamelii*	Carminic acid, 95–99%	Kermesic and Flavokermesic acids, 1.0–4.2%
Lac dye[20,22]	Insect	*Kerria lacca*	Laccaic Acids, 71–96%	Kermesic and Flavokermesic acids, 3.6–9.0%
Alizarin[21]	Plant	*Rubia tinctorum* (Madder plant)	Alizarin, 55–62%	Purpurin, 24–31%

A closer look at this table is very instructive. For example, the coloring matter of Aztec Red and Armenian Cochineal is virtually 100% carminic acid, whereas Kermes Cochineal and Polish Cochineal contain fairly high percentages of flavokermesic acid, which is yellow. This explains why the Aztec and Armenian varieties were more highly esteemed than the kermes and the Polish, which were reputed to have a much weaker red color – they actually did! Chemical analysis has revealed the reason. Not only that, the amount of the yellow flavokermesic acid could vary by very large amounts in both the Polish and the Kermes varieties, leading to uncertainties in their coloring power from batch to batch.

Since cochineal is soluble in hot water, filtration or sedimentation from a hot aqueous solution will dispense of the body parts. To prepare the pigment, called cochineal carmine, from the dye extract, addition of aluminum or calcium salts will precipitate out a **lake** pigment. By "fine-tuning" the carminic acid/aluminum ratio, it is possible to produce colors ranging from pale strawberry to blackcurrant.[24] Precipitation with other metal salts produces other **lake** colors: lead salts yield a deep purple color and tin salts a bright scarlet color (Drebbel's Scarlet; **Chapter 11.2**). This is because the metals form **chelate** complexes with the dye.[25,26] The **chelates** give a much more intense color than just the carminic acid alone. To complicate matters, it is also possible to change the color by controlling the pH, *i.e.*, the acidity or basicity of the colorant; carminic acid is straw yellow in low pH (high acidity) solutions, but deep red to purple in high pH (low acidity) solutions.

10.3.5 Cochineal Carmine Arrives on the Artists' Palette

Though Aztec Red arrived on European shores in 1523 and was immediately snapped up by dyers as the best red dye that money could buy, its journey to the artists' palette was glacial – partly because of the secrecy of its processing, uncertainty of its permanence, ignorance of its true nature, and possibly because of its great expense due to heavy import duties.

At first, the New World cochineal pigment was prepared by the indirect method by which the more ancient cochineals were made: wool shearings or pieces of cloth, called clothlets, were immersed in the dyebath until all the dye had been absorbed, through multiple dippings and dryings. Then the dye was extracted from the dyed textile shearings by adding alkali (often calcium or sodium salts as carbonates) and alum (potassium aluminum sulfate) to precipitate the colorant onto a substrate of aluminum oxide, forming the **lake** pigment. This New World product seems to make its first appearance in works by Joachim Beuckelaer (1533–1574)[27] and Paolo Veronese (1528–1588) in the 1570s.

Not long after that, in the 1580s, a fundamentally different method to directly prepare cochineal carmine came into play by accident. The story goes that a Franciscan monk and pharmacist in Pisa inadvertently added some acid to a mixture of cochineal extract and cream of tartar (potassium hydrogen tartrate). Instead of tossing this useless mix away, he set it aside – and lo! After a few days, he discovered a red precipitate. A painter found this pigment much to his liking and asked for more.[28] In the seventeenth century, it has been identified in works by Tintoretto (1518–1594),

Figure 10.3   Giovanni Antonio Canaletto. A Regatta on the Grand Canal, *ca.* 1740. Bowes Museum, Barnard Castle. Reproduced from ref. 31.

Peter Paul Rubens (1577–1640),[29] Rembrandt (1606–1669),[30] Guido Reni (1575–1642), Diego Velázquez (1599–1660), Hendrick ter Bruggen (1588–1629), Anthony van Dyke (1599–1641) and Pierre Mignard (1612–1695). By the eighteenth century, it had become a permanent resident on the palettes of Canaletto (1697–1768), Joshua Reynolds (1723–1792), Thomas Gainsborough (1727–1788) and J.M.W. Turner (1775–1851). Eventually, Aztec Red became the essential red pigment among most easel painters until the end of the nineteenth century. Figure 10.3,[31] Canaletto's "A Regatta on the Grand Canal," displays red draperies on the left identified as New World cochineal.[32]

10.3.6   Aztec Red's Life Beyond the Palette

As the twentieth century dawned, cochineal carmine's use declined for several reasons. Artists were becoming aware of its fugitive nature. It was observed to be susceptible to light-induced discoloration exacerbated by relative humidity and atmospheric pollution.[33,34] At the same time, synthetic pigments derived from coal tar and petroleum were catching the notice of artists ever since the first commercially viable one, mauve, appeared in 1859[35] (**Chapter 12.2**). No doubt, the practicing artist took heed of British chemist, A.H. Church, who admonished: "No artist who cares for his work and hopes for permanency should employ [it]."[36]

As cochineal's use among artists waned, a more ancient pigment, madder, recovered its slow but sure ascendancy. We know that it was used as a pigment in the Middle Ages, as certified by the word "rubric" – in its Latin name, *Rubia tinctorum*. Presently meaning "direction" or "fixed protocol," its root comes from "*rubia*," red, because the directions that preceded prayer rituals in missals were written in red madder. Its storied use by Alexander the Great as military camouflage involved adroitly marking his soldiers at appropriate sites on their bodies with wound marks in red madder to fool the enemy into thinking that the army was on its last legs. England's Henry II (1133–1189) decreed that all fox hunters and military officers had to have jackets dyed in red madder. Some seven centuries later in 1840, France's King Louis Philippe (1773–1850) decreed that French soldiers' caps and trousers were to be dyed madder red.[17, p. 13] Madder's use as a pigment has been poorly documented by ancient sources; evidence of its commercial use appears only at the beginning of the nineteenth century when an apparatus was invented to continuously extract it from madder root under pressure.[37] When the soluble extract, alizarin (Table 10.1) is precipitated as a **lake**, it can offer a rainbow of colors: rose red with aluminum, bluish red with calcium, red-violet with tin, black-violet with iron and brown-violet or red-brown with chromium.[38] Today, there is a limited but steady demand for madder as a natural pigment of superior quality.[17, p. 17]

Meanwhile, cochineal migrated into the realms of cosmetics, foods and medical applications. It can be found in lipsticks, blushes and other cosmetics requiring a red color. It also finds use in baked goods, meats, dairy products and in some pharmaceutical products. Although it is a natural product and as such is exempt from US Food and Drug Administration (FDA) regulation, any food or cosmetic that contains it must declare either "cochineal extract" or "carmine" on the label because a significant subset of the population might display an allergenic reaction to its presence.[39] As a result, most ice cream products now use beet juice, which contains non-allergenic water-soluble red-violet betacyanins and yellow betaxanthins.

Not everyone is happy about the prospect of finding *Dactylopius coccus* in their favorite beverage either. In 2012, in the interests of trending to natural colorants, Starbucks announced that they were switching to cochineal extract to color their Soy Strawberry Crème Frappuccino and several other offerings.[40] The outcry was loud and vigorous. Vegans and vegetarians united to demand a change – and it took little more than a month for Starbucks to rescind the order (19 April 2012), substituting lycopene, the red coloring matter in tomatoes, for the offending bug. Cochineal is also anathema to certain groups for religious reasons.

The synthetic pathway to produce carminic acid in the laboratory is not commercially viable, so carmine is a natural product all the way. This status has its drawbacks, however, because of lack of control over the supply chain. A case in point arose back in 1971 when Massachusetts General Hospital experienced an outbreak of *Salmonella cubana* among its patients. Intense detective work over several months finally pinpointed the culprit: cochineal carmine capsules used as a dye marker for transit time tests.[41] The dye in question was imported from Peru where cochineal production relies heavily on a cottage industry. Individuals and families tend nopal cactus in their back yards and then dry the harvested insects on the ground or in doorways in close proximity to playing children and frolicking animals – leaving them wide open to all kinds of infection, among them, *Salmonella*.

Today, you can identify carmine extract or cochineal carmine by its *Colour Index* designation as C.I. Natural Red 4. As a constituent in foods in the European Union, it is labeled E120. Peru produces 80% of the natural product, which is the substance that appears on your table. Advice: Read your labels!

While red could stand for blood, in some ancient cultures it also stood for sacrifice. In ancient Israel, James Michener (1907–1997) describes such a fictional scene (but with historic roots): "With a red dye from the seashore they stained the wrists of Urbaal's son and then directed the farmer to halt the screaming of his wife....Urbaal spent that night by himself in the room of the four Astartes, and there he entered upon the full conflict of death and life, for in a cradle in a corner slept his son with red-marked wrists, unaware of the ritual which he would sanctify next day..."[42] Then Moloch opened his great flaming belly, and the victim, when scarcely at the edge of the cavity, disappeared like a drop of water on a red-hot plate, and white smoke rose amid the great scarlet color.[43]

Meanwhile, across the Atlantic in the land of the ancient Maya, blue stood in for red.

10.4 INTRODUCING MAYA BLUE

Between 1904 and 1911, the American archaeologist Edward Thompson (1857–1935) dredged up artifacts from the bottom of a natural geological sinkhole near the major Maya city of Chichen Itzá. Since many of the items were made of precious gold and jade, Thompson concluded that they had been thrown into the hole, called the Sacred Cenote, as sacrifices to appease the gods. Digging deeper, he raised up a grisly collection of 127 human skeletons that had sunk into a 14-foot layer of blue sediment at the bottom of the well, confirming his suspicions.[44]

Blue, to the ancient Maya, was the color of sacrifice and the color of the rain god, Chaak. When the bright blue firmament spanning the usually arid Maya world remained blue for too long, the only solution was to appeal to Chaak for a restorative rainfall by sending him a sacrificial victim. Everything about the ritual was blue, the color of the desired water. The priests, the victim, the clothing, the tools, the altar, even the weaving spindles were all painted entirely blue.[45]

10.4.1 Maya Blue Discovered

Apparently, Thompson's original description of the sinkhole full of blue did not register with either scholars or archaeologists until years later. This extraordinary colorant, one of the most intriguing and technologically savvy of the ancient pigments, had to wait until 1931 to emerge on the international scene. Herbert E. Merwin "discovered" it on a wall painting in the Temple of the Warriors in Chichen Itzá, Yucatán, Mexico.[46] Discoveries of other spectacular wall paintings soon followed, such as that at the Chiapas site of Bonampak, the oldest accurately dated murals containing Maya Blue[47] (Figure 10.4).[48] When information about the site was published in 1955, it attracted immense interest because of the remarkable durability of its fresco paintings despite the hot humid climate to which they were exposed for approximately 1200 years.[49] Images of these murals became famous around the world.

Figure 10.4   Detail of Mural from the Bonampak Excavation Site: Room 1, East Wall. Procession of Musicians, lower register. 8th century CE. Reproduced from ref. 48.

Dubbed "provisionally" Maya Blue in 1942 by Gettens and Stout,[50] there is evidence of its use throughout Mexico and Central America, beginning at about 300 BCE or a bit later, attesting to its misnomer: its presence was far more widespread than just the expanse of Mayan influence. In addition to wall paintings, it was the most common blue applied to pottery, sculpture and pre-Columbian codices.[51]

10.4.2   The Nature of Maya Blue: A Multi-decade Conundrum

Twenty years after he "baptized" Maya Blue, in a 1962 paper Gettens admitted to being stumped by its mysterious nature. This pigment, he writes, "is different from any blue that has ever been identified on the ancient and medieval paintings from Europe or Asia."[52] He further affirms its uniqueness by citing its great stability to heat and chemicals and lack of any "chromogenic" element but iron. Then he throws up his hands by declaring "but the source of the blue color principle remains unknown."

The ancient Maya probably exploited the massive deposits of attapulgite that are being mined to this day in the villages of Sacalum and Ticul in Yucatán to manufacture their signature blue pigment.[53] However, since none of the usual metallic **chromophores** – copper, cobalt or iron – were detected in the samples in any appreciable amount, numerous suggestions were put forward until the true Maya Blue was found to be **indigo** (from the Mexican indigo bush, *Indigofera suffruticosa*) intimately mixed with attapulgite, or another supporting clay structure such as sepiolite. Reasoning that indigo was the most stable blue pigment universally available in Mesoamerica, H. Van Olphen studied attapulgite–indigo and sepiolite–indigo complex formation and their stability upon heating at various temperatures. He hypothesized that both types of clay had crystal structures featuring parallel channels of molecular dimensions in which the indigo molecules could lodge.[54] The sequel to this finding and the entire fascinating story is told by Delamare (Box 10.1).[55]

Over the course of sixty years of research, the characteristics of this remarkable pigment came to light: it is a hybrid **nanomaterial** with an indigo **chromophore** interwoven as **nanoparticles** in a clay matrix. Universally used in Central America and relatively easy to prepare, it is impervious to the harshest treatment as well as to some of the most degrading climatic conditions.[58] The greatest difficulty in determining the cause of Maya Blue's permanence was its relationship to the clay support. Was the dye bonded to the clay surface, or did it enter the channels inherent to the clay structure?

Box 10.1   Maya Blue Discovery and Synthesis: A Scenario

Attapulgite (also known as palygorskite) is a rather rare grayish white magnesium hydroxysilicate clay with the formula $(Mg,Al)_2Si_4O_{10}(OH)\cdot4(H_2O)$. It is named after the town of Attapulgus, Georgia, whereas palygorskite takes its name from its discovery site in the Ural Mountains. In addition to a related clay, sepiolite, it differs structurally from other clays in that it is fibrous, containing tubules that could harbor guest molecules within or at the entrances of these channels. This structural peculiarity along with the natural abundance of attapulgite in Mesoamerica was one of the necessary conditions for the development of Maya Blue technology. The Maya valued attapulgite for its medicinal properties; it was used for bad burns, infections, stomach problems, goiter and diarrhea.[56] The other ingredient, indigo, was viewed by the Maya as a medicinal plant with antispasmodic, diuretic, and digestive properties.[57] If these two medicinal components were burned together as incense for the gods, it is not too difficult to envision the eventual discovery and technological development of a blue substance impervious to external stresses that could be used as a durable pigment.

10.4.3   How Do the Pieces Fit Together?

A member of the Gettens team, Tae Young Lee, first identified the **indigo** pigment; thence followed a series of hypotheses and experimental trials to determine how the dye was fixed onto or into the clay base. Chiari and co-workers demonstrated by molecular modeling that the indigo could fit snugly into palygorskite molecular channels without impediment and could form strong **hydrogen bonds** between the carbonyl group (–C=O) of the dye and the structural water of the clay.[59] The definitive solution came from the work of Sonia Ovarlez and co-workers in 2006 by mixing indigo with sepiolite and then heating the mixtures at varying temperatures.[60] Below 130 °C, the indigo remained on the clay surface and behaved like ordinary indigo. Heat applied at temperature ranges between 130 and 300 °C was sufficient to separate aggregated indigo molecules from one another, forming individual indigo particles small enough to enter the sepiolite's ion channels. There, indigo's nitrogen atoms were able to bond with sepiolite's magnesium ions, forming a stable indigo/sepiolite complex. At even higher temperatures (up to 550 °C), indigo's carbonyl group could also bond with the host clay forming an even more stable complex, retaining its structure to temperatures as high as 800 °C. Figure 10.5 illustrates these structural changes.[61] Note that Figure 10.5 (left)[62] shows an erroneous model with the indigo tucked well inside the clay body. Recent thermogravimetric analyses by G. Chiari[63] show that the true picture is actually Figure 10.5 (right) where the indigo molecules are more tenuously held in the structure.

Further work by the Giustetto group demonstrated that the nature of the host clay material affects Maya Blue's stability. Clays with narrow ion channels allow the indigo to **hydrogen bond** to both sides of the channel, whereas channels with larger diameters can support bonding only to one side or the other, dramatically reducing host/guest interaction.[64] Additional work on Maya Blue formation has led to the suggestion that incorporation of indigo's oxidized form, dehydroindigo, confers even greater stability to the pigment because these molecules can penetrate more deeply into the clay support.[65]

An excellent literature review[66] summarizes the various models for the formation of Maya Blue, experimental details of syntheses, and results of analytical methods to characterize the

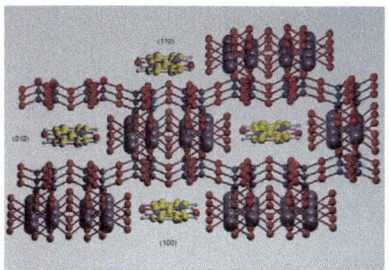 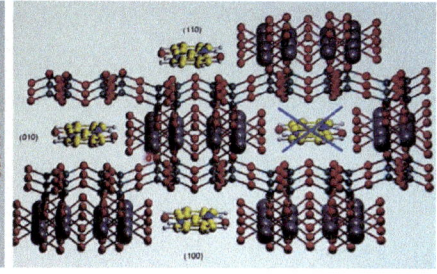

Figure 10.5   Left: Atomic model of Maya Blue Formation. Crystal mesh of palygorskite; water molecules are replaced by indigo molecules on heating to 150 °C; indigo is completely absorbed into the ion channels; Right: Corrected atomic model of Maya Blue Formation. Crystal mesh of palygorskite; water molecules are replaced by indigo molecules on heating to 150 °C; indigo is only partially absorbed and does not fill all the ion channels (note the crossed-out molecule). Left: Reproduced from ref. 62 with permission from Springer Nature, Copyright © 2007, J. Paul Getty Trust. Right: Reproduced from ref. 63, image courtesy of Giacomo Chiari.

synthesized products. It concludes that (a) there is great support for the dehydroindigo model in which the ratio of dehydroindigo/**indigo** has been observed to increase with increasing temperature and appears to be a significant factor in determining the color of Maya Blue, and (b) the nature of the clay support is significant in terms of indigo trapping ability and in facilitating the oxidation of **indigo** to dehydroindigo.

Since Maya Blue is considered the first ever hybrid organic-inorganic pigment, a **nanomaterial** centuries ahead of its time, fascination with it will not go away for a long, long time. It will more than likely serve as an inspiration and model for the development of new hybrid materials.[67] It has certainly already inspired chemical educators to emulate our Mesoamerican forbears: having students synthesize it as an "amazing topic to engage students into scientific literacy."[68]

"Who knows where this interest and research will take us in the development of new host/guest hybrid materials? The Native Americans of Mesoamerica hardly knew that their extraordinary discovery would both baffle and inspire scientists around the world for centuries to come."[69]

10.5 CONCLUSION

Building on traditions forged over millennia, the ancient peoples of Mesoamerica developed two indigenous colors that now span the globe and excite awe and spawn research even today. Though possibly discovered by trial and error, their production was carefully cultivated and rigidly controlled. We shall see in the next chapter that palette expansion can take place in precisely the opposite mode, by accident and by serendipity, but with no less noteworthy results.

REFERENCES

1. M.-L. Bernadac and P. du Bouchet, *Picasso, Master of the New*, Thames and Hudson, London, 1993, p. 142.
2. A. N. Haney, Red, *Louisiana Literature*, 2019, **36**(1), 100–101.
3. Dactylopius coccus growing in Barlovento, La Palma, Canary Islands, https://commons.wikimedia.org/w/index.php?title=File:Dactylopius_coccus_(Barlovento)_04_ies.jpg&oldid=473021301, accessed June 2021.
4. Dactylopius coccus on Opuntia ficus-indica, https://commons.wikimedia.org/w/index.php?title=File:Dactylopius_coccus_on_Opuntia_ficus-indica_(003).jpg&oldid=110132385, accessed June 2021.
5. Lucretius, *De rerum natura, Book IV*, 1987.
6. H. Schweppe and H. Roosen-Runge, Carmine, in *Artists' Pigments: A Handbook of Their History and Chracteristics*, ed. R. L. Feller, National Gallery of Art, Washington, DC, 1986, vol. 1, pp. 255–284.
7. C. M. Salinas, Mexican cochineal, local technologies and the rise of global trade from the sixteenth to the nineteenth centuries, in *Global History and New Polycentric Approaches*, ed. M. P. Garcia and L. de Sousa, Palgrave Macmillan, Singapore, 2018, pp. 255–274.
8. E. Phipps, Global colors: dyes and the dye trade, in *The Interwoven Globe: Worldwide Textile Trade 1500–1800*, ed. A. Reck, Metropolitan Museum of Art, New York, 2013, pp. 120–135.
9. F. B. de Sahagún, *Historia general de las cosas de la Nueva España*, A. Valdés, México, 1829–1830.

10. F. Bernardino de Sahagún, *General History of the Things of New Spain (Florentine Codex)*, Book XI. Natural Things. On Colors, 1577, p. 216v, https://www.wdl.org/en/item/10622/view/1/434/, accessed June 2021.

11. F. Bernardino de Sahagún. *General History of the Things of New Spain (Florentine Codex)*, Book XI. Natural Things. On Colors. 1577, p. 221v, https://www.wdl.org/en/item/10622/view/1/441/, accessed June 2021.

12. D. A. Kerpel, Painters of the New World: The process of making the Florentine Codex, in *Colors between Two Worlds: the Florentine Codex of Bernardino de Sahagún*, ed. G. Wolf, J. Connors and L. A. Waldman, Florence: Kunsthistorisches Institut in Florenz, Max-Planck-Institut: Villa I Tatti, the Harvard University Center for Italian Renaissance Sbtudies, Cambridge, MA: Worldwide distribution by Harvard University Press, 2011, pp. 46–76.

13. P. Baglioni, R. Giorgi, M. C. Arroyo, D. Chelazzi, F. Ridi and D. M. Kerpel, On the nature of the pigments of the general history of the things in New Spain: The Florentine Codex, in *Colors between Two Worlds: the Florentine Codex of Bernardino de Sahagún*, ed. G. Wolf, J. Connors and L. A. Waldman, Florenc: Kunsthistorisches Institut in Florenz, Max-Planck-Institut: Villa I Tatti, the Harvard University Center for Italian Renaissance Studies, Cambridge, MA: Worldwide distribution by Harvard University Press, 2011, pp. 78–105.

14. A. B. Greenfield, *A Perfect Red*, Harper Collins, New York, 2005, p. 75.

15. A. van Leeuwenhoek, *Phil. Trans.*, 1704, **24**(292), 1614–1628.

16. J. R. Partington, *A History of Chemistry*, Macmillan London, 1961, vol. 2, pp. 451–454.

17. W. F. Leggett, *Ancient and Medieval Dyes*, Chemical Publishing Co., Brooklyn, 1944, p. 88.

18. Strange Science, *The Rocky Road to Modern Paleontology and Botany*, https://www.strangescience.net/linn.htm, accessed November 2020.

19. M. L. Vázquez de Ágredos Pascual, M. T. Doménech Carbó, D. J. Yusá-Marco, S. Vicente Palomino and L. Fuster López, *Arché*, 2007, 2, 131–136.

20. V. Finlay, *Color, a Natural History of the Palette*, Random House, New York, 2004, pp. 150–160.

21. A. Serrano, A. van den Doel, M. van Bommel, J. Hallett, I. Joosten and K. J. van den Berg, *Anal. Chim. Acta*, 2015, **897**, 116–127.
22. I. Boldiszár, Á. László-Bencsik, Z. Szűcs and B. Dános, *Acta Pharm. Hung.*, 2004, **74**, 142–148.
23. J. Wouters and A. Verhecken, *Stud. Conserv.*, 1989, **34**, 189–200.
24. A. G. Lloyd, *Food Chem.*, 1980, **5**, 91–107.
25. G. Favaro, C. Miliani, A. Romani and M. Vagnini, *J. Chem. Soc., Perkin Trans.*, 2002, **2**, 192–197.
26. M. V. Orna, *The Chemical History of Color*, Springer, Heidelberg, 2013, pp. 40–41.
27. J. Kirby, M. Spring and C. Higgitt, *National Gallery Technical Bulletin*, 2007, vol. 28, pp. 69–95.
28. P. Schützenberger, *Traité des matières colorantes comprenant leurs applications à la teinture et à l'impression et des notices sur les fibres textiles, les épaississants et les mordants, 2 vols*, V. Masson, Paris, 1867, vol. 2, p. 356.
29. B. C. Anderson, *J. Interdiscip. Hist.*, 2015, **45**(3), 337–366.
30. E. Phipps, *Cochineal Red: The Art History of a Color*, Metropolitan Museum of Art, New York, 2010, p. 35.
31. La Régate sur le Grand Canal, https://commons.wikimedia.org/w/index.php?title=File:La_R%C3%A9gate_sur_le_Grand_Canal.jpg&oldid=354888646 , accessed June 2021.
32. J. Kirby and R. White, *National Gallery Technical Bulletin*, 1996, vol. 17, pp. 56–80.
33. D. Saunders and J. Kirby, *National Gallery Technical Bulletin*, 1994, vol. 15, pp. 79–97.
34. D. Saunders and J. Kirby, *National Gallery Technical Bulletin*, 2004, vol. 25, pp. 62–72.
35. S. Garfield, *Mauve*, W. W. Norton, New York, 2001.
36. A. H. Church, *The Chemistry of Paints and Painting*, Seely & Co., London, 1901, p. 166.
37. R. Harley, *Artists' Pigments C, 1600–1835*, Butterworths, London, 1970, pp. 127–132.
38. D. De Santis and M. Moresi, *Ind. Crops Prod.*, 2007, **26**, 151–162.
39. Federal Register, https://www.federalregister.gov/documents/2009/03/11/E9-5286/listing-of-color-additives-exempt-from-certification-food-drug-and-cosmetic-labeling-cochineal, accessed November 2020.

40. *USA Today*, Thursday March 12, 2012.
41. B. Roueché, *New Yorker*, 4 September 1971, pp. 66–81.
42. J. Michener, *The Source*, Penguin Books, New York, 1965, pp. 127–128.
43. Paraphrased from G. Flaubert, *Salammbo*, Mondial, New York, 2006, p. 186.
44. K. Chang, *The New York Times*, 29 Feb. 2008, https://www.nytimes.com/2008/02/29/science/29bluew.html, accessed November 2020.
45. S. G. Morley, *The Ancient Maya*, Stanford University Press, Stanford, 1956, 3rd edn, p. 171.
46. H. E. Merwin, Chemical analysis of pigments, in *Temple of the Warriors at Chitchen-Itzá, Yucatan*, ed. E. H. Morris, J. Charlot and A. A. Morris, Carnegie Institution of Washington, Washington, DC, 1931, pp. 355–356.
47. C. Reyes Valerio, *De Bonampak al Templo Mayor. El azul maya en Mesoamerica: Colección America Nuestra. v. 40*: Siglo XXI Editores, Mexico D. F, 1993.
48. Bonampak mural. Room 1. Musicians and dancers, https://commons.wikimedia.org/w/index.php?title=File:Bonam-pak_mural._Room_1._Musicians_and_dancers.jpg&ol-did=218701980, accessed June 2021.
49. S. G. Morley, G. W. Brainerd and R. J. Sharer, *Bonampak, Chiapas, Mexico. Carnegie Institution of Washington Publication 602*, Carnegie Institution of Washington, Washington, DC, 1955.
50. R. J. Gettens and G. L. Stout, *Painting Materials: A Short Encyclopaedia*, Dover Publications, New York 1966, p. 130.
51. D. Buti, D. Domenici, C. Miliani, C. García Sáiz, T. Gómez Espinoza and F. Jímenez Villalba, *et al.*, *J. Archaeol. Sci.*, 2014, **42**, 166–178.
52. R. J. Gettens, *American Antiquity*, 1962, **27**(4), 557–564.
53. D. E. Arnold, Maya Blue, in *Encyclopaedia of the History of Science, Technology, and Medicine in Non-western Cultures*, ed. H. Selin, Springer, Dordrecht, 2014. https://doi.org/10.1007/978-94-007-3934-5_10170-1, accessed November 2020.
54. H. Van Olphen, *Science*, 1966, **154**(3749), 645–646.
55. F. Delamare, *Blue Pigments: 5000 Years of Art and Industry*, Archetype Publications, London, 2013, pp. 306–313.
56. D. E. Arnold and B. F. Bohor, *Archaeology*, 1975, **28**(1), 23–29.

57. D. E. Arnold, *Antropología y Técnica*, 1987, **2**, 53–84.
58. M. Sánchez del Río, A. Doménech, M. T. Doménech-Carbó, M. L. Vásquez de Ágredos Pascual, M. Suárez and E. García-Romero, The Maya blue pigment, in *Developments in Palygorskite and Sepiolite Research: A New Outlook on These Nanomaterials; Developments in Clay Science*, ed. E. Galan and A. Singer, Elsevier, Amsterdam, 2011, vol. 3, pp. 453–481.
59. G. Chiari, R. Giustetto and G. Ricchiardi, *Eur. J. Mineral.*, 2003, **15**(1), 21–33.
60. S. Ovarlez, A.-M. Chaze, F. Giulieri and F. Delamare, *C. R. Chim.*, 2006, **9**, 1243–1248.
61. Figures adapted from R. Giustetto, F. X. Llabrés i Xamena, G. Ricchiardi, S. Bordiga, A. Damin, R. Gobetto and M. R. Chierotti, *J. Phys. Chem. B*, 2005, **109**(12), 19360–19368.
62. G. Chiari, R. Giustetto, J. Druzik, E. Doehne and G. Ricchiardi, *Appl. Phys. A*, 2008, **90**(1), 3–7.
63. G. Chiari, *Measurement Techniques to Uncover Mysteries in Art, Plenary Lecture Delivered on Monday, 10 December 2018 at the 3rd International Conference on Techniques, Measurements and Materials in Art and Archaeology*, Jerusalem, Israel, 9–13 December 2018. Book of Abstracts, p. 21.
64. R. Giustetto, K. Seenivasan, F. Bonino, G. Ricchiardi, S. Bordiga, M. R. Chierotti and R. Gobetto, *J. Phys. Chem. C*, 2011, **115**(34), 16764–16776.
65. R. Rondão, J. Sérgio Seixas de Melo, V. D. B. Bonifácio and M. J. Melo, *J. Phys. Chem. A*, 2010, **114**(4), 1699–1708.
66. A. Doménech-Carbó, S. Holmwood, F. Di Turo, N. Montoya, F. M. Valle-Algarra, H. G. M. Edwards and M. T. Doménech-Carbó, *J. Phys. Chem. C*, 2019, **123**(1), 770–782.
67. C. Dejoie, P. Martinetto, N. Tamura, M. Kunz, F. Porcher and P. Bordat, *et al.*, *J. Phys. Chem. C*, 2014, **118**(48), 28032–28042.
68. I. M. V. Leitão and J. Sérgio Seixas de Melo, *J. Chem. Educ.*, 2013, **90**(11), 1493–1497.
69. M. V. Orna, Historic Mineral Pigments: Colorful Benchmarks of Ancient Civilizations, in *Chemical Technology in Antiquity*, ed. S. C. Rasmussen, American Chemical Society, Washington, DC, 2015, pp. 16–69, at 55.
70. J. Keats, Sonnet written in answer to a sonnet by J. H. Reynolds, in *The Complete Works of John Keats, Vol. II, Posthumous Poems to 1818*, ed. H. B. Forman, Thomas Y. Crowell & Co., New York, 1900, pp. 198–199.

CHAPTER 11

Alchemical Anomalies: Accidents Will Happen†

Name the greatest of all the inventors. Accident.[1]

Mark Twain

11.1 INTRODUCTION

The chemistry in this chapter is both serendipitous and revolutionary. Born of some lucky accidents and guided by quantitative principles, a great explosion of new pigments appeared at almost the same time as modern chemistry.

11.2 A REAL LIVE PROSPERO

Imagine an impresario who, in the early 1600s, boasted that he could create storms, make the Great Hall of Westminster so cold that the king would take to flight, build a telescope that could make a manuscript legible from one mile, build a submarine that could traverse the Thames underwater, and construct a perpetual motion clock that ran on an invisible form of energy.[2,3]

---

†Glossary entries and chapter cross-references are in boldface.

March of the Pigments: Color History, Science and Impact
By Mary Virginia Orna
© Mary Virginia Orna 2022
Published by the Royal Society of Chemistry, www.rsc.org

Figure 11.1   Left: The barometric clock of Cornelis Drebbel patented in 1598 and then known as "perpetuum mobile." Print by Hiesserle von Choda (1557–1665). Reproduced from ref. 3. Right: Gobelin tapestries in the Linderhof Palace, Upper Bavaria, Germany, *ca.* 1895. Reproduced from ref. 10.

Reminiscent of Prospero's claim in Shakespeare's "The Tempest," who declared that "Graves at my command/have wak'd their sleepers, op'd, and let them forth/By my so potent art,"[4] (Figure 11.1, left) Cornelis Drebbel (1572–1633), inventor and member of James I's (1566–1625) court, asserted that he could do the same. In fact, he claimed that he could call up some of them, such as Alexander the Great, by name.[5] Since William Shakespeare (1564–1616) and Drebbel were contemporaries and members of the same court, it is not unlikely that Shakespeare, who wrote "The Tempest," his last play, in 1610, was cognizant of Drebbel and borrowed some of his "magic art" to create his character, Prospero.[2]

11.2.1   Drebbel the Alchemist

Drebbel, a native of the Netherlands, burst upon the London scene in 1605, and seems to have worked on numerous (and bewildering) special projects for a succession of monarchs. Some of his noted contemporaries like Constantijn Huygens (1596–1687), Ben Jonson (1572–1637) and Francis Bacon (1561–1626) recognized his great talent and inventive spirit. He may be the inventor of the compound microscope; he gifted one to Huyghens who gave it to his physicist son Christiaan, who put it to very good use indeed. Many of Drebbel's inventions are featured in Bacon's "The New Atlantis," a visionary novel published posthumously in 1626.[6] Drebbel was clearly influenced by Paracelsus and practiced a form of alchemy that relied not so much on an appeal to the supernatural but rather to the practical manipulation of nature through chemistry and mathematics. He might even be credited as the first discoverer of oxygen since he provided a bottled form of it for the operators of his submarine by burning saltpeter, KNO_3.

Assessments by his contemporaries, including the celebrated painter Peter Paul Rubens (1577–1640), seem to indicate that that Drebbel was considered a shade below the natural philosophers of his time. He was a man engaged in practical matters, more concerned with how things worked than with why they worked. One might venture to say that he was the first of a new type, an early engineer.[6]

11.2.2 Alchemical Accident #1: Drebbel's Scarlet

Though Drebbel and his many ingenious inventions have receded into the shadows of history, due largely to his own penchant for secrecy, his one accidental discovery has not. There are at least three stories about the birth of Drebbel's Scarlet, a new color that enchanted the European world. One has it that he placed an extract of cochineal red on his windowsill intending to fill a glass tube with the solution to make a thermometer. But a broken phial of *aqua regia* (a mix of nitric and hydrochloric acids) interfered by accidentally dripping into it and colored the dye a remarkable intense scarlet. Further inquiry revealed that the *aqua regia* had hit the tin window frame first, and that it was the presence of the dissolved tin that endowed the color. Indeed, another story relates that he added pewter filings to his mix. (Pewter is a tin alloy.) In the Science Museum, South Kensington, London, there once were two specimens of silk dyed with cochineal, one done with alum and the other with tin salt. The fine color of the one where tin was used was striking.[5] As chemists, we now know that besides tidy laboratories being liabilities to invention, many metals such as tin, aluminum and copper are capable of forming **chelate** complexes with **anthraquinoid** dye molecules, a class of compounds to which cochineal belongs (**Chapter 10.3.4**). According to one expert on color, Drebbel's conscious addition of tin to the cochineal formulation, and subsequent improvements, amounted to the initiation of the age of scientific dyeing (Box 11.1).[7]

Drebbel's two canny sons-in-law, the Kuffler brothers Abraham and Johannes Sibertus, set up a dye works in Stratford-Bow, naming their product "Bow Dye" which quickly became the rage of London – though it remained a family secret for many years. They subsequently expanded to the continent and became famous when the prestigious Gobelin tapestry works contracted for the dye, now sold as Holland Dye, to be incorporated into their wall hangings. From then on, the tapestries were no longer considered hangings, but woven pictures (Figure 11.1, right).[10]

A variation on Drebbel's recipe came into play in Gloucestershire. Called Stroudwater Scarlet, the Stroud Valley's most celebrated export, it was recognized worldwide as the colorant of the famous British Redcoats. However, one of the ingredients had some very stringent requirements. In his 1823 *Practical Treatise on Dying* (*sic*), William Partridge lays them out plainly:

Box 11.1   Alchemical Aside

Palette expansion could not have proceeded unless accompanied by the alchemical activity that led to the isolation and purification of many elements and compounds. Key to the necessary chemical reactions was the discovery of strong acids some time in the early Middle Ages. The so-called False Geber (*ca.* 1300 CE) (as opposed to the Arabic alchemist, Jabir ibn Hayyam (d. 815?)) is credited with discovering sulfuric acid by oxidizing green vitriol, $FeSO_4$, to form $Fe_2(SO_4)_3$ which, on further heating, decomposes to Fe_2O_3 and SO_3. Dissolving the latter in water produces sulfuric acid, H_2SO_4. From there it is a short step to forming hydrochloric acid, HCl, from chloride salts such as NaCl, and nitric acid from nitrate salts such as KNO_3. The alchemist's magic tool, *aqua regia*, or "royal water," is a mixture of nitric and hydrochloric acids, powerful enough to dissolve even intractable gold.[8] Early pigment manufacture depended upon being able to dissolve metals and precipitate insoluble compounds using techniques such as distillation, sublimation, filtration and recrystallization.[9]

"Urine that is fresh voided will not scour well. That from persons on a plain diet is stronger and better than that from luxurious livers. The cider and gin drinkers are considered to give the worst, the beer drinker the best. When urine is collected it should be kept in close (*sic*) vessels until it has completely undergone those changes by which ammonia is developed."[11] In addition to this stale urine, other necessary ingredients to make the Redcoat brilliant are cutbear (lichen found on sea rocks on the Atlantic coast), argol (a deposit from fermented wine), and incidentally, cochineal, madder, alum and tin liquor. Not exactly a recipe for the contemporary tie-dyeing workshop![12]

Up until this point, we have seen that each pigment seems to have a certain unique "personality," exhibiting unexpected behavior. Each is one-of-a-kind, and some elude classification as **inorganic** or **organic** in nature. Drebbel's Scarlet was clearly organic until the uninvited tin fell in. Prussian Blue, our next topic, not only falls into step as a one-off reality, but it also straddles the line between **inorganic** and **organic**, but not so clearly as Maya blue in the previous chapter. To meet this next alchemical accident, we must step into the next century.

11.3 ALCHEMICAL ACCIDENT #2

This time, imagine yourself almost a century later in 1707 Berlin. We are visiting the laboratory shared by Johann Konrad Dippel (1673–1734), a shady alchemist, and Johann Jacob Diesbach, a bumbling color-maker. Their quarters may have looked something like Figure 11.2,[13] but perhaps a bit more disorderly.

A setting like this is an accident waiting to happen, and happen it did. Diesbach was hard at work making his signature red pigment that he called Florence Lake, having nothing to do with Florence, as it contained cochineal, alum and English vitriol (iron(II) sulfate) mixed in an alkaline medium. Running out of alkali, he asked for some waste potash from Dippel and, as might be suspected given the slovenly character of the place, the potash happened to be contaminated with bull's blood. Adding red to red, the contents of the pot turned a glorious deep blue, an alchemical miracle that both men realized could make their fortunes.

Figure 11.2   *Interior with an Alchemist.* Thomas Wijck. Oil on Canvas. Wellcome Collection. Reproduced from ref. 13 under the terms of the CC BY 4.0 license, https://creativecommons.org/licenses/by/4.0/deed.en.

11.3.1   Prussian Blue Concealed and Revealed

Diesbach immediately set himself up in business and managed to keep his recipe hush-hush for almost twenty years. Meanwhile, Dippel absconded with his knowledge of how to make Prussian Blue, as it was sometimes called, and marketed it until at least 1714 in the Netherlands.[14]

In 1724, an Englishman named John Woodward (1657–1728) outed Diesbach's secret by publishing the recipe in Latin, as he had received it in a personal communication from Germany.[15] Harley summarized the recipe thus: "To an alkali calcined with bullock's blood, dissolved and brought to boiling point, a solution of alum and ferrous sulfate was added while also boiling. During the effervescence which followed, the mixture turned green, and, after it had been allowed to stand, it was strained. The residual greenish precipitate turned blue as soon as spirit of salt (HCl) was poured on it."[16] Given the bloody ingredient, she classified it as an organic pigment.

11.3.2   A Complex Chemistry

The actual chemistry behind Prussian Blue's synthesis remained a mystery for more than two centuries. Carl Wilhelm Scheele (1742–1786), in Germany, found that he could produce a strange-smelling gas by heating it, and Joseph Louis Gay-Lussac (1778–1859) identified the gas as highly toxic hydrogen cyanide, HCN. It is fortunate that neither Scheele nor Gay-Lussac sniffed that gas for too long. Cyanide is lethal in 10 minutes at 27 ppm. Subsequent analyses showed that the pigment contained iron, potassium and cyanide in the ratio of $KFe_2(CN)_6$. We can deduce in hindsight that the necessary ingredients were the ferrous ion provided by the "English vitriol" and the cyanide ion arising in some way from the calcined bullock's blood. (Later experimenters eventually found that any kind of animal flesh would do the trick of providing the necessary material to make the pigment.) The essential, but then-unknown ingredient, cyanide, was more than likely produced by the degradation of molecules containing the C–N bond such as hemoglobin.[17] However, it was not until 1977 that the actual structure and chemical composition were determined.[18] Accepted formulas are $Fe_4[Fe(CN)_6]_3 \cdot xH_2O$, bearing the chemical name iron(III) hexacyanoferrate(II), $KFe[Fe(CN)_6]$ and $NH_4Fe[Fe(CN)_6]$, the latter introduced in the United States after World War I since potassium had become very expensive.

   Later formulations of Prussian Blue found that animal wastes and alum were not needed; the required nitrogen could be supplied by atmospheric nitrogen or by ammonia, much to the relief of those who resided in the vicinity of Prussian Blue factories. The noxious odors spewed forth by the industry were both unpleasant and unhealthy.

   So, an earlier classification about the organic nature of Prussian Blue had to be revised. If it can be synthesized from mineral carbonates and atmospheric nitrogen, it should be classified as inorganic.

A modern formulation of Prussian Blue, $M^I Fe^{III} Fe^{II}(CN)_6 \cdot nH_2O$, where M^I may be K^+, Na^+ or NH_4^+ and $n = 14-16$[19] reveals the true nature of the pigment as a mixed valence compound, which accounts for its intense blue color *via* the charge transfer mechanism. Translation: There are two iron atoms in the compound, each with a different valence, Fe(III) and Fe(II). The fun is that Fe(II) tosses its remaining valence electron to the Fe(III), thus reducing it, while it itself becomes oxidized. This atomic game of back-and-forth catch can be symbolized thus:

$$\text{Fe(III)} + \text{Fe(II)} \rightarrow \text{Fe(II)} + \text{Fe(III)} \text{ or } \text{Fe(III)} + e^- \rightarrow \text{Fe(II)}; \text{Fe(II)} \rightarrow \text{Fe(III)} + e^-$$

$$(11.1)$$

The energy involved in accomplishing this electron transfer results in absorption of almost all the wavelengths of visible light except those in the blue region, which then get reflected to your eye. A brief article by Andreas Ludi, who was a member of the team responsible for determining the crystal structure of Prussian Blue, discusses its mixed valence nature, gives several recipes for its production, and a brief summary of its major uses.[20]

11.3.3 An Artists' Pigment and a Workhorse

Prussian Blue, which also eventually went under other names such as Berlin Blue, Milori Blue, and iron blue, was an instant success. Artists and colorists immediately added it to their palettes. Kraft[14] reports that its earliest recorded use was by the Dutch painter, Adriaen van der Werff (1659–1722) in 1709, no doubt supplied by Dippel. By the mid-18th century, it had become the blue pigment of choice for oil painting. Delamare[21] documents its presence in the paint box of Jean-Honoré Fragonard (1732–1806) in two different formulations and in the work of Jean-Baptiste Perronneau (1715–1783)[22] and the Japanese artist, Katsushika Hokusai (1760–1849)[23] (Figure 11.3).

Prussian Blue also found use in house paints, wallpaper, laundry bluing, and textile dyeing. Toward the end of the millennium, it was the most commonly used blue pigment. Annual production exceeded 50 000 metric tons until about 1970 when

Figure 11.3 Left: Jean-Baptiste Perronneau. *A Girl with a Kitten.* 1745. National Gallery, London. Reproduced from ref. 22. Right: Katsushika Hokusai. *The Great Wave of Kanagawa, ca.* 1830. Metropolitan Museum of Art. Reproduced from ref. 23.

phthalocyanine blue superseded it. However, all through its career, Prussian Blue suffered criticism regarding its permanence: some claimed that it was absolutely permanent, and others recommended that it be avoided. These contradictory views could be reconciled when one realizes that as the pure pigment, it really was permanent, but that when mixed with others, or placed in compromising environments, it could be rated as only moderately durable.[19] It certainly was fugitive in fresco since the alkaline medium rapidly decomposes the pigment to form brown hydrated ferric oxide.[24] George Field, in his 1885 treatise *Chromatography*,[25] called it a vacillating pigment that "has the singular property of fluctuating under certain conditions, according to whether the surroundings are favourable for it to acquire or relinquish oxygen." He said it also darkens in damp and polluted air such as that found in his "test compartment," a shut-up privy.

11.3.4 Alchemy's Last Gift: Medicine and More

Delamare[21] dubs Prussian Blue "a last gift from alchemy," which it truly is. He also lists some unusual properties that stretch the limits of its use as an artists' pigment: (a) even though it harbors the lethal cyanide ion in its formula, the nature of its bonding within the compound renders it non-toxic (although it can release lethal HCN under mildly acidic conditions); (b) its monovalent counterion, be it K^+ or NH_4^+, can easily exchange with other mon-ovalent positive ions, making it an excellent sequestering, *i.e.* immobilizing, agent for toxic ions like thallium, Tl^+, and cesium, Cs^+. This property is due to its open crystal lattice which is capable of harboring ions as large as 182 pm (picometers; there are one trillion pm in a meter) in exchange for its normal potassium counterion. (The ionic radii of Tl^+ and Cs^+ are 164 and 181 pm respectively.)[17] Prussian Blue is marketed as a poison antidote under such brand names as *Radiogardase, Antidotum*, and *Thallii-Heyl* in capsule form or powder form. The only side effect noted is the production of spectacularly blue feces!

Victoria Finlay[26] describes another property of Prussian Blue that gave the world the "blueprint": "...Prussian Blue became the basis of something that changed the face of architecture and design forever: the first ever industrial photocopying process." She goes on to describe how Sir John Herschel (1792–1871), a chemist and astronomer, was the first Englishman to take up photography. She says, "In 1842 he realized that if he held a pattern drawn on tracing paper over photo-sensitive paper and shone a lamp over both of them, the bits that were not protected by dark lines would change their chemical formula in the light. They would shift subtly from ammonium ferric citrate to ammonium ferrous citrate." The photo-sensitive material is essentially the Fe(III) (ferric) ion that, on exposure to ultraviolet light from the sun or from some other source, is photo-reduced to Fe(II) (ferrous) ion. Other photo-sensitive salts, such as ferric ammonium oxalate, can also be used. Once the paper has been activated, it can then be sprayed with or immersed in a solution of potassium ferricyanide which then reacts to form Prussian Blue:

$$Fe^{2+} + K_3Fe(CN)_6 \rightarrow KFe[Fe(CN)_6] + 2K^+ \quad (11.2)$$

This is a two-step process that Finlay describes, but it is also possible to combine both the ammonium ferric citrate and the potassium ferricyanide, impregnate a paper with the mixture, and carry out the exposure in a single step. The paper, in either case, should then be rinsed to remove unreacted ammonium ferric citrate. Finlay continues, "The resulting ghostly pattern of white lines on blue paper was called a 'blueprint,' and the word's meaning has shifted subtly to mean any design for the future, whether or not it is in photocopiable form." This process, called cyanotype photography, has found many uses since its invention by Herschel.[27]

If one follows the ups and downs of 18th and 19th century chemists as they tried to make Prussian Blue, it soon becomes clear that each step along the way was totally empirical – they tried various methods to improve the hue and the yield, but no one knew what they were working with. A certain aura of "witchcraft" developed around the art of its synthesis, mainly driven by the prospect of financial rewards, especially in France.[21] In the end, it came to be viewed mostly as a workhorse pigment, but certainly not a suitable substitute for the regal ultramarine.

In a new 21st century application, Prussian Blue sheds its identity as a pigment and is poised to solve a problem that has always plagued the energy-on-the-run folks who build our batteries. Because it can transfer electrons very rapidly, virtually friction-free, from one form of iron to another, a Prussian Blue-based battery presents a very low explosion risk. Capable of discharging in two minutes and recharging in eight, it is an ideal substitute for applications that don't require heavy duty energy density such as small appliances, data centers and small electrical grids. Entrepreneur Colin Wessells, following a suggestion of his doctoral mentor Robert Huggins (b. 1929), is finding ways to use Prussian Blue as a possible energy reservoir system – storing and releasing energy in an everlasting game of electron-catch between Fe^{II} and Fe^{III}. Watch for news about his nascent company, Natron Energy, as it moves forward.[28] While Prussian Blue batteries may never replace lithium ion batteries, they will certainly be invaluable as low-cost light duty power sources.

Nature did not endow the Paleolithic hunter-gatherers who painted their caves and grottoes many millennia ago with any pigments in the blue region of the spectrum. It remained for enterprising miners and later, for proto-chemists who worked with glassy substances, to find the blues that we so treasure today. Over a period of about five thousand years, we have not contributed very many new blue members to this exclusive club: on average, one would appear every 300 years or so, but not on a regular basis. Prussian Blue was a great surprise and a welcome addition to the palette. More surprises were in store as we progressed into the 20th and 21st centuries. Manganese blue, though neither alchemical nor an accident, was a brief stepping stone to be noted – a discontinued pigment. The two following were most definitely accidents, but lacking alchemical roots.

11.4 MANGANESE BLUE (1935–1995)

This pigment, like Halley's Comet, appeared and then quickly disappeared from the artistic scene, but it will not return. Though known since 1869, it was first patented by I. G. Farbenindustrie AG in 1935. Strongly heating mixtures of barium nitrate, sodium sulfate, and potassium permanganate yields a blue barium manganate, $BaMnO_4$, fixed on a barium sulfate base.[29] Chemically inert to acids, bases, light, and heat, it nevertheless has poor hiding power. Its major use was coloring cement, particularly the bottoms of swimming pools. Because of toxicity and environmental concerns, its use has been discontinued.[30]

11.5 ACCIDENT #3: PHTHALOCYANINE BLUE

"Sightings" of what came to be called phthalocyanines (etymology: rock oil dark blue) happened as early as 1907, when A. Braun and J. Tcherniac, working at Strasbourg, heated a not so common compound, 2-cyanobenzamide $(C_6H_4(CN)(CONH_2))$[31] and produced phthalonitrile $(C_6H_4(CN)_2)$. But the happy accident that led to phthalocyanine blue did not come to fruition until 1935. A step along the way was the work of a group of chemists at the University of Fribourg, Switzerland, who noticed indigo-like blue colors develop when *ortho*-dibromobenzene was heated together with copper(I) cyanide and pyridine.[32] Subsequently, Scottish Dye Works (later Imperial Chemical Industries, ICI) patented a blue color that had been in development since 1927 after their researchers noticed a greenish color in a reaction vessel that should have contained a colorless product. They tracked down where the color was centered: at a chip in the enamel reactor that had exposed the iron underneath.‡ When the ICI group replaced the iron with copper, they got a very intense blue solid – the object

---

‡An interesting first-person narrative documenting this discovery, along with the research leader's falsely accusing the foreman of contaminating the reaction vessel before discovering the faulty reactor, can be found in ref. 66.

of the 1935 patent. The product, copper phthalocyanine, was very easy to make: one simply heated a $4:1$ mixture of phthalonitrile and elemental copper to 200 °C. In determining its empirical formula $(C_{32}H_{16}CuN_8)$ and structure,[§] they found that the metal ion was linked by four nitrogen atoms in a planar configuration, very similar to a spider (the metal) sitting in the center of its web (the organic porphyrin-like part).[33] The extended **conjugated system** of alternating single and double carbon–carbon and nitrogen–carbon bonds gives rise to the observed color. From this accidental discovery arose the pigment that almost completely dominates the blues market today.[21] The *New York Times* was so excited about its discovery that the 25 November 1935 issue carried the banner: BLUE PIGMENT DISCOVERY IS THE FIRST IN 100 YEARS.[34] A 1963 ACS monograph, complete with the discoverers' remarks in the original German and French, documents blow-by-blow the discovery's 18-year history.[35]

Copper phthalocyanine is also the primary blue pigment used in artists' materials as well. Other blues and greens can be obtained by substituting various other groups onto the ring system and by changing the central metal ion. They have been marketed by color names such as Peacock Blue, Viridian Hue and Permanent Blue, among others, leaving artists in the dark as to their true nature.[36] It has been identified in works by Yves Klein (1928–1962) (Monochrome Vert 1957), Bernard Rancillac (b. 1931) (La suite Américaine 1970),[37] Josef Albers (1888–1976) (Tautonym B, 2944),[38] Paul Delvaux (1897–1994) (Faubourg),[39] and in works of many other 20th century artists including Roy Lichtenstein (1923–1997) and Wassily Kandinsky (1866–1944). About 90% of phthalocyanine production goes into pigments for paints, printing inks, plastics, coatings, textile, paper, leather and spin dyeing industries. The predominant phthalocyanine derivatives are formed with chlorine (green pigments), and sulfuric acid derivatives (dyes). Their chemical stability and other desirable properties have expanded applications in fields like lasers, electronics, catalysts, semiconductors and lubricants.[40]

Figure 11.4 shows the basic structure of phthalocyanine blue, a **macrocyclic compound** containing four pyrrole units.[41,42]

[§]R.P. Linstead of London's Imperial Institute of Science and Technology gave a preliminary report on the structure at the September 1933 meeting of the British Association for the Advancement of Science and the publication appeared in 1937 (see ref. 67).

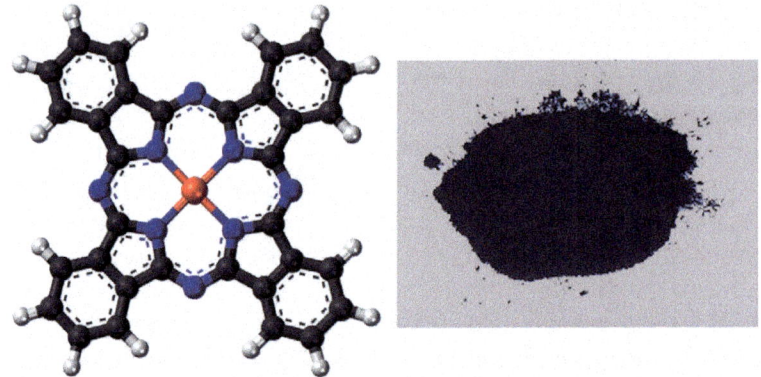

Figure 11.4 Left: Structure of phthalocyanine blue showing copper (red), nitrogen (blue), carbon (black), and hydrogen (white). Reproduced from ref. 41 under the terms of the CC0 1.0 license, https://creativecommons.org/publicdomain/zero/1.0/deed.en. Right: Phthalocyanine Blue, Pigment Blue 15. Reproduced from ref. 42.

Substituting cobalt and nickel for the central copper atom produces violet and dark turquoise pigments respectively. About four dozen other metals from almost every group in the periodic table can form additional metal phthalocyanines. This structure is in common with compounds important for life. When magnesium, Mg, substitutes for the copper, we have the basic structure of the chlorophylls; when iron replaces the copper, the molecule is an analog of heme, an essential component of hemoglobin.

We are not finished with accidents by a long shot. They extend into the 21st century and, as we shall see in the next chapter, the supreme accident of all time launched the synthetic dye industry, an event that spawned a fallout which we are still trying to deal with today. However, we still have a related accident to deal with now: another inadvertent blue pigment.

11.6   ACCIDENT #4: YInMn BLUE

A chance discovery at the University of Oregon gave rise to the 21st century's first blue pigment. While working with white $YInO_3$ and black $YMnO_3$, a research team headed by M. A. Subramanian wondered if a mixed oxide would exhibit properties that might be useful in the electronics industry. They started to dope $YInO_3$ with manganese, Mn^{3+}, making oxides with the formula $YIn_{1-x}Mn_xO_3$, and much to their surprise, they created an intensely blue compound even when x was as low as 0.05, and the intense blue color persisted over much of the solid solution range. The team concluded that the blue color was due to electronic energy transitions in the manganese.[43] The authors also realized that the presence of manganese in any blue pigments to be prepared in the future would not be the environmental obstacle experienced with manganese blue – in that case, the toxicity was due to the presence of barium.[44]

In 2019, the group announced the discovery of yet another blue pigment, hibonite blue, based on the structure of the mineral, hibonite, $CaM_{12}O_{19}$ (M = Al, Co, Ti). Although the **chromophore** is carcinogenic Co^{2+}, the Co concentration in hibonite blue is less than 15% of commercially available spinel cobalt blue $(CoAl_2O_4)$.[45]

The accidental discovery of Prussian Blue bookended the start of the 18th century; the serendipitous discovery of a new element with a colorful bent forms the bookend for the conclusion of the century. Read on – this is a marvelous tale that encompasses the birth of chemistry as we know it today.

11.7   A REVOLUTIONARY SERENDIPITOUS DISCOVERY

In 1789, Antoine-Laurent Lavoisier (1743–1794) was the first to publish a modern listing of the then-known "simple substances," the first to define the law of conservation of matter, and the first to propose a consistent explanation of combustion. Together with Claude Berthollet (1748–1822), L.-B. Guyton de Morveau (1737–1816) and Antoine-François Fourcroy (1755–1809), he devised a method for standardized chemical nomenclature. All future chemical endeavor was based upon this beginning which has come to be called the "Chemical Revolution."[46] In the same year, another revolution of great consequence was underway when, on 14 July, a mob stormed and took over the infamous Paris Bastille.

Both Revolutions continued on parallel tracks while a young protégé of Fourcroy's, hailing from a tiny Norman village, Nicolas-Louis Vauquelin (1763–1829), continued to pursue his study of chemistry, physics and philosophy. Both Vauquelin and Lavoisier got caught up in the political upheaval and, at the height of the Reign of Terror in late 1793, Vauquelin had to flee Paris,[¶,47] and in the spring of 1794, Lavoisier lost his head.

Once a semblance of normalcy returned, Vauquelin, now Citizen Vauquelin, took up a professorship at the Paris School of Mines. In this capacity, a beautiful red crystalline mineral called Siberian red lead (present name: crocoite) caught his eye. Though examined many times previously, the elemental composition of this very scarce material remained unknown. Vauquelin felt equal to the task, managing to produce beautiful red, yellow and green precipitates in the course of his analysis.[48] They were all due to a hitherto unknown simple substance that he called chromium, based on the Greek term for color, *chroma* (χρώμα).

11.7.1 Chromium: A Many-splendored Thing

For lack of more samples, Vauquelin suspended his chromium studies until a deposit came to light in France in 1809. We now know that chromium ranks 21st in crustal abundance; it is not as rare as previously thought. Chromite, chromium's chief ore, abounds in the United States, Turkey, South Africa and Central Asia.

Meanwhile, Vauquelin found time to discover yet another element, beryllium, and to apply his analytical skills to a variety of projects that earned him the reputation of being one of the greatest analytical chemists of all time. He even found that when he analyzed an emerald from Peru that its green color was due to the presence of trace amounts of his element, chromium. Trace amounts of chromium in otherwise colorless corundum are responsible for the exquisite color of ruby.[49] Chromium is also an essential nutrient in animal metabolism.[50]

When Vauquelin resumed his chromium investigations, he promoted the oxide's merits, waxing eloquent over the results that the Royal Porcelain Works at Sèvres and at Limoges had with this new color: "It gives an extremely beautiful green, which we had never been able to obtain with other metals."[51]

¶This was not the first time that Vauquelin had to flee under inauspicious circumstances. While, as a teenager, he was apprenticed to a pharmacist in Rouen, it seems that he was subjected to some bullying on the part of his fellow apprentices due to his avid study of chemistry. His master thought it best that he transfer to Paris, and it was there that he came under Fourcroy's benevolent influence.

It was only a matter of time before a bewildering bouquet of chromium colors began to grace not only porcelain but also the artists' palette and school buses. Chrome yellow ($PbCrO_4$) is found on both. Then we have chrome red ($PbCrO_4 \cdot Pb(OH)_2$); chrome orange ($PbCrO_4 \cdot PbO$); chromium oxide green, opaque (Cr_2O_3); viridian, Guignet's Green or hydrated chromium oxide green transparent ($Cr_2O_3 \cdot 2H_2O$); zinc yellow ($K_2O \cdot 4ZnCrO_4 \cdot 3H_2O$); basic zinc yellow ($ZnCrO_4 \cdot 4Zn(OH)_2$); molybdate orange ($PbCrO_4 \cdot PbMoO_4 \cdot PbSO_4$); lemon yellow or strontium yellow (sometimes misnamed Ultramarine Yellow, SrCrO4); light lemon yellow or barium yellow ($BaCrO_4$). Other chromates not often encountered or used as pigments are light yellow calcium chromate ($CaCrO_4$), basic cadmium chromate, basic iron chromate, red mercuric chromate ($HgCrO_4$), and bismuth chromate. Chrome Green or Zinnober Green is a mixture of Prussian Blue and chrome yellow, as are some of the other chromates mentioned. A popular website lists all the colored compounds of chromium,[52] and several excellent references highlight the artists' pigments (Box 11.2).[53,54]

Box 11.2   Background to Spectroscopy

While Vauquelin relied upon wet chemical methods for his discovery of chromium, things were about to change radically in the next century: an analytical method for element discovery that had been gestating for almost 200 years. In 1666, Isaac Newton (1643–1727) proposed his color theory based upon his prismatic observation of the solar spectrum. William Hyde Wollaston (1766–1828) viewed the spectrum through a narrow slit and observed dark lines that turned out to be images of the slit itself. Joseph von Fraunhofer (1787–1826) identified 570 of these lines, meticulously mapped them out and measured their frequencies, but had no idea of their significance. In 1822, Sir John Herschel (1792–1871), while studying the visible spectra of colored flames, remarked: "The colours thus contributed by different objects to flame afford...a ready and neat way of detecting extremely minute quantities of them."[55] Almost four decades later, Gustav Kirchhoff (1824–1887) made the connection: he realized that the frequencies of Herschel's bright flame colors corresponded exactly with the dark line frequencies reported by Fraunhofer, and that they were fingerprints to identify the elements.[56] Within a few months, he and his colleague, Robert Wilhelm Bunsen (1811–1899) succeeded in building a device that would form spectra of these colors using a prism: the earliest spectroscope. With it they soon discovered two new elements and set in motion the growth of a whole new field, spectroscopy. Rapid advances, still going on today, have given rise to a plethora of spectrophotometric instruments coupled with digital computers and appropriate light sources, like lasers, that accomplish rapid determinations with great sensitivity.[57]

11.7.2   Chromium Pigments: History of Use

Chrome yellow appeared in paintings almost immediately after Vauquelin followed up on his discovery. It has been identified in several paintings done in 1810–1812 located in the Bayerische Staatsgemäldesammlungen.[54] The earliest known mention of the pigment in artists' literature comes from the 18 September 1815 diary entry of Danish painter, C. W. Eckersberg (1783–1853), who bought one ounce of "jaune de crome" in Rome.[58] By the second quarter of the 19th century, chrome yellow was a fixture in water-colors and in oils. It has since become one of the major commercial pigments for highway markings, signs, and vehicles, mainly because black on yellow is one of the easiest color combinations to see.

Chromium oxide green has definitely been identified in paint-ings dating from 1845, but it may have been used by J. M. W. Turner as early as 1811 in a painting called *Somer Hill, Tonbridge*.[53] It is the most permanent commercially available green pigment, a property touted by Vauquelin at the very beginning.

Strontium chromate yellow was widely used throughout the 19th and 20th centuries. Among noted artists employing it were Henri Matisse (1869–1954), Pierre Bonnard (1867–1947), Fernand Cormon (1845–1924), Jean-Gabriel Domergue (1889–1962) and Diogène Maillart (1840–1926).

Chrome orange was one of the compounds that Vauquelin (Figure 11.5)[59] prepared during his initial studies of chromium compounds in 1809. It was rarely used commercially or by art-ists during the 19th century. One notable exception was Pierre-Auguste Renoir (1841–1919) who dazzled the eye with his rendition of an orange skiff against a bright cobalt blue background (Figure 11.5).[60]

11.8   THE EXPANDED PALETTE: 1775–1850

By the end of the 18th century, the Mars pigments, synthetic iron oxides, were being exploited by Joshua Reynolds (1723–1792), who also delighted in daubing with Indian Yellow. Reynolds' use of this exotic pigment can be traced from 1781. Bolstered by the large number of chromium-based pigments produced in the early 19th century, the artists' palette virtually doubled during the lifetime of one key artist, J. M. W. Turner (1771–1851). And we know that he used almost all of them (Box 11.3).[61]

Figure 11.5   Left: Statue of Nicolas-Louis Vauquelin. Sculptor: Pierre Hébert (1804–1859). Faculté de Pharmacie, Rue de l'Observatoire, Paris. Vauquelin is holding a phial labeled "chrome" in his left hand. Reproduced from ref. 59 with permission from American Chemical Society, Copyright 2014. Right: Pierre-Auguste Renoir, *The Seine at Asnières* (1875), The National Gallery London. Reproduced from ref. 60.

Box 11.3   Exotic and Novel Pigments

Perhaps, aside from mummy (**Chapter 4.4.1**), the most exotic (or even outlandish) artists' pigment was *Indian Yellow*. At first of mysterious provenance, its origin was purportedly traced to a small group of cattle owners in Monghyr (near Calcutta) who fed their cows exclusively on a diet of mango leaves, harvested their urine, evaporated it and handpacked the precipitate into balls before drying.[16] The pigment content is mainly magnesium and calcium salts of euxanthic acid. It was presumably banned and unavailable commercially after 1921.[62] Intrepid, as always, in her primary research, color sleuth Victoria Finlay journeyed to Monghyr only to find no trace of this dreadful practice. She mused, "I will always wonder whether the explanation that I have heard is reality or merely a reflection of reality, and whether this story is simply an example of somebody gently, and literally, taking the piss."[26] *Chinese White*, at the time of its introduction, was a novel substitute for toxic lead white. Its chemical content is zinc oxide, ZnO, commercially available for the first time around 1834, marketed as Zinc White.[30] *Mars colors*, introduced in the late 18th century, are synthetic iron ochres (iron = Mars, Roman god of war) precipitated from ferrous and aluminum sulfates in an alkaline medium. The initial product is a yellow ferric hydroxide; the yellow shade is controlled by the proportion of aluminum used. To obtain other Mars colors, Mars yellow is heated, the degree of which yields various hues of orange, red, brown and violet.[29]

Turner, in addition to all of the chromium-based pigments, was profligate in the use of cobalt blue, emerald green, synthetic ultramarine, Chinese white, orange vermilion, and a host of organic yellows, reds and greens.[63] Unfortunately, Turner and his confrères were also prone to experiment in techniques of dubious value such as gelled mediums, bituminous paints and fugitive pigments.[64] Strangely, another major painter, James McNeill Whistler (1834–1903), was not known to have used any of the new chromium pigments. Though the availability of these pigments was a great boon to creativity at the time, the modern museum curator is much more aware of an ongoing problem: the fugitive nature of many of the new offerings. A little tale from *The Turner Society News* highlights this problem:[65] "Mr Winsor (of the color manufacturing company, Winsor and Newton) during his life was pretty intimate with the great artist, Turner, who used to get his colours from us and noticing from time to time the very fugitive colours that Turner bought from us, plucked up courage one day to remonstrate with him for so doing. Turner's answer, in spite of the friendship between the two men, was somewhat uncompromising; he is alleged to have said, 'Your business Winsor is to make colours for Artists, mine is to use them.' I am afraid that poor Mr Winsor had nothing to say in reply."

Nor do we. To many an artist, color is an end in itself, and to worry about its permanence is someone else's job.

11.9   CONCLUSION

Other artists, concerned about legacy, take every means to ensure that their works will stand the test of time. The generation of artists that followed Turner was very much aware of the deterioration in quality of art materials and they began to insist on standards of excellence and pigments of reliable quality. In fact, Winsor and Newton were among the earliest vendors to publish information on the permanence and composition of their colors, labeling each with its *Colour Index* number when that publication became available in 1924.[65]

But, often, pigments durable in an ideal environment find themselves compromised by excessive light exposure, atmospheric pollution, high humidity, and incompatible pigment neighbors. So whether the fugitive nature of a pigment is intrinsic or induced, the problem came to the fore in the 20th century following an unprecedented explosion of synthetic colors never before seen or tested. This is the story of the next chapter which opens with the rebirth of ancient Tyrian purple as a modern dye – by accident! Stay tuned.

REFERENCES

1. K. Chowder, *Eureka! Smithsonian Magazine*, September 2003, https://www.smithsonianmag.com/science-nature/eureka-89414180/, accessed May 2021.
2. R. Grudin, *Huntingt. Libr. Q.*, 1991, **54**(3), 181–206.
3. The barometric clock of Cornelis Drebbel patented in 1598 and then known as "perpetuum mobile." Print by Hiesserle von Choda (1557–1665), https://commons.wikimedia.org/w/index.php?title=File:Drebbel-Clock.jpg&oldid=483537748, accessed June 2021.
4. W. Shakespeare, *The Tempest* V, I, pp. 48–50.
5. R. Colie, *Huntingt. Libr. Q.*, 1955, **18**(3), 245–260.
6. G. Tierie, *Cornelis Drebbel, 1572–1633*, H. J. Paris, Amsterdam, 1932, pp. 17–18.
7. D. Cardon, *Le Monde des Teintures Naturelles*, Belin, Paris, 2003, pp. 51–52.
8. L. Campanella, F. Cardone and G. Oliveti, *Chimica e Storia dell'Arte*, UniversItalia, Roma, 2012, pp. 52–53.
9. A. Abbott, *Sci. Prog.*, 2013, **96**(4), 398–416.
10. Gobelin Tapestries, Linderhof Palace, Upper Bavaria, https://commons.wikimedia.org/w/index.php?title=File:The_Gobelin_Tapestries,_Linderhof_Palace,_Upper_Bavaria,_Germany,_ca._1895_(1).jpg&oldid=284154895, accessed June 2021.
11. W. Partridge, *Practical Treatise on Dying (Sic)*, H. Wallis, New York, 1823, p. 47.
12. E. H. Edwards, *Military Illustrated: Past and Present*, 1995, vol. 81, pp. 15–16.
13. Interior with an alchemist. Oil painting by Thomas Wijck. Wellcome L0027193, https://commons.wikimedia.org/w/index.php?title=File:Interior_with_an_alchemist._Oil_painting_by_Thomas_Wijck._Wellcome_L0027193.jpg&oldid=454284782, accessed June 2021.
14. A. Kraft, *Bull. Hist. Chem.*, 2008, **33**(2), 61–67.
15. J. Woodward, *Phil. Trans.*, 1724, **33**, 15–17; English translation: *Phil. Trans.* Abridged series 7, 1809, 4–6; doi.org/10.1098/rstl.1724.0005, accessed January 2019..

16. R. Harley, *Artists' Pigments C 1600–1835*, Butterworths, London, 1970.
17. M. Ware, *J. Chem. Educ.*, 2008, **85**(8), 612–621.
18. H. J. Buser, D. Schwarzenbach, W. Petter and A. Ludi, *Inorg. Chem.*, 1977, **16**(11), 2704–2710.
19. B. H. Berrie, Prussian Blue, in *Artists' Pigments: A Handbook of Their History and Characteristics*, ed. E. W. FitzHugh, National Gallery of Art, Washington, DC, 1997, vol. 3, pp. 191–217.
20. A. Ludi, *J. Chem. Educ.*, 1981, **58**(12), 1013.
21. F. Delamare, *Blue Pigments: 5000 Years of Art and Industry*, Archetype Publications, London, 2013.
22. Jean-Baptiste Perronneau – A Girl with a Kitten, https://commons.wikimedia.org/w/index.php?title=File:Jean-Baptiste_Perronneau_-%20_A_Girl_with_a_Kitten_-_WGA17212.jpg&oldid=374424194, accessed June 2021.
23. Katsushika Hokusai: The Great Wave off Kanagawa, https://commons.wikimedia.org/w/index.php?title=File:Tsunami_by_hokusai_19th_century.jpg&oldid=511578528, accessed June 2021.
24. J. Kirby, *National Gallery Technical Bulletin*, 1993, vol. 14, pp. 62–71.
25. G. Field, *Chromatography; or, A Treatise on Colours and Pigments and of Their Powers in Painting*, Winsor & Newton, London, 1885, p. 149.
26. V. Finlay, *Color: A Natural History of the Palette*, Random House, New York, 2004, p. 313.
27. J. W. F. Herschel, *Philos. Trans. R. Soc. London*, 1842, **132**, 181–214.
28. M. A. Watson, *Chem. Eng. News*, 2020, **98**(44), 40–41.
29. R. J. Gettens and G. L. Stout, *Painting Materials: A Short Encyclopedia*, Dover Books, New York, 1966, p. 129.
30. G. Accorsi, G. Verri, A. Acocella, F. Zerbetto, G. Lerario, G. Gigli, D. Saunders and R. Billinge, *Chem. Commun.*, 2014, **50**, 15297.
31. A. Braun and J. Tcherniac, *Eur. J. Inorg. Chem.*, 1907, **40**(2), 2709–2714.
32. H. de Diesbach and E. von der Weid, *Helv. Chim. Acta*, 1927, **10**(1), 886–888.
33. M. Dahlen, *Ind. Eng. Chem.*, 1935, **31**, 839–841.
34. *New York Times*, 25 November 1935, p. 12.
35. F. H. Moser and A. L. Thomas, *Phthalocyanine Compounds*, Reinhold, New York, 1963, pp. 1–5.

36. S. Q. Lomax, *Stud. Conserv.*, 2005, **50**(suppl. 1), 19–29, Rev. Conserv. 6.

37. N. Sonoda, J.-P. Rioux and A. R. Duval, *Stud. Conserv.*, 1993, **38**(2), 99–127.

38. G. Poldi, C. Anselmi, A. Daveri and M. Vagnini, Josef Albers' use of 20th century pigments: a non-invasive analytical approach, in *Science and Art: The Contemporary Painted Surface*, ed. A. Sgamellotti, B. G. Brunetti and C. Miliani, Royal Society of Chemistry, London, 2020, pp. 75–76.

39. P. Vandenabeele, A. Hardy, H. G. M. Edwards and L. Moens, *Appl. Spectrosc.*, 2001, **55**(5), 525–533.

40. G. Loebbert, Phthalocyanine compounds, *Kirk-Othmer Encyclopedia of Chemical Technology*, Wiley Online Library, 2000, DOI: 10.1002/0471238961.1608200812150502.a01, accessed December 2020.

41. Phthalo-blue-3D-balls, https://commons.wikimedia.org/w/index.php?title=File:Phthalo-blue-3D-balls.png&oldid=457172234, accessed June 2021.

42. Copper Phthalocyanine Blue, https://en.wikipedia.org/w/index.php?title=File:Copper_Phtalocyanine_Blue.JPG&oldid=609952659, accessed June 2021.

43. A. E. Smith, H. Mizoguchi, K. Delaney, N. A. Spaldin, A. W. Sleight and M. A. Subramanian, *J. Am. Chem. Soc.*, 2009, **131**, 17084–17086.

44. D. Reinen, T. C. Brunold, H. O. Guedel and N. D. Yordanov, *Z. Anorg. Allg. Chem.*, 1998, **624**, 438–442.

45. B. A. Duell, J. Li and M. A. Subramanian, *ACS Omega*, 2019, **4**(26), 22114–22118.

46. The Chemical Revolution of Antoine-Laurent Lavoisier, https://www.acs.org/content/acs/en/education/whatischemistry/landmarks/lavoisier.html, accessed November 2020.

47. O. Lafont, *Ann. Pharm. Fr.*, 2014, **72**, 221–228.

48. L.-N. Vauquelin, *Nicholson's J.*, 1798, **2**, 387–393.

49. M. E. Weeks, *Discovery of the Elements*, Easton, PA, *J. Chem. Educ.*, 7th edn, 1968, pp. 278–280.

50. W. Mertz, *Biol. Trace Elem. Res.*, 1992, **32**, 3–8.

51. L.-N. Vauquelin, *Ann. Chim.*, 1809, **70**, 70–94.

52. R. J. Lancashire, *Chromium Chemistry*, http://wwwchem.uwimona.edu.jm/courses/chromium.html, accessed November 2020.

53. R. Newman, Chromium oxide greens, in *Artists Pigments: A Handbook of Their History and Characteristics*, ed. E. W. FitzHugh, National Gallery of Art, Washington, DC, 1997, vol. 3, pp. 273–293.

54. H. Kühn and M. Curran, Chrome yellow and other chromate pigments, in *Artists Pigments: A Handbook of Their History and Characteristics*, ed. R. L. Feller, National Gallery of Art, Washington, DC, 1986, vol. 1, pp. 187–218.

55. J. Lewis, *Spectroscopy in Science and Industry*, Blackie and Son, Glasgow, 1936, p. 4.

56. G. Kirchhoff, *Ann. Phys.*, 1860, **185**(2), 275–301.

57. M. V. Orna, *The Chemical History of Color*, Springer, Heidelberg, 2013, pp. 99–107.

58. A. Raft, Eckersberg's Farbekob I Rom, *Meddelelser fra Thorvaldsen's Museum*, 1973, vol. 15–17, pp. 152–154.

59. M. V. Orna, Paris: A Scientific Theme Park, in *Science History: A Traveler's Guide*, ed. M. V. Orna, American Chemical Society, Washington, DC, 2014, pp. 136–158.

60. Pierre-Auguste Renoir: The Skiff (La Yole), https://commons.wikimedia.org/w/index.php?title=File:Renoir_-_the-seine-at-asnieres-the-skiff-1879.jpg!PinterestLarge.jpg&oldid=527083056, accessed June 2021.

61. J. H. Townsend, Painting techniques and materials of Turner and other british artists 1775–1875, in *Historical Painting Techniques, Materials and Studio Practice*, preprints of a Symposium held at the University of Leiden, Netherlands, 26–29 June 1995, ed. A. Wallert, E. Hermens and M. Peek, pp. 176–185.

62. N. S. Baer, A. Joel, R. L. Feller and N. Indictor, Indian Yellow, in *Artists' Pigments: A Handbook of Their History and Characteristics*, ed. R. L. Feller, National Gallery of Art, Washington, DC, 1986, vol. 1, pp. 17–36.

63. J. H. Townsend, *Stud. Conserv.*, 1993, **38**, 231–254.

64. M. R. Katz, William Holman Hunt and the "Pre-Raphaelite Technique", in *Historical Painting Techniques, Materials and Studio Practice*, preprints of a Symposium held at the University of Leiden, Netherlands, 26–29 June 1995, ed. A. Wallert, E. Hermens and M. Peek, pp. 158–165.

65. P. Bower, *Turner Society News*, 1989, **53**, 5.

66. P. Gregory, *J. Porphyrins Phthalocyanines*, 1999, **3**, 468–476.

67. A. H. Cook and R. P. Linstead, *J. Chem. Soc.*, 1937, 929–933.

CHAPTER 12

Out of the Depths: Synthetic Colors From the Coal Tar Industry[†]

Time is the best appraiser of scientific work, and I am aware that an industrial discovery rarely produces all of its fruit in the hands of its first inventor.[1]

Louis Pasteur

12.1 INTRODUCTION

In the previous chapter, we saw that several lucky accidents and the advent of the chemical revolution accelerated **inorganic** pigment development over the first half of the 19th century. Now came **organic** chemistry's turn to contribute another lucky accident. It, together with the maturation of structural chemistry, unleashed many thousands of systematically synthesized dyes and pigments into the ever-growing stream of colorants. This development also contributed to reconfiguring the political, social, economic and educational structures of the modern world.

[†]Glossary entries and chapter cross-references are in boldface.

March of the Pigments: Color History, Science and Impact
By Mary Virginia Orna
© Mary Virginia Orna 2022
Published by the Royal Society of Chemistry, www.rsc.org

12.2 A MOST PROPITIOUS ACCIDENT

Easter came early in 1856, the 23 of March, to be exact. The weather report was dismal for the preceding holiday week: a dry northerly with persistent high winds and very cold temperatures. Young William Henry Perkin (1838–1907), on vacation from the Royal College of Chemistry, sat huddled in his makeshift laboratory, an unheated garret open to the fierce North Sea winds that came tearing down the Thames Estuary. For the second try that afternoon, his awkward fingers trembling with the cold, he drizzled a solution of potassium dichromate into a jug of aqueous aniline sulfate. Another bust! A black unyielding gunk. And for not the first time in history, alcohol was the liquid that loosened things up to witness a purple personality emerge from the mess. Another accidental find!

12.2.1 "Chance Favors the Prepared Mind". L. Pasteur

An accident is one thing. What you do with it is something else. Louis Pasteur (1822–1895) would have been proud of Perkin for following his advice[‡] and abandoning his teacher's playbook.[2] August Wilhelm Hofmann (1818–1892) was arguably the world expert on **organic** synthesis and an excellent research director. He sent his student on a hunt for a desperately needed commodity, quinine, the only known effective medicine for the killer disease, malaria. He advised Perkin to start with allyltoluidine, $C_{10}H_{13}N$ – the formula needed only an oxygen atom to square up with quinine, so a powerful oxidizing agent like dichromate was just the ticket. What we know today is that quinine is only one of 815 other possible products of this oxidation sequence: the formula gives only the element ratio, but not a clue as to how the molecule is put together. Perkin's genius move no. 1 was his departure from the playbook to try another starting material, aniline. Genius move no. 2 was to clean up the failed experiment with alcohol. Genius move no. 3 was to recognize a good thing when he saw

---

[‡]In 1860, Pasteur and Perkin briefly corresponded regarding a paratartaric acid preparation from succinic acid.

it. Genius move no. 4 was to realize that this beautiful purple color could make (or break) his fortune.

Perkin had enrolled in chemistry initially against his father's wishes. George Fowler Perkin (1802–1864), a carpenter, was not at all keen on chemistry as a choice of a profession, viewing it as a dead-end field. But young William had been fascinated in his early teens by the possibility of discovering something new through research and particularly by observing the process of crystallization to form regular geometric solids.[3]

Enchanted by his new color, especially when it produced a brilliant sheen on silk, William and his older brother, Thomas Dix Perkin, toyed with producing a larger batch of the dye to determine more properties. Their friend and schoolmate, Arthur Church, encouraged them to seek advice from an actual dyer, whom they identified as Robert Pullar of Perth, to whom they sent samples. Following an encouraging letter from Pullar, William filed his first patent in August of 1856; thence followed some visits to Perth to demonstrate and try out his new color. By this time it was evident that he had made a decision, perhaps rashly, to try to commercialize his discovery, which he called Tyrian purple, evocative of that long-lost and longed-for Phoenician commodity. He was already deeply indebted to his father for support: patents and travel do not come cheaply.[2]

In October, when he approached Hofmann to inform him of his intention to leave school to commercialize his discovery, Hofmann was visibly annoyed. Ostensibly he voiced his opinion that this was not a good career move. However, could his objection have been tinged with a bit of jealousy or even the future intent to launch into a commercial venture on his own? Perhaps it was Perkin's very naiveté that moved him to forge ahead, but it boggles the mind that he could convince his father to sink his entire means into the venture. After a long search, the Perkins purchased about six acres of property at Greenford Green, a few miles northwest of London. George and Thomas, who abandoned his promising architectural career, set about building the plant while William worked on methods of scaling up and cutting costs (Box 12.1).

Box 12.1   August Wilhelm von Hofmann

Hofmann landed on the ground floor of organic chemistry, so to speak, through his education under Justus von Liebig (1803–1873) at Giessen. After taking his degree, he left for Bonn as a Privatdozent but, in 1845, Prince Albert himself recruited him to head academics at the newly founded Royal College of Chemistry in London. Hofmann turned out to be the star of the show. Not only did he create the leading British school of organic chemistry, but he also developed ammonia Type formulas, advocated Stanislao Cannizzaro's (1826–1910) atomic weights, and elucidated the structure of aniline dyes. He had a marvelous power of stimulating his students and communicating his own enthusiasm; he took the strongest personal interest in their work, conferring with each three or four times during the week. His directorship of research was powerful and remarkable. His mode of instruction was informal, individual, practical – given to the student *via* an impromptu lecture right at the bench. The magnetic force of his personality, his ability to get to the heart of a student's work with a few searching questions, and his tendency to drive them to reach success were characteristics of his pedagogy. Hofmann was equally good at developing networks among his peers which led to his leadership in forming Germany's chemical power-base when he returned to his home country in 1867.[4]

12.2.2   Perkin & Sons, Dye Works

By late 1857, the factory was ready to roll out its first batch of aniline purple, or mauve. Perkin's color was not the first coal tar dye. As early as 1845, M. Guinon in Lyon had been producing yellow picric acid from the phenol fraction of the tar by nitration. Mauve's introduction raised greater awareness of the immense potential offered by coal tar as the feedstock from which benzene, and thence, aniline, could be obtained. But development of a beaker-full of color was quite different from commercial scale-up, a daunting task for an untried teenager.[5]

However, Hofmann had given Perkin a sound basic training in chemistry, particularly a knowledge of the substituted **aromatic amines** and how to prepare them. It was this foundation that Perkin built upon in developing and scaling up his process. Though he was not the first scientist to study the oxidation of aniline and its derivatives, he was the first to produce a useful dyestuff in large amounts and successfully bring it to market.[6] And perhaps Perkin's head for business was even more responsible for his success than his chemical knowledge: at the time, there was no set theory nor was there any knowledge of structural formulas. Perkin was literally working in the dark with molecular formulas that led to essentially trial-and-error reactions. That he succeeded at all is no small miracle. He improved Nikolay Zinin's (1812–1880) synthetic pathway to aniline,[7,8] a key step in coal tar dye production.[9] Perkin also developed ways to make the new dye adhere to cotton as well as silk; he set a price that would bring him a profit; he staved off competition from a number of quarters, and not least, he survived preparation of large amounts of his dye in primitive and hazardous conditions using reagents like fuming nitric acid and hydrogen sulfide.

Though early on he called his product "Tyrian" purple or aniline purple, the name was changed to mauve a year into production as a marketing scheme to link the violet color with Parisian fashion. Perkin's great good luck was twofold: there was high demand for this stunning color in the fashion world, and it was adopted on a large scale by one of the largest dyeworks in the British Isles. This effectively quashed two competitors that derived their product from natural sources.

The first was Roman purple, sometimes called murexide, that gave it the aura of having come directly from the *Murex* snail, whereas its inglorious source was uric acid extracted from guano imported from Peru.[10,11] Murexide was known since C. W. Scheele's (1742–1786) synthesis from uric acid in 1776; in 1818, William Prout (1785–1850) succeeded in making it from *boa constrictor* excrement and in 1853, Jules-Albert Schlumberger (1804–1892) from pigeon and chicken excrement,[12] not terribly viable sources for an industrial-scale product. In his 1975 classic, *The Periodic Table*, Primo Levi (1919–1987) considered using python droppings as a source for uric acid for making alloxan as a starting material for cosmetics. However, the director of the Turin zoo turned him down because the zoo already had lucrative contracts with pharmaceutical companies, so he turned to chickens as his source of uric acid. Levi philosophized: "The fact that alloxan, destined to embellish ladies' lips, would come from the excrement of chickens or pythons was a thought which didn't trouble me for a moment. The trade of chemist (fortified, in my case, by the experience of Auschwitz) teaches you to overcome, indeed to ignore, certain revulsions that are neither necessary or congenial: matter is matter, neither noble nor vile, infinitely transformable, and its proximate origin is of no importance whatsoever."[13,14]

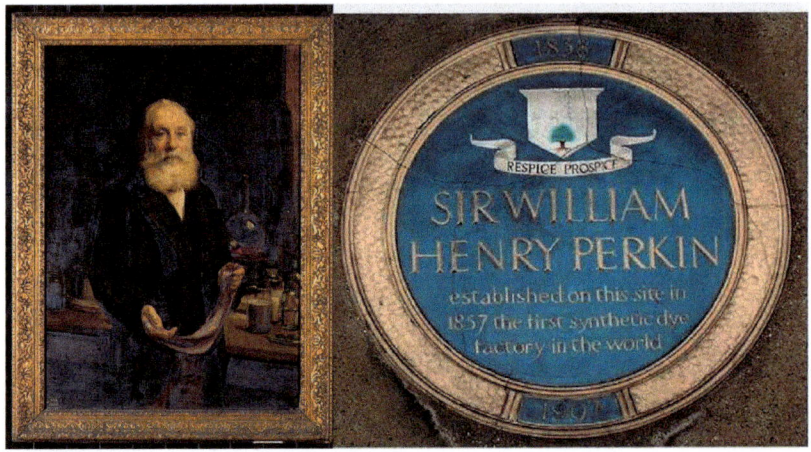

Figure 12.1   Left: Sir William Perkin. Wellcome Iconographic Collection. Reproduced from ref. 15 under the terms of the CC BY 4.0 license, https://creativecommons.org/licenses/by/4.0/deed.en. Right: Blue Plaque near the Grand Union Canal, Greenford, 2006. Reproduced from ref. 16 under the terms of the CC B-SA 2.5 license, https://creativecommons.org/licenses/by-sa/2.5/deed.en.

The second was French purple (*pourpre Française*), made by the action of ammonia and air on various species of lichens. It, and Perkin's mauve (Figure 12.1),[15,16] appeared at the same time, like identical twins in both shade and behavior. They were often confused with one another. In 1862, F. Crace Calvert, a prominent coal tar chemist, said "It is a curious coincidence that after many years of anxious search, two purples from widely different sources should have been first discovered in the same year in different countries (Box 12.2)."[17]

12.2.3   The Aniline Dye Revolution

Mauve mania was not the only circumstance that drove Perkin's success. As they say, timing is everything in life; thus pronounced C. J. T. Cronshaw of Imperial Chemical Industries: "If a Faery Godmother had given Perkin the chance of choosing the precise

---

Box 12.2   Two Bizarre Purples

A new purple literally exploded on the scene in the work of a 16th century alchemist-monk with the pseudonym of Basil Valentine. Called *"aurum ful-minans,"* Valentine made it by dissolving a gold ducat in *aqua regia* containing ammonium ions to produce gold(I) hydrazide, $AuNHNH_2$. This highly unstable but beautiful purple compound was commonly, if ticklishly, used by ceramicists and jewelers from at least the 15th century. The color was ultimately due to the formation of gold **nanoparticles**.[18] In the case of the more well-known pigment Purple of Cassius, reducing gold trichloride, $AuCl_3$, with stannous chloride, $SnCl_2$ formed the nanoparticles. The resulting purple suspension was widely used in the production of ruby glass and as enamel color on ceramics and is still in use today. Its nature confounded some of the most prestigious chemists of the day such as J.J. Berzelius (1779–1848) and J.L. Gay Lussac (1778–1850), until Richard Zsigmondy (1865–1929) finally elucidated it as part of the body of work for which he received the 1925 Nobel Prize in Chemistry.[19]

---

moment for his discovery, he could not have selected a more appropriate or auspicious time."[20]

However, a third circumstance that really drove Perkin & Sons over the top was a new concept in the chemical industry: problem-solving, support and service. Perkin knew he had to convince dyers to buy his product. If they had problems, he had to solve them or they would revert to the natural dyes they knew so well.[21]

Buoyed by the continuation of mauve mania in the fashion world and his firm's monopoly in Britain, along with his ingenuity in technical service, Perkin, by the age of 36, became a wealthy man. But mauve, in a sense, was a flash in the pan, with production ceasing in 1873, only 17 years after it came on the scene. It was a victim of its own success in that it encouraged so many competitors trying to find better, less expensive dyes.[2] However, it was not just this one dye that changed the world – it was the train of events that followed it, especially the international competition it engendered. Essentially, Perkin's "decision to get out on the road and solve problems of application, using the methods of both the laboratory and the colorist, and the sweaty laborious work on production processes for intermediates...were all critical components that enabled the aniline dye to replace its lichen-derived competitor."[22] Perkin did not place all his bets on mauve either. He continued dye research, discovering several more blockbuster dyes, including Britannia Violet and Perkin's Green. He also set about producing large quantities of synthetic alizarin (a derivative of **anthraquinone**) as soon as an economical pathway to that ancient standby opened up.[23]

Eventually, Perkin & Sons developed **lake**-making procedures, thus enabling the company to introduce aniline dyes to printers of wallpapers and to makers of lithographic and other printing inks. The De La Rue Company, the first printers to make use of inks containing synthetic coloring matters, used **lakes** from mauve and Britannia Violet to produce the British Postage and Revenue Stamps.[24]

When Perkin began his research, he could calculate the molecular formulas of the substances he was working with. It would be another decade before even the faint stirrings of the ideas of Kekulé, Couper and others would lead to a realization of the importance of structural formulas in understanding organic chemical reactions. Perkin's success was not based on his serendipitous discovery of a beautiful color, but on both good fortune and hard work, mostly the latter. He needed to show that the color could be fixed not only on "animal" fibers, but also would be good for cotton and for calico printing. Cost was a big factor: his dye needed to be more beautiful and more permanent, but also less expensive. To accomplish this, Perkin had to learn more chemistry by his bootstraps – realizing that his starting materials had to change from aniline to benzene, which was much cheaper. The stepwise synthesis of aniline from benzene was critical to forming the dye, which turned out to be a condensation product of several aniline molecules according to eqn (12.1).

$$\text{Benzene} \xrightarrow[\text{Nitration}]{HNO_3} \text{Nitrobenzene} \xrightarrow[\text{Reduction}]{Fe}$$
$$\text{Aniline} \xrightarrow[\text{Oxidation}]{\text{Dichromate}} \text{Mauve} \qquad (12.1)$$

This multistep process presented a more cost-effective method because of the inexpensive starting materials, but it was hazardous in many other ways, especially when scaled up to fulfill increased demand. He was just plain lucky that these starting materials rearranged themselves into something useful, and he was very astute in recognizing that fact. He also refined his process with starting materials that gave the best results: aniline mixed with two forms of toluidine.[22] But to his dying day, he never knew exactly what mauve was – strangely enough, it was not until 1994 that the correct formulas were finally determined and there turned out to be at least seven of them in the mixture.[25] How to distinguish one from the other was impossible without the theoretical framework that developed during the following decade.

Perkin's mauve drove three world powers – Britain, France, and Germany – to seek domination in the business of producing colors. Within five years of mauve's appearance, there were already 28 dye manufacturers, not only in the "big three," but also in Austria and Switzerland – many of them destined to become the industrial giants that we know today by such familiar acronyms as AGFA and BASF.[26]

12.3 STRUCTURAL STIRRINGS

Flashback a few years to the early to mid-1850s when the English chemist, Edward Frankland (1825–1899) began to speak of atoms having a certain binding capacity. His idea eventually evolved into our modern concept of an element's combining power, or valence. Pondering this idea, in 1865, Friedrich August Kekulé von Stradonitz (1829–1896) had a dream. At least, the tale that is told and the myth that arose, whether actual fact or not,[27,28] is that Kekulé dozed off while wrestling with the problem of benzene's puzzling molecular formula, C_6H_6. Later on, he would write: "[T]he atoms were gamboling before my eyes...all twining and twisting in snake-like motion. But look! What was that! One of the snakes had seized hold of its own tail, and the form whirled mockingly before my eyes. As if by a flash of lightning I awoke; and this time also I spent the rest of the night working out the consequences of the hypothesis."[29]

Kekulé had made the rounds of the hotspots of chemical ideas in Europe, studying first at Giessen under Justus von Liebig's (1803–1873) influence, later in London where he met William Odling (1829–1921), and still later with Robert Bunsen (1811–1899) at Heidelberg. He moved to a professorship in Ghent in 1858 and it was there that he made his breakthrough regarding the structure of benzene. However, as with anything else, we stand on the shoulders of those who preceded us, or at least, we move together, side by side, in developing new ideas. Kekulé's structural ideas owe much to his association with Odling in London and Charles Frédéric Gerhardt (1816–1856) in Paris. In 1858, he and Archibald Scott Couper (1831–1892), a young Scot working in Charles-Adolphe Wurtz's (1817–1884) Paris laboratory, (Figure 12.2),[30,31] came to the same conclusion simultaneously:

- that carbon atoms were tetravalent (had a bonding capacity of four); and
- that they had the ability to bond with one another.

As this idea began to take hold, chemists began to think and speak about the structure of molecules, realizing that how the atoms related to one another in a molecule was a key to

Figure 12.2 Left: Kekulé Monument in Bonn; Sculptor: Hans Everding, 1903. Reproduced from ref. 30 under the terms of the CC BY-SA 2.0 DE license, https://creativecommons.org/licenses/by-sa/2.0/de/deed. en. Right: Archibald Scott Couper. Reproduced from ref. 31.

understanding their identity and their properties. Eventually it would give rise to theory-based synthesis of colorants rather than the dominant hit-or-miss method of earlier years.[32] This new approach bore so much fruit that by 1892, a comprehensive publication[33] came out of the University of Basel that gave the major classifications of dyestuffs that had been produced up until that time. Although the volume includes the natural colors, the book describes over 650 dyestuffs created in the 35 years since mauve's debut. This flowering of the scientific approach to synthesis would grow into the thousands in the decades to come; many of these new compounds would land as pigments on the artist's palette – for weal or for woe.

Given the fact that all of Europe was in the throes of welcoming the new synthetic dyes, it may seem counterintuitive to investigate synthesizing the baseline natural dyes in the laboratory. However, alizarin, the dye derived from madder root was still very much in demand. Its firm reputation in the world of textile dyeing was unshakable, as was that of its close relative, **indigo**. The chemist who could find an inexpensive pathway to these two colors would have his reputation and his fortune made. Tune in to see how two relatively simple chemical reactions changed the world.

12.4   THE JOURNEY FROM MADDER TO ALIZARIN

In the game of cricket, a skilled bowler can deliver a ball with a high velocity and elusive spin. For a batsman trying to make contact, the only visual cue for when and where to swing the bat is a split-second flash of the glossy red cricket ball. To make the ball more visible, early cricket ball makers relied on rose madder to impart a bright red color to the ball's leather cover. Although rose madder is no longer used, the synthetic analogs now favored by cricket ball manufacturers owe their existence to research on this important dye. In 1826, French chemists Jean Jacques Colin (1784–1865) and Pierre Jean Robiquet (1780–1840) isolated the primary red coloring agents in rose madder, alizarin and purpurin. The problem with winning the natural dye from the roots of *Rubia tinctorum* is that on average, alizarin ranges from 9.1 to 21.8 mg g^{-1} out of a total **anthraquinone** content of 15.6 to 39.4 mg g^{-1} in *Rubia tinctorum* L. roots. This amounts to

Figure 12.3   Left: Harvested madder crop of 2 year-old plants, prior to washing; Right: Dried homegrown madder root after a 12 hour soak.
Image courtesy of Michelle LaBerge.[34]

about 55–62% alizarin colorant, but note the tiny gross amount
in the roots. A little goes a long way in chromatic strength, but
all you get is a little out of each root[34] (Figure 12.3).[35] A synthetic
pathway was just waiting to be found.

12.4.1   A Synthetic First

Forty-two years after alizarin was chemically isolated, in 1868, the
German chemists Carl Graebe (1841–1927) and Carl Liebermann
(1842–1914), under the direction of Adolf von Baeyer (1835–1917),
prepared alizarin[§] from anthracene and were the first to produce a synthetic substitute for a naturally occurring dye.[36] This
was no mean feat: alizarin is highly complex and also had enormous commercial value.[22] However, the method necessitated a
bromination followed by a debromination, two very expensive
steps. A year later, they managed to devise a more direct synthesis, filed a patent, and interestingly, beat Perkin to the punch by
one day.[37] The German application was delayed awaiting further
experimentation, while Perkin's was railroaded through. Graebe
and Liebermann, with good reason, felt that Perkin had been
given special treatment. They finally negotiated an agreement to

---

[§]Alizarin is the red structural "first cousin" of colorless **anthraquinone**, the parent
compound for a whole host of other red to yellow colorants.

respect each other's patents and to divide the market. This provided Perkin & Sons with new life for a few years, but in 1873, he sold the business and retired a very wealthy young man.[22]

Valued from ancient times, madder was an important agricultural product throughout the world, principally in India, Turkey, and many parts of Europe, particularly Holland. Naturally occurring madder actually consists of two principal colorant components: alizarin and pseudopurpurin, both belonging to the group of compounds known as anthraquinones after the parent compound, **anthraquinone**. Alizarin is the major component; purpurin does not occur in the freshly cut root, but only in roots that have been placed in storage, because of the degradation of pseudopurpurin.[38] Although purpurin is a dye in its own right, it is usually considered an undesirable contaminant of alizarin extracted from the madder root.[39] Among the almost 30 naturally occurring anthraquinones that have been identified,[40] alizarin is by far the most commercially viable and therefore the compound for which a viable synthetic route was sought in the 1860s.

12.4.2   Alizarin Pluses and Minuses

Alizarin is a **mordant** dye, that is, it does not have a strong chemical affinity for textile fibers and therefore requires a fixing agent, or mordant. Alizarin, when mordanted with aluminum, tin, chromium or copper, will form red, pink, puce brown and yellow brown complexes respectively.[41] Among the colors possible with other **mordants** are purple, black, and even chocolate. Furthermore, the compounds present in the coloring matter are unusually stable, so they can be treated by various agents to improve or modify the shade.[42]

Given its versatility, and the fact that it could impart brilliant colors to cotton and wool, alizarin was a prize worth synthesizing. However, although successful synthesis of natural dyes of historic interest was viewed as a chemical feat of the first order, no naturally occurring dye could stand up to the stringent requirements eventually set by the dye industry: consistent and controllable hue, lightfastness, and fastness to washing with soaps and other alkali agents, and later on, detergents. Alizarin was no exception. It was recognized early on that the natural material, madder, had some excellent qualities but also some serious drawbacks.

On the plus side, madder was more fast than most other natural colorants, it gave strong bright colors, and it was relatively inexpensive because Europeans could grow it themselves. But in an industry that required great precision, the big problem was that the madder roots varied greatly in quality. The dye was also sensitive to both pH and temperature and had a tendency to give an orange shade.[43] Synthetic alizarin could obviate some of these problems, but others were inherent in the nature of the dyestuff itself. It is no wonder, then, that chemists seeking new and better dyestuffs would first try to characterize and synthesize the natural colorants, but then use their structures as models for developing new colorants with more desirable properties. Nevertheless, the advent of synthetic alizarin had disastrous consequences for the madder growers and carmine-cochineal cultivators. Almost overnight, the *Dactylopius coccus* growers of the Canary Islands were bankrupted; the madder farmers of Southern France had to convert their fields to potatoes and (perhaps happily for some) grapevines. And most important of all, the center of gravity of the dye industry shifted once and for all from England to Germany, the major patent-holder for the new technology.

12.5 MOOD INDIGO

The next field to conquer was the other major natural dye, **indigo**. Adolf von Baeyer was up to the task: he won the 1905 Nobel Prize in Chemistry for its successful synthesis.[44] We must remember that, for Europeans, **indigo** was an expensive imported luxury up until the latter part of the 16th century. Largely reserved for inks and paints, it made up a portion of the exotic spices that made their long hazardous journey from Asia and the Middle East. As an item that demanded laborious processing in addition to importation, it often cost double the price of other luxury commodities.[45]

12.5.1 Indigo the Pigment

Along with madder, **indigo** was a fixture on the Renaissance artist's palette from as early as the 13th century, mainly in fresco paintings.[46] Analyses have shown **indigo**'s presence in European paintings from the high Middle Ages (13th century) through the 19th century. Some notable examples are Russian icons,

Figure 12.4   Left: Cosimo Tura (1430–1495): A Muse, *ca.* 1460. The bodice of the figure is painted with indigo. National Gallery, London; Reproduced from ref. 48. Right: A resist-dyeing display, Goldberger Textile Museum, Budapest. The hangings from rear to front show heavily dyed, lightly dyed and undyed printed textiles.

Leonardo da Vinci's *Last Supper*, and works by Rubens, Steen, Vermeer, and Turner[47] (Figure 12.4).[48,49] However, it does not do well in oils, often giving a flat and stilted look.[50] There is also evidence that it was used as a pigment in printed hand-colored manuscripts well into the 18th century.[51]

12.5.2   Jeans Blue

Indigo, or indigotin (its proper common chemical name) is probably the oldest coloring matter known and is, to this day, one of the few naturally occurring dyes in wide use. Most of the 40 000 tons produced annually go into coloring blue jeans alone. As its name suggests, indigo has its origin in India, the root of the word derived specifically from the Indus River; the peoples living along the Indus were the first to use indigo as a dye. Marco Polo encountered indigo on his travels and was the first to describe its production in India from the plant material.

To obtain the blue dye from the plant is not an easy matter. Its precursor, colorless indoxyl, bonded to a sugar molecule, is the starting material regardless of the plant species used. Processing the raw material required fermentation for days in an alkaline (non-acidic) solution (typically urine, which exudes ammonia on standing) with periodic maceration (bare feet were recommended). Chemically, this treatment initiated enzymatic cleavage of the sugar-indoxyl bond. The **enzyme**, also produced by the indigo plant, was only brought into contact with the indoxyl in the vat mix when the leaves' cell walls were mechanically broken down.[52,53] If sufficiently alkaline, the fermentation broth is nearly colorless and ready for dyeing. The textiles to be dyed are soaked in it for a time, and then hung up to air dry (Figure 12.4). The indoxyl spontaneously air oxidizes to indigo which adheres to the textile fibers as small particles.

It should be noted that the formation of solid **indigo** is completely reversible. When it is transported and sold as solid blue cakes, it must be **vat dyed**, that is, placed in an alkaline solution in a vat so that it goes into solution as the near-colorless "leuco" form (from the Greek *leukos* = white).

One of the 19th century's greatest chemists, Adolf von Baeyer, devoted a considerable portion of his professional life to indigo because it was the major industrial colorant throughout the century. His goal was to elucidate the structure of indigo and thereby deduce a pathway toward its laboratory synthesis in order to bypass the time-consuming and laborious processes of its extraction. As early as 1868, he used the Humpty-Dumpty effect: he pulled the molecule apart step by step, found its structural building block, indole, C_8H_7N, and then spent the next 15 years trying to put it together again. By 1883, he managed, but without all the king's horses and all the king's men.[22] It took another seven years before von Baeyer was able to devise a commercially feasible pathway to its synthesis. Another chemist, Karl Heumann (1851–1894), cleared the way by discovering how to produce anthranilic acid from cheap and plentiful naphthalene for use as a cost-effective starting material. This breakthrough led directly to the first commercially viable process for indigo synthesis in 1897. In that same year, BASF and Hoechst in Germany began to manufacture synthetic indigo (Box 12.3).

Box 12.3   The Madness of George III and Bright Blue Urine

While the precursors of indigo exist in some species of plants, its presence in human urine has been noted as early as 1767. The most famous specimen was that of King George III (1738–1820), found at the height of his insanity. While the king may have been suffering from congenital porphyria, it seems to be unrelated to his blue urine.[54] Further investigations showed that a possibly constipation-related overload of dietary tryptophan, an essential amino acid, can result in indigo-blue urine. The mechanism involves conversion of the tryptophan to indole by bacteria in the gut and subsequent oxidation of the indole in the liver to indoxyl, the indigo precursor. A similar pathway has been noted in certain pathologies such as Crohn's disease.[55]

12.5.3   Consequences of Synthetic Indigo

The cataclysm was 30 years in the making, but it arrived at last. The effect of this synthesis on the natural product industry was, on the scale of nations, catastrophic and immediate. Indigo plantations all over the world experienced a shock wave similar to a modern tsunami. In India alone, especially in northern India, this virtual single-crop industry was literally decimated. In the time frame of only 17 years, from 1897 to the outbreak of World War I in 1914, the number of indigo-planted hectares had nosedived from 700 000 to 121 000, an astonishing and debilitating 83% decrease.[56] Measured in tonnage, the figures were even drearier: an 89% drop; but even worse – when measured in pounds sterling, the drop was 94%.[57]

Indigo plantations in India had been a bone of contention for centuries. The British, under the aegis of the East India Company, began their Indian indigo industry in 1777, virtually enslaving large swaths of the native population by forcing them to grow indigo instead of food and then buying the crop at a fixed price that kept the farmers in starvation mode. Following a peasant revolt in 1860, matters became worse, and such was the situation in 1897 when the industry began to decline. Mohandas Gandhi (1869–1948) is sometimes credited for sparking the Indian independence movement when, in 1917, he proposed a massive nonviolent protest against the oppressive conditions on the indigo plantations.[58] However, this seems to be a stretch since the industry was already virtually dead, and independence did not arrive until 30 years later.

Since the indigo supply was cut off due to World War I, Switzerland, the United States, and other countries developed their own synthetic indigo industries; Britain did as well, once it overcame a patent stranglehold,[59] and by the end of the war, naturally occurring indigo was an artifact of history. Furthermore, today's synthetic methods, requiring high temperatures and utilizing a feedstock of petroleum-based aniline, large amounts of formaldehyde, cyanide and corrosive reducing agents, may be consigned to the chemical waste-bin in favor of greener and more sustainable, enzyme-driven processes.[60]

The world is full of stories of unintended consequences. Rabbits introduced as a food source into Australia have had a devastating impact on agriculture; smokestacks designed to reduce air pollution have only exacerbated the problem by releasing particulates. As we have already seen in the cases of madder and indigo, chemical synthesis of the same commodities can destroy local industries and upend the economy. As the synthetic dye industry developed, a chain of utterly unforeseen events led to changing the entire history of the 20th century.

12.6 HOW THE SYNTHETIC DYE INDUSTRY CHANGED THE WORLD

We can distinguish three steps in the development of dyes that set history on a course of no return. The first, from about 1850, was the large-scale growth of the dye industry in terms of volume and multiplicity of new colors. The second was A. W. Hofmann's growing influence among British colorists and his nurturing of a network of German visitors who came to absorb his knowledge and expertise. The third was Hofmann's return to Germany in 1865 as the acknowledged leader in aniline dye synthesis which provided the impetus for the unprecedented growth of German businesses and universities.[61]

12.6.1 An Organizational Coup

Two years after the successful synthesis of alizarin in the laboratory, another major breakthrough occurred in Germany, but this time of an organizational, not a technical, nature. In 1867, presided over by none other than A. W. von Hofmann, the German

Chemical Society (*Deutsche chemische Gesellschaft*) was founded. This society was to have far-reaching effects on two major fronts: (1) through its journal, the *Berichte*, it became a vehicle for the spread of modern chemical ideas and the recording of the enormous amount of experimental work leading to the rapid determination of the structural formulas for thousands of chemical compounds, and (2) it fostered the development of in-house industrial research laboratories leading to the rise of an innovative, science-based chemical industry in Germany and industrial-academic collaboration.[62] By 1873, Germany had put in place a comprehensive patent law system for protecting chemical inventions, and by 1900, Germany controlled most of the world's dye market.[63] The industry was embedded in an economic system that, in enticing greater consumption, led to frantic research, which led in turn to an increased pace of organic chemical discovery and to innovations particularly in processes.[64]

12.6.2 Areas Transformed by the Dye Industry

Fashion. Prior to aniline dyes, color austerity was the order of the day: only the rich could afford the rainbow of colors available in the natural dyes. But the ordinary person's desire for the riot of color, now inexpensive and available to all, coupled with a clarity, variety, and fastness not found in the natural dyes, greatly stimulated dye research.

The educational structure of industrial society. The needs of the dye industry stimulated educational reforms that were already under way at the university level so that at the end of the nineteenth century, Germany had the finest system of scientific and engineering education worldwide – and this system influenced higher learning in all parts of the world.

The social structure. By employing large numbers of academically trained chemists in its plants and by growing a white-collar managerial proletariat, the dye industry displaced the independent tradesman and raised the educational level of the masses.

Political action. The dye industry organized and supported lobbies, exerted strong influence on legislation, and helped establish a model patent system.

Industrial research. The dye industry, led by Germany, originated and developed the industrial research laboratory and the research team. Driven by patent law changes and the acceleration of discovery and invention in the fields of organic chemistry and dye chemistry, this organizational structure prevails to this day.[65]

Power. Through its hegemony in the dye industry, Germany became the foremost industrial power in Europe, leading to it becoming the foremost power in Europe. World War I was a battle of technologies that promoted growth in other branches of the chemical industry, while the dye industry itself declined in importance.[66]

Not all of these developments were unequivocally positive. Expertise in the chemistry of dyes segues naturally into expertise in the chemistry of high explosives since both categories are structurally similar. Thus, through the dye industry, Germany became a giant war machine, manufacturing explosives, poison gases, photographic film, drugs, natural product substitutes, and a whole host of other products born of its technology. The industrial juggernaut also led to massive air and water pollution; above all, it led to an unprecedented increase in chronic diseases, deaths by heavy metal poisoning, birth defects, and so much more. It led to the power to instigate two World Wars, taking an incalculable toll on human life. The rivers that ran red with industrial effluent also ran red with blood. The developments in the European dye industry of the late nineteenth century arguably steered the course of 20th century history right up to and beyond World War II.[¶]

12.7   CONCLUSION

While this chapter was devoted to the development of the synthetic dye industry and its legacy,[‖] the importance of these events to the addition of innumerable pigments to the artists' palette cannot be overestimated. During the 19th century, about a dozen new **inorganic** colors and a single discovery were to inundate the art world with a whole new movement. As the 20th century dawned, hundreds more colors made from coal tar were to dominate the art scene for generations to come – for good and for ill. The story continues in the next chapter.

[¶]Parts of this chapter are adapted from ref. 23, pp. 90–91 by permission from Springer.
[‖]Colorants Industry History, www.colorantshistory.org (accessed May 2021) is an excellent resource for further information on the development of the industry.

REFERENCES

1. L. Pasteur, *Studies on Fermentation*, Macmillan, London, 1879, p. viii.
2. S. Garfield, *Mauve: How One Man Invented a Color that Changed the World*, W.W. Norton, New York, 2001.
3. R. Brightman, *Nature*, 1956, **177**(4514), 815–821.
4. J. Bentley, *Ambix*, 1970, **17**, 153–181.
5. W. J. Hornix, *Br. J. Hist. Sci.*, 1992, **25**(1), 65–90.
6. C. Heichert and H. Hartmann, *Z. Naturforsch.*, 2009, **64b**, 747–755.
7. N. Zinin, *Ann. Chem. Pharm.*, 1842, **44**(3), 283–287.
8. N. Zinin, *J. Prakt. Chem.*, 1842, **27**, 140–153.
9. N. M. Brooks, *Bull. Hist. Chem.*, 2002, **27**(1), 26–36.
10. E. Homburg, *J. Soc. Dyers Colour.*, 1983, **99**, 325–333.
11. D. Hollett, *More Precious than Gold: The Story of the Peruvian Guano Trade*, Fairleigh Dickinson University Press, Madison, 2008.
12. J.-A. Schlumberger, *Bull. Soc. Ind. Mulhouse*, 1853, **25**, 242–263.
13. P. Levi, *The Periodic Table*, Schocken Books, New York, 1984, pp. 183–184.
14. P. Ball, The other periodic table, https://www.chemistry-world.com/features/primo-levi-and-the-other-periodic-table/3010512.article, accessed March 2021.
15. Sir William Perkin, Oil Painting, https://commons.wiki-media.org/w/index.php?title=File:Sir_William_Perkin._Oil_painting._Wellcome_V0018025.jpg&oldid=462942212, accessed June 2021.
16. Perkin factory, Henry Rzepa (Rzepa at English Wikipedia), https://commons.wikimedia.org/wiki/File:Perkin_factory.jpg, accessed June 2021.
17. F. C. Calvert, *J. R. Soc. Arts*, 1862, **10**(481), 169–180p. 171.
18. N. Zumbulyadis, *Bull. Hist. Chem.*, 2014, **39**(1), 7–17.
19. L. B. Hunt, *Endeavour, New Series*, 1981, **5**, 61–67.
20. C. J. T. Cronshaw, *Chem. Ind.*, 1935, **54**, 515 and 547.
21. W. V. Farrar, *Endeavour*, 1974, **33**, 149–155.
22. A. S. Travis, *The Rainbow Makers: The Origins of the Synthetic Dyestuff Industry in Western Europe*, Lehigh University Press, Bethlehem, 1993.

23. M. V. Orna, *The Chemical History of Color*, Springer, Heidelberg, 2013, pp. 73–75.
24. M. R. Fox, *Dye-makers of Great Britain 1856–1976*, Imperial Chemical Industries, Manchester, 1987, p. 97.
25. O. Meth-Cohn and M. Smith, *J. Chem. Soc., Perkin Trans.*, 1994, **1**, 5–7.
26. S. B. McGrayne, *Prometheans in the Lab: Chemistry and the Making of the Modern World*, McGraw-Hill, New York, 2001, p. 22.
27. J. H. Wotiz and S. Rudofsky, *Chem. Br.*, 1984, **20**, 720–723.
28. A. J. Rocke, *Angew. Chem., Int. Ed.*, 2015, **54**, 46–50.
29. F. R. Japp, *J. Chem. Soc.*, 1898, **78**, 97–138.
30. Kekulé-Denkmal in Bonn, https://commons.wikimedia.org/w/index.php?title=File:FA_Kekule.jpg&oldid=452383777, accessed June 2021.
31. Couper Archibald Scott, https://commons.wikimedia.org/w/index.php?title=File:Couper_Archibald_Scott.jpg&oldid=497646837, accessed June 2021.
32. A. J. Ihde, *The Development of Modern Chemistry*, Dover Books, New York, 1984, p. 218 ff.
33. R. Nietzki, trans. A. Collin and W. Richardson, *Chemistry of the Organic Dyestuffs*, Gurney and Jackson, London, 1892.
34. I. Boldiszár, Á. László-Bencsik, Z. Szűcs and B. Dános, *Acta Pharm. Hung.*, 2004, **74**, 142–148.
35. M. LaBerge, The Heart of the Madder: An Important Prehistoric Pigment and its Botanical and Cultural Roots, MS thesis, University of Wisconsin – Milwaukee, 2018, p. 70 and 85.
36. C. Gräbe and C. Liebermann, *Ber. Dtsch. Chem. Ges.*, 1868, **1**(1), 104–106.
37. Henry Rzepa's Blog, The history of alizarin (and madder), https://www.ch.imperial.ac.uk/rzepa/blog/?p=20333, accessed December 2020.
38. R. Hill and D. Richter, *J. Chem. Soc.*, 1936, 1714–1719.
39. L. Wacker, *J. Prakt. Chem.*, 1896, **54**, 88–94.
40. H. Schweppe and J. Winter, Madder and alizarin, in *Artists' Pigments: A Handbook of Their History and Characteristics*, ed. E. W. FitzHugh, National Gallery of Art, Washington, DC, 1997, vol. 3, pp. 109–142.
41. R. H. Peters, *Textile Chemistry, vol. III: The Physical Chemistry of Dyeing*, Elsevier, New York, 1975, p. 649.

42. E. Schunck, *Q. J. Chem. Soc.*, 1860, **12**, 198–221.
43. A. B. Greenfield, *A Perfect Red*, HarperCollins, New York, 2005, p. 28.
44. F. Steinmüller, Adolf von Baeyer, in *Nobel Laureates in Chemistry, 1901–1992*, ed. L. K. James, American Chemical Society, Washington, DC, 1993, pp. 30–34.
45. J. Balfour-Paul, *Indigo – Egyptian Mummies to Blue Jeans*, Firefly Books, London, 2011, p. 27.
46. R. H. Michel, J. Lazar and P. E. McGovern, *Archaeomaterials*, 1992, **6**(1), 69–83.
47. H. Schweppe, Indigo and woad, in *Artists' Pigments: A Handbook of Their History and Characteristics*, ed. E. W. FitzHugh, National Gallery of Art, Washington, DC, 1997, vol. 3, pp. 80–107.
48. Cosimo Tura: A Muse (Calliope?), https://commons.wikimedia.org/w/index.php?title=File:Studiolo_di_belfiore,_calliope_di_cosm%C3%A8_tura,_national_gallery_di_londra.jpg&oldid=378928954, accessed June 2021.
49. Dyebaths on display, Goldberger Textile Museum, Budapest.
50. A. P. Laurie, *The Materials of the Painter's Craft*, T.N. Foulis, London, 1910, p. 357.
51. P. Zannini, P. Baraldi, M. Aceto, A. Agostino, G. Fenoglio and D. Bersani, *et al.*, *J. Raman Spectrosc.*, 2012, **43**, 1722–1728.
52. S. Struckmaier, *Chem. Unserer Zeit*, 2003, **37**(6), 402.
53. W. Maier, B. Schumann and D. Gröger, *Phytochemistry*, 1990, **29**(3), 817–819.
54. A. Dronsfield and C. Cooksey, *George III, Indigo and the Blue Ring Test*, https://edu.rsc.org/feature/george-iii-indigo-and-the-blue-ring-test/2020154.article, accessed December 2020.
55. A. H. Jackson, R. T. Jenkins, M. Grinstein, A.-M. Ferramola de Sancovich and H. A. Sancovich, *Clin. Chim. Acta*, 1988, **172**, 245–252.
56. B. Schaefer, *Natural Products in the Chemical Industry*, Springer, New York and Heidelberg, 2014, pp. 23–28.
57. Royal Economic Society, Asiaticus, *Econ. J.*, 1912, **22**(86), 237.
58. B. Campos Seijo, Indigo and Indian Independence, *Chemistry World*, The Royal Society of Chemistry, https://www.chemistryworld.com/opinion/indigo-and-indian-independence/7294.article, accessed March 2021.
59. P. Reed, *Br. J. Hist. Sci.*, 1992, **25**(1), 113–125.

60. E. Landhuis, *Chem. Eng. News*, 2019, **97**(44), 11 November, 22–25.
61. F. Caron, *Dynamics of Innovation: The Expansion of Technology in Modern Times*, Berghahn Books, New York, 2013, p. 110.
62. J. A. Johnson, Germany: Discipline-Industry-Profession. German Chemical Organizations, 1867-1914, in *Creating Networks in Chemistry: The Founding and Early History of Chemical Societies in Europe*, ed. A. K. Nielsen and S. Štrbáňová, RSC Publishing, Cambridge, 2008, pp. 113–138.
63. A. S. Travis, *Bull. Hist. Chem.*, 2008, **33**(1), 1–11.
64. J. J. Beer, *Isis*, 1958, **49**(2), 123–131.
65. E. Homburg, *Br. J. Hist. Sci.*, 1992, **25**(1), 91–111.
66. J. J. Beer, *The Emergence of the German Dye Industry*, University of Illinois, Urbana, 1981, pp. 149–151.

CHAPTER 13

Monet's Garden: Impressionist Innovation and Beyond[†]

Color has a logic of its own, no less strict than that of form.[1]

Pierre Bonnard

13.1 INTRODUCTION

Almost by default, Paris became the locus of not one, not two, but three seismic revolutions during the 18th and 19th centuries. First came the chemical revolution personified in Antoine Laurent Lavoisier; next came the French Revolution that upended the political map of Europe; finally came the artistic revolution in which color superseded form and the artist moved out of the studio into the open air: the birth of Impressionism.

13.2 MONET REFUSES THE (CATARACT) OPERATION[2]

Doctor, you say that there are no halos around the street-lights in Paris and what I see is an aberration caused by old age, an affliction.

[†]Glossary entries and chapter cross-references are in boldface.

March of the Pigments: Color History, Science and Impact
By Mary Virginia Orna
© Mary Virginia Orna 2022
Published by the Royal Society of Chemistry, www.rsc.org

I tell you it has taken me all my life to arrive at the vision of gas lamps as angels, to soften and blur and finally banish the edges you regret I don't see...

Fifty-four years before I could see Rouen cathedral[3] is built of parallel shafts of sun, and now you want to restore my youthful errors...

What can I say to convince you the Houses of Parliament[4] dissolve night after night to become the fluid dream of the Thames?

I will not return to a universe of objects that don't know each other, as if islands were not the lost children of one great continent. The world is flux, and light becomes what it touches,...lilac and mauve and yellow...

Doctor, if only you could see...

Yes, if only you could see![‡] Color was to become the watchword of late 19th century artists, while the attention to drawing and form demanded by the French *Académie* dissolved away like Monet's *Houses of Parliament* (Figure 13.1).

From 1775 to 1875, punctuated by Vauquelin's discovery of chromium in 1797, (**Chapter 11.7**) a rainbow of no fewer than 30 new manufactured artists' colors appeared like wildflowers on the artistic scene. Derived mainly from the three Cs, chromium, cadmium and cobalt, they were seized upon by the Romanticist painters, particularly by Eugène Delacroix (1798–1863) and Gustave Courbet (1819–1877), and found their way into every Impressionist painting that followed these precursors. But not without a struggle of monumental proportions.

The establishment, bounded by convention and opposition to change, leaving no pathway open to accommodate originality

---

[‡]In 1912, Claude Monet was diagnosed with cataracts which caused increasing visual impairment until he finally submitted to eye surgery in 1923.

Figure 13.1   Claude Monet (1840–1926). Left: Rouen Cathedral, West Façade. Oil on Canvas, 1894, National Gallery of Art, Washington, DC. Reproduced from ref. 3. Right: The Houses of Parliament, Sunset. Oil on Canvas, 1903. Chester Dale Collection, National Gallery of Art, Washington, DC. Reproduced from ref. 4.

and individual expression, was dominated by the "mighty, indisputable, absolute despot"[5] of academic painting, Jean-Auguste-Dominique Ingres (1780–1867). Not only was his dictum "drawing is everything, the whole of art lies there"[6] directly opposed to the new fascination with color, but also was rendered passé by the advent of photography. Even Delacroix, a respected member of the establishment, ran afoul of Ingres with his adherence to free techniques; between them lay a mortal enmity. This struggle opened the way to a new movement in search of truth and sincerity in art – Impressionism.

13.3   BIRTH OF IMPRESSIONISM

Although the spirit of informality and directness, hallmarks of Impressionism, were boiling away under the surface for quite some time, it all emerged with two works by Édouard Manet (1832–1883). Realizing that he could accomplish shading and modeling with color alone, his 1866 *The Fife Player* speaks this new language of painting so that "every brush stroke, every color patch is equally 'real' no matter what it stands for in nature."[7] Émile Zola (1840–1902) said of paintings like that, "they break the wall."[8] Manet's next work completed the revolution and earned him the title of "The First Impressionist;" his *In the Boat* (1874) is flooded with color, outdoor sunlight, brightness, all motion, a quick glimpse, an "impression" of what two people are doing with reference to nothing else (Figure 13.2 right).[9]

Obviously, this approach outraged the critics. They labeled such works as unfinished sketches, not serious or profound, just a "slice of life," with too much raw, violent color.[10] Breaking away at the same time were some of the household names we associate with Impressionism today: Monet, Pierre-Auguste Renoir (1841–1919), Camille Pissarro (1830–1903), Edgar Degas (1834–1917), Berthe Morisot (1841–1895), Alfred Sisley (1839–1899). Routinely rejected by the great annual Salon of the École des Beaux Arts, which could make or break a budding career, most of these artists chose to exhibit on their own in eight Impressionist exhibits between in 1874 and 1886.

Appropriately enough, Monet, who is considered the iconic Impressionist by many, inadvertently baptized the whole movement with the title of a work he exhibited in 1874 at the first independent Impressionist show, *viz.*, *Impression, Sunrise*, now hanging at Paris's Musée Marmottan Monet (**Chapter 16, Figure 16.2A**).

Figure 13.2   Left: Édouard Manet. The Fife Player, 1866. Oil on canvas. Musée d'Orsay. Reproduced from ref. 7. Right: Édouard Manet. In the Boat, 1874. Oil on canvas. The H. O. Havemeyer Collection, Metropolitan Museum of Art. Reproduced from ref. 9.

The term "Impressionism" became current, and predictably, the art critic Louis Leroy, who coined the term to mean "unfinished," satirized the show.[11]

Although painting out of doors, *plein air*, was not exactly an Impressionist innovation, it was a practice central to their goal of achieving truth and sincerity in their works. Catching the fugitive effects of changes in weather and sunlight, even minute by minute, stimulated the development of their color and brushwork techniques.[11] But at a price! Swarming, gesticulating observers plagued Berthe Morisot; police accused other easel painters of starting riots because of the crowds they attracted. Monet was probably one of the most courageous outdoor painters, braving the vagaries of weather: sleet, snow, ice, wind, rain, and even a threatening incoming tide.[11] He often worked with 15 or more canvases at once for a few minutes at a time because he said that no painter could paint more than a half-hour on any outdoor effect and keep the picture true to nature.[12] But perhaps the greatest difficulty of all was handling the many pigment and paint containers that made up a bulky paint box prior to the invention of portable collapsible metal paint tubes. In fact, Jean Renoir quotes his father as saying: "Without paints in tubes, there would have been no Cézanne, no Monet, no Sisley or Pissarro, nothing of what the journalists were later to call Impressionism (Box 13.1)."[13]

13.4   RÉVOLUTION À LA FRANÇAISE

Why did France, and particularly Paris, become the hub of this revolutionary approach? Perhaps because France was the seat of revolutions and knew how to handle them. As we saw in **Chapter 11.7**, the chemical revolution took place in the midst of the political revolution, with dire consequences for Lavoisier. There was a quasi-revolution in pigment development since the vast majority of new **inorganic** colors were discovered in France, aided in part by the expanding French metallurgical industry: the chrome pigments, iodine scarlet, synthetic ultramarine and first manufacture of cobalt blue.[15] Much of this development was due to the work of French chemists who outlined the theoretical framework for the discovery of new elements and compounds and those

---

Box 13.1   Bladder Control

What does fine art have to do with slaughterhouses? Commonplace among 18th century pigment vendors was their advertisement of bladder colors on their trade labels. Among artists, no explanation was needed. The bladder was indeed that anatomical necessity for most animals: watertight, flexible, strong, an ideal container for liquids. Pig bladders were just the right size for many applications: inflated they became children's balls or balloons, resonators for some types of musical instruments, rattles when filled with peas or seeds. Uninflated, they were the ideal container for artists' pigments. Artists could cut pieces of bladder, place their pigments ground in oil media on top, gather the corners and secure the whole with twine or a clamp. Or they could buy ready-made colors stored that way. To access the colors, they pierced the bladder with a tack, squeezed out the desired amount of paint, and replaced the tack firmly. Some painters, like J. M. W. Turner, substituted the pig bladder with parchment packets, but they were not airtight and the pigments solidified quite rapidly. Pig's bladder was reputed to keep the paints fresh for decades.

Once set up in the studio, the artist was ready to go to work. But imagine an artist carting an assembly of pigs' bladders outdoors along with all the other paraphernalia needed to do a day's work *plein air*. In 1841, an enterprising American painter solved the problem. John G. Rand's (1801–1873) answer – a collapsible metal tube with a screw-cap closure – was both revolutionary and momentous. It immediately relieved the artist of the burden of bulky, awkward containers requiring two hands for manipulation. From then on, painting with a squeeze was a breeze.[14]

---

who were at the forefront of laboratory research at the time, *e.g.*, Vauquelin, Louis-Jacques Thénard (1777–1857) and Bernard Courtois (1777–1838).

Another revolution was the contribution of chemist Michel Eugène Chevreul (1786–1889) to color theory. His laws of the simultaneous contrast of colors, studied assiduously by the Impressionist artists, had a profound influence on the development of painting in France. As Director of Dyeing at the fabled Royal Gobelins tapestry concern, part of his job was to find out why some of the tapestries exhibited unexpected colors. The story, as told in Ogden Rood's (1831–1902) masterpiece on color theory and color mixing, *Modern Chromatics*,[16] goes like this: some dealers requested that some plain red, violet-blue and blue textiles be overprinted with black patterns. When the dealers received the goods back, they complained that the patterns were not black, but that the red was overprinted with green, the violet with dark greenish yellow and the blue with a copper color. Chevreul found that the overprint was indeed black; the observed effects were due entirely to contrast. He found that apparent brightness in colors depended not so much on the colors themselves but on the ones in close proximity. Complementary colors, that is, any two colors that complete the visible spectrum, (**Figure 1.3**) will produce the optical illusion of shining more brightly when placed next to one another. Two such pairs are orange and blue, used to great effect by Renoir in *The Seine at Asnières* (**Figure 11.5**): the differences between the cobalt blue and chrome orange in hue, color saturation and brightness are at their greatest when touching.[11,17]

Parenthetically, one of the most widely held myths, that the Impressionists exploited a process of "optical mixing" by using separate brushstrokes (*taches*) of pure prismatic colors that fused in the eye of the beholder, is simply not true – nor is it possible. "The continual appearance of this error is doubtless due as much to superficial observation of the pictures as to incomplete knowledge of the facts of color, since the pictures never confirm it."[18] The dabs are too large for retinal fusion to take place at normal viewing distances. Rather, they used this technique of "a thousand little dancing strokes... in vital competition for the whole impression"[19] to suggest form and light by clearly differentiating their colors.[11] Two young French artists, among the first of the "post-Impressionists," Georges Seurat (1859–1891) and Paul Signac (1863–1935), were to minimize the "little dancing strokes" of the Impressionists to the dimensions of "points" in order to – ahem!, make their point. That point, to some art historians, is "retinal fusion:" additive color mixing to achieve a mix. But the points, the smallest of which in their paintings is about 1/16 in., are too small to see when viewing from the required distance of fusion, which is 12–14 feet away. Signac and Seurat carried Chevreul's ideas to their ultimate conclusion, founding what came to be known as "Pointillism," but achieved retinal fusion in only a very minute area of their work, if at all.[18]

An "American Impressionist," Winslow Homer (1836–1910), was very much aware of Chevreul's physiological optics as reported by Rood.[16] Homer was in the forefront of his artistic contemporaries to actually study modern color theory and to apply it in his work. He once stated: "You can't get along without a knowledge of the principles and rules governing the effects of one color upon another. A mechanic might as well try to get along without tools."[20] Many clear examples of proper application of color theory can be found in all periods of his watercolor work.[21]

However, despite Homer's adherence to Chevreul's science, he chose to ignore another bit of inconvenient science: Winsor & Newton's warnings that many of the new colors were impermanent. Throughout his career, he used light-sensitive red **lakes.** As a result, some of his works are out of color balance because of the fading of the reds. In some works, he used Hooker's Green, a mix of Prussian Blue and fugitive yellow gamboge (**Chapter 14.4**), thus shifting his greens toward the blue. His layered purple swaths are now almost entirely faded toward the blue with the disappearance of his red **lake** and cochineal formulations.[21] Another victim was Seurat, betrayed by an inorganic pigment, chrome yellow, that turned his greens brown[22] due to a photochemical reaction with visible and ultraviolet light.[23]

Vincent van Gogh (1853–1890) came to his color theory *via* the influence of Paul Signac and a book by theorist Charles Blanc.[24,25] For example, in a letter to a friend he wrote: "What I should like to find out is the effect of an intenser blue in the sky... There is no blue without yellow and without orange, and if you put in blue, then you must put in yellow, and orange too..."[26]

Despite external pressures, Homer managed to keep control of his own pigments, whereas van Gogh did not. Most of Homer's mixtures were of his own making. Such was not the case for many Impressionists and those who followed, as we will see in the next section.

13.5   THINGS ARE SELDOM WHAT THEY'RE CALLED

Living on the edge, many Impressionist artists had regular color merchants, such as Julien (Père) Tanguey (1825–1894), who would provide pigments and paints in exchange for canvases if the artist could not pay for the material.§ Far more difficult, however, was the gnawing uncertainty regarding the identities and qualities of the paints the artists purchased from their suppliers. This was a far cry from artistic practice in the 17th century when

---

§When Tanguey died, numerous unsold Impressionist paintings were found at the rear of his shop; they were auctioned off by his heirs.

painters ground and prepared their own materials. By the late 18th century, scientific and industrial production methods had begun to distance painters from an understanding of their paint. By the second half of the 19th century, the industrialization of the painter's craft was complete.[14]

This parting of the ways led to a practice already rife among vendors of food and cosmetics – fraudulent adulteration. Divorced from preparing their own materials, artists, too, were at the mercy of unscrupulous colormen who not only added, but sometimes completely substituted, cheaper pigments and media for the genuine article. For example, by the mid-19th century, artists were being duped into buying a spurious "Naples yellow" made of inexpensive cadmium yellow mixed with lead or zinc white, in contrast to the genuine article, lead antimonate, $Pb_2Sb_2O_7$ (**Chapter 9.3.2**). Resins and oils suffered the same fate.[27] Such practices inadvertently caught art forgers in their trap. When the Dutch forger of Vermeers (1632–1675), Han van Meegeren (1889–1947), was brought to trial, one of the proofs of his perfidy was the presence of cobalt blue in one of his purported "genuine Vermeer paintings."[28] In this case, the con man was hoodwinked in turn: he had paid full price for genuine ultramarine to lavish on Christ's robe in *Christ with the Woman Taken in Adultery*. What he got from his dodgy color dealer was an adulterated sample containing cobalt blue, a pigment not available in Vermeer's time.

By the mid-20th century, the United States and many European countries had passed legislation regulating adulteration of food and cosmetics (**Chapter 15.4**); unhappily, artists' pigments were not included in the list. The standards that exist are voluntary. Ideally all tubes of oil paint should display not only the *Colour Index* names and number, but the common color name, the percentage of pigment to oil, the specific drying oil, and waxes or fillers used,[29] but this is seldom the case. It should be noted that tubes labeled as "hue" colors such as Alizarin Crimson Hue or Sap Green Hue are rarely single colors, but unspecified mixtures. It is also useful to distinguish among pigments (colored powders), paints (pigments mixed with liquid media) and colors (marketing names given to paints).[30]

When possible, it is always better to purchase single pigment paints. However, not even they were all advisable, as we have seen with respect to Winslow Homer's choices. Jehan-Georges Vibert (1840–1902), a "minor artist" who had "little patience for the new Impressionist style of painting,"[31] played the role of watchdog by making weighty recommendations for stable artists' colors. For many years a leading member of the commission in charge of art restoration in the museums of France, he predicted that "many pictures of the present day will fade into insignificance before they are 50 years old."[32] His short list of 13 stable recommended colors (four of which are traditional; nine are the products of nineteenth-century technology)[11] is accompanied by a much longer list of bad, or very bad, colors[33] (Table 13.1). The ones that merit unqualified condemnation are the new aniline colors along with Prussian Blue and most lead, arsenic and copper compounds.

Of the acceptable pigments, the first four are traditional pigments used from antiquity; the remaining nine were all products of 19th-century technology and were underwritten as permanent by scientific investigation.[11]

Table 13.1 Vibert's list of acceptable and unacceptable artists' pigments.

Acceptable pigments	Unacceptable pigments
Earths, ochres, marls (oxides of iron, if properly labeled)	All copper carbonate compounds such as verdigris and malachite
Lead white (basic lead carbonate) if used with caution	All other lead-containing pigments such as minium (red lead) and lead iodide
Vermilion (mercury(II) sulfide, HgS)	All arsenic compounds such as orpiment, realgar, and the arsenic-containing greens
Madder **lakes**	Other **organic** reds such as carmine cochineal
Zinc white (zinc oxide, ZnO)	Prussian Blue
Strontium chromate	Barium chromate (lemon yellow) and calcium chromate
Cadmium yellow (cadmium sulfide)	Chrome yellow (lead chromate)
Cobalt blue (cobalt aluminate)	Egyptian blue (called "useless")
Cobalt green (cobalt oxide + zinc oxide)	Smalt
Chromium oxide green	All vegetable yellows such as gamboge
Cobalt violet (cobalt phosphate + silicate)	Brazilwood
Mineral violet (manganese phosphate)	Campeachy
Synthetic ultramarine	All aniline colors

Other pigments that regularly graced the Impressionist palette were zinc white (zinc oxide), Naples yellow (lead antimonate), chrome orange (basic lead chromate), Scheele's green (copper arsenite), emerald green (copper acetoarsenite), viridian (hydrated chromic oxide), cerulean blue (cobalt stannate), chrome green (Prussian Blue + chrome yellow), and occasionally ivory black (boneblack) even though Vibert proscribed some of them. Vauquelin's iodine scarlet (HgI_2) glowed with a matchless brilliance, but was recognized as unreliable from the start.[34] We'll take a look at how Monet used these pigments in the next section.

13.6 THE IMPRESSIONIST PALETTE IN PRACTICE

Claude Monet's use of some of these pigments, both approved and condemned, illustrates a typical Impressionist approach in Figure 13.3.[35] His palette is not very extensive. The subject matter is evanescent – travelers and trains come and go. Although this painting, one of four painted by Monet of the interior of Paris's Gare Saint-Lazare, seems to be out of step with the *plein air* Impressionist technique, only the painter himself is under a roof in this scene. What is really odd is the subject matter: we are used

Figure 13.3   Claude Monet. The Gare Saint-Lazare, 1877. Oil on canvas. National Gallery, London. Reproduced from ref. 35.

to seeing scenes in parks and on water or in gardens and forests, but not the spectacle of engine clangor, hissing steam, and disorderly crowds – all part of the 19th century industrial revolution. But, as art historian Ashok Roy remarks, it is certainly a "challenging subject to explore the technical limits of the new method of painting."[36]

Indeed, Monet stretched these limits to the full: he requested that all the trains be halted while he painted. Judging from the **energy dispersive X-ray spectroscopy** data and the way the pigments were handled, he recorded the scene very rapidly, no doubt under the hostile glare of the station personnel. Monet used eight principal pigments, half of which are on Vibert's acceptable list: lead white, cobalt blue, synthetic ultramarine and vermilion. The others are cerulean blue, emerald green, viridian and a red **lake** pigment. What is remarkable, given the dark tone of the painting, is the virtual absence of black pigment. (Only one or two stray black particles were found and identified as possibly ivory black in some samples.) This is entirely in line, however, with the Impressionist "campaign" to eliminate black from the palette ("There is no black in nature," said they) as well as all earth colors (not bright enough). Monet achieves his "optical blacks" by intricate mixtures of cerulean blue, cobalt blue, ultramarine, viridian and emerald green, dropping in traces of lead white, red **lake** and vermilion. The light mauve of the steam is chiefly lead white with minor amounts of cobalt blue and vermilion, and traces of cerulean blue, viridian, red **lake** and possibly a particle or two of ivory black. Cerulean blue was only newly available in 1877; most of the blue paint samples contain it in mixtures with ultramarine and cobalt blue.[37] Given the evidence that most of these mixtures were done directly on the canvas itself, Monet's rapidity of execution of complex paint mixtures places him in a virtuoso class of experimentalists.[36]

Impressionism as a movement was but a fleeting instant on the art historical trajectory, but its influence was both lasting and profound. As thousands of new synthetic colors began to flood the market toward the end of the 19th century, artists began to take advantage of the spectral expansion they afforded and, predictably, indiscriminately.

13.7 BEYOND IMPRESSIONISM: EVOLUTION OF THE PALETTE

The Victorian era saw color history turned on its ear by the advent of William Henry Perkin's mauve at the very moment that Europe was experiencing massive industrial expansion. Taken together with the paradigm shift in the way the world was viewed, the new coal tar colors gradually superseded traditional artists' colors. With a groundbreaking discovery of **azo** pigments in 1884, there was no turning back.[38] Traditionally, **organic** pigments had to be co-precipitated with an inorganic substrate to form a so-called **lake** pigment – otherwise, they were too transparent to paint with. But the salts of azo colorants, those containing the azo group, –N=N–, did not require laking – they could be used as standalone pigments. Thus, the first azo pigment, tartrazine yellow (Pigment Yellow 100), patented in 1884 and still used in artists' paints, took a bow. From the early 1970s, about half of the market share of artists' pigments consisted of synthetic **organic** pigments (SOPs) and of those, about 80% were azos, largely because they have a wide color range and are easy to prepare. Furthermore, most azos are nontoxic and suitable as additives in food and cosmetics. However, most of them are light-sensitive and many are sensitive to solvents, important clues for art conservation protocols (Box 13.2).[39]

In addition to azo pigments, as of the mid-2000s, almost 20 classes of SOPs were on the market. An SOP is a manufactured colorant that has no counterpart in nature. It is a true pigment, not a dye, characterized by a carbon ring structure with mobile electron clouds, called a **chromophore**, that imparts the color. Other atomic groupings may be present that can change the appearance of the primary color; these are called auxochromes. SOPs now dominate the colorant market, offering the entire spectral range of hues. Their applications, aside from artists' paints, include industrial, architectural and automotive paints, and colorants for plastics, rubber, textiles and printing. The azo pigments took the lead between 1890 and 1910, followed by the phthalocyanines in the 1930s (**Chapter 11.5**), the **quinacridones** in the 1950s and high performance azo pigments in the 1960s.[43]

Box 13.2   Titanium White

While discovery of dozens of SOPs was moving apace, the first quarter of the 20th century witnessed the discovery of a new white **inorganic** pigment, titanium white (titanium dioxide, TiO_2, or any white pigment containing a compound of titanium). White is not considered a "glamourous" pigment like luscious iodine scarlet that can excite like no other red ever could. The traditional darling of all white pigments, lead white, fell out of favor because of its toxicity, and titanium white was a welcome substitute. The element, titanium, was discovered by the Rev. William Gregor (1761–1817) in 1791 and named in 1795 by **Martin Heinrich Klaproth** (1743–1817), who recognized it in its oxide. However, although titanium oxide exists in nature (actually in three crystalline forms), it was never exploited as a pigment because it was highly contaminated with impurities. The child of early 20th century chemical technology, a viable process enabled commercial production of titanium oxide by 1918.[40] Nontoxic, opaque and stable, it is now an $18 billion business and the most widely used pigment of all time except for carbon black. It can be found in rubber, plastics, enamel, glass, ceramics, inks, textiles, cosmetics, leather, paper, and in artists' pigments. Among the first artists to incorporate it into their works were Jean Arp (1886–1966) by 1924 and Georgia O'Keeffe (1887–1986) by 1928.[41] "It took a long time for people to figure out how to mass-produce the stuff, but once they did, it changed the world."[42]

The **azo pigments** frequently found in artists' pigments are pyrazolone, naphthol, β-naphthol, arylide (Hansa), diarylide, nickel-azo yellow and benzimidazolone.[39] Phthalocyanine pigments go under many manufacturers' aliases including Sap green, Peacock Blue and Permanent Hooker's Green; their paints are sold under trade names like Fastogen, Hostaperm, Irgalite, Thalo, Monastral and Sunfast. **Quinacridones** are sold in artists' paints under names like Quinacridone Violet, Permanent Magenta, Mauve Red and Cobalt Violet.[44] Note that these names are not descriptive of the contents but of quality or hue. It is always best to purchase pigments *via* their *Colour Index* (CI) numbers (**Chapter 1.7.3**) to know what you are really getting. A handy database is the group of 170 organic pigments relevant to artists' paints collected from historic collections and modern manufacturers that were characterized in 2009. Flow charts based on color, pigment class, group and individual CI numbers were presented to help identification of unknowns and mixed paint samples.[45]

The last major group of pigments to be introduced in the 20th century was the DPP (diketopyrrolopyrrole) pigments, discovered at Ciba-Geigy in 1974.[46] After a highly secret production route had been established, they were patented in 1983.[47] They have a red-orange hue range and have heat, bleed and chemical resistance which make them desirable despite their high price.[38] The most important DPP pigment is CI Pigment Red 254.

13.8 IMPRESSIONISM'S LEGACY

Expansion, discovery and invention were the key words, applied to both style and palette, of the post-Impressionist era artists. The first of these, often called the "first post-Impressionist," was Paul Cézanne (1839–1906), the father of cubism, the integral link to where modern art begins.[8] He paved the way but always kept one foot in the traditional past. This was especially true of his palette which did not differ in any way from that of the Impressionists.

Another household word of the early post-Impressionist era is Vincent van Gogh. His prolific output of paintings plus his equally prolific documentation *via* frequent letters make him an ideal artist to study regarding pigment usage and degradation. The urgency and intensity of van Gogh's style, soon to be called Expressionism, was matched by his early use of the newly available SOPs. One very exuberant red, eosin, beckoned to him, and by 1907 had betrayed him by fading badly. Figure 13.4 shows an example of an original van Gogh (above)[48] in which the eosin has faded badly, and a partially reconstructed image (below) based on analytical data. Van Gogh made multiple orders for eosin paint under the commercial name of *laque géranium*, or geranium **lake**, as soon as he arrived in Arles and continued to use it until the end of his life.[49] His use of other red lakes, such as cochineal and brazilwood, fared just as badly as documented by three major museums that hold his works.[50] The fading of these pigments has been attributed to oxidation of –OH groups on adjacent carbon atoms (catechol groups), transforming the molecules into colorless species. Knowledge of this mechanism is helpful in determining protocols for conservation or restoration.[51]

The red lakes aren't the only pigments that two-timed van Gogh. Red lead, or minium (Pb_3O_4) whitened in many of his works, but only since the use of X-ray powder diffraction tomography (see **X-ray diffraction** in the Glossary) were they able to determine that an exotic lead mineral, plumbonacrite

Figure 13.4   Comparison of Vincent van Gogh's Undergrowth with Two
Figures, oil on canvas, 1890, Cincinnati Art Museum (above)
with a merged reproduction (below) in which the left half is
unrestored and the right half is digitally restored. The digital
reconstruction rejuvenates a pink hue to 38% of the white flowers. Please see ref. 49 for details. Top: Reproduced from ref. 48,
Bottom: Image courtesy of Jeffrey E. Fieberg and Gregory D.
Smith.

$(3PbCO_3 \cdot Pb(OH)_2 \cdot PbO)$, forms on the surface of the minium
prior to the formation of the more common lead degradation products, hydrocerussite $(2PbCO_3 \cdot Pb(OH)_2)$ and cerussite
$(PbCO_3)$. This analysis allowed chemists to trace the progress of
the photochemical degradation of minium, an important clue
for conservation.[52] Another tricky pigment that changed the
appearance of van Gogh's sunflowers forever was yellow lead
chromate, $PbCrO_4$, that turns brown when the hexavalent chromium is reduced to Cr^{3+}.[53]

An early Expressionist, Edvard Munch (1863–1944), famously
expressed better than anyone else how World War II engulfed
humanity by his *The Scream*. But now the yellow cadmium sulfide
in the 1910 version at the Munch Museum in Oslo is degrading

to colorless cadmium sulfate. An international team of conservationists have found that the degradation is not light-induced but as yet there is no conclusive explanation.[54]

Cubism was crucial to the development of other art forms like abstract art, Dadaism and surrealism. In addition to the many new SOPs available, artists could choose among a range of new media such as **acrylics**, enamels, oil pastels, aerosol sprays, wax crayons, felt tipped pens and colored pencils. Many of the SOPs themselves are highly fugitive, as dramatically exemplified by Mark Rothko's (1903–1970) use of Lithol Red (Pigment Red 49), one of the earliest SOPs, for his Harvard and the Tate's Seagram Murals.[55]

In 2015, a team of chemists and conservation scientists performed a multi-analytical study on the photochemical degradation of some selected SOPs shown in Table 13.2 by artificial aging techniques.[56]

Table 13.2 List of selected synthetic organic pigments rated for lightfastness. Adapted from ref. 56 with permission from Elsevier, Copyright 2015.[a]

Colour index usage number	Manufacturer	Pigment class or name	ASTM lightfastness
Pigment Blue 1	EC Pigments	Triarylcarbonium/ Victoria Blue	V (gouache)
Pigment Orange 5	Winsor & Newton	B Naphthol/Azo Orange	II (oil)
Pigment Orange 16	Sun	Diarylide Yellow	V (gouache)
Pigment Orange 46	Sun	B Naphthol Ba/ Sunset Yellow	IV
Pigment Red 3	Clariant	B Naphthol Toluidine	IV (oil)
Pigment Red 53 : 1	Lansco	B Naphthol Ba/**Lake** Red	V (oil)
Pigment Red 83	Winsor & Newton	Anthraquinone/ Alizarin Crimson	III (oil)
Pigment Red 170	Winsor & Newton	Naphthol Red AS	II (oil)
Pigment Violet 1	Magruder	Rhodamine Violet	V (gouache)
Pigment Yellow 3	Golden	Monoazo Yellow	II (oil)
Pigment Yellow 100	H. Kohnstamm	Monoazopyrazolone Al	V (watercolor)

[a]ASTM (American Society for Testing and Materials) Lightfastness Categories: I: Excellent, II: Good, III: Fair, IV: Poor, V: Very Poor.

This study was the first systematic and comprehensive survey on the lightfastness properties of some pure synthetic **organic** artists' pigments. The team evaluated color change *via* colorimetric measurements and rationalized some of the degradation pathways using pyrolysis/**gas chromatography/mass spectrometry (GC/MS) (pyr/GC/MS)**. Since many of these pigments are extensively utilized in paint formulations, these results are of the utmost importance for artists, manufacturers, conservators and anyone involved with cultural heritage and preservation of works of art.

Artists of today have a much greater selection of materials as well as a growing wealth of information regarding the qualities of their pigments. As a result, art conservationists have a much more difficult job keeping up with them. We have seen that the innovative spirit of the Impressionists together with the skill of synthetic chemists enabled the evolution of this prolific and exciting world of modern art to enrich our lives.

13.9   CONCLUSION

Judging from the crowds that throng the Impressionist galleries of almost every major museum Sunday after Sunday, the public has ruled that Impressionism is good art. And if good art, it plays the role of the cosmic myth – it gives us a sense of meaning, of belonging, and enables us to take a personal part in it. But cosmic myth has a geography, one that places gardens and forests at the center of consciousness. The author of *Genesis* locates human origins in a garden; Robert Frost[57] marks the forest a Valley of Decision (Joel 3 : 14). In the next chapter, we will see that the denizens of the forest, in their struggle for survival, provide us with some legendary pigments and a pathway to major scientific breakthroughs through "color writing."

REFERENCES

1. A. Terrasse, *Bonnard: Biographical and Critical Study*, World Publishing, Cleveland, 1964, pp. 76–78.
2. L. Mueller, *Monet Refuses the Operation, Alive Together: New and Selected Poems*, Louisiana State University Press, Baton Rouge, 1996, pp. 186–187.

3. Claude Monet: Rouen Cathedral Portal, https://commons.wikimedia.org/wiki/File:Claude_Monet_-_Rouen_Cathedral,_West_Façade_-_Google_Art_Project.jpg, accessed June 2021.
4. Claude Monet: The Houses of Parliament Sunset, https://commons.wikimedia.org/wiki/File:Claude_Monet_-_The_Houses_of_Parliament,_Sunset.jpg, accessed June 2021.
5. C. Baudelaire, *The Mirror of Art*, Doubleday Anchor Books, New York, 1956, p. 209.
6. C. Lloyd and R. Thomson, *Impressionist Drawings: from British Public and Private Collections*, Phaidon Press, Oxford, 1986, p. 9.
7. Édouard Manet: The Fife, https://commons.wikimedia.org/wiki/File:Manet,_Edouard_-_Young_Flautist,_or_The_Fifer,_1866_(2).jpg, accessed June 2021.
8. As quoted in J. Barnes, *Keeping an Eye Open*, Knopf, New York, 2015, p. 75.
9. Édouard Manet: Boating, https://commons.wikimedia.org/wiki/File:Edouard_Manet_Boating.jpg, accessed June 2021.
10. H. W. Janson and D. J. Janson, *The Picture History of Painting*, Harry N. Abrams, New York, 1957, pp. 249–251.
11. D. Bomford, J. Leighton, J. Kirby and A. Roy. *Impressionism: Art in the Making*, National Gallery, London, 1991.
12. L. C. Perry, *The American Magazine of Art*, 1927, **18**(3), 119–126.
13. J. Renoir, *My Father*, Collins, London, 1962, p. 73.
14. J. Ayres, *The Artist's Craft: a History of Tools, Techniques, and Materials*, Phaidon Press, New York, 1985.
15. J. H. Townsend, *Stud. Conserv.*, 1993, **38**, 231–254.
16. O. N. Rood, *Students' Text-book of Color: or, Modern Chromatics, with Applications to Art and Industry*, D. Appleton, New York, 1881, pp. 243–244.
17. R. G. Kuehni and A. Schwarz. *Color Ordered: A Survey of Color Order Systems from Antiquity to the Present*, Oxford University Press, New York, 2008, pp. 84–85.
18. J. C. Webster, *Coll. Art J.*, 1944, **IV**(1), 3–22.
19. J. Laforgue, *L'Impressionisme, review of an exhibition at the Gurlitt Gallery, Berlin*, 1883, reprinted in ed. L. Nochlin, *Impressionism and Post-impressionism 1874–1904: Sources and Documents*, Prentice-Hall, Englewood Cliffs, 1996, pp. 16–17.
20. L. Goodrich and J. W. Beatty, *Winslow Homer,* Macmillan, New York, 1944, p. 222.

21. B. Berrie, F. Casadio, K. Dahm, Y. Strumfels, M. Tedeschi and J. Walsh, A Vibrant Surface: Investigating Color, Texture and Transparency in Winslow Homer's Watercolors, in *Science and Art: the Painted Surface*, ed. A. Sgamellotti, B. G. Brunetti and C. Miliani, Royal Society of Chemistry, Cambridge, 2014, pp. 404–428.

22. M. Pastoureau, *Green: the History of a Color,* Princeton University Press, Princeton, 2014, p. 188.

23. R. Haug, *Dtsch. Farben-Z.*, 1951, **5**, 343–348.

24. C. Blanc, *Grammaire des arts du dessin*, Librairie Renouard, Paris, 1867.

25. B. Berrie, The Alchemy of Artists: From Pigments to Painting, *14th Color and Imaging Conference Final Program and Proceedings*, Society for Imaging Science and Technology, Springfield, VA, 2006, pp. 273–273.

26. As quoted in H. B. Chipp, *Theories of Modern Art: a Sourcebook by Artists and Critics*, University of California Press, Berkeley, 1968, p. 33.

27. L. Carlyle, *The Conservator*, 1993, **17**, 56–60.

28. F. Wynne, *I Was Vermeer: the Rise and Fall of the Twentieth Century's Greatest Forger*, Bloomsbury, New York, 2006, p. 183.

29. S. Saitzyk, *Art Hardware: The Definitive Guide to Artists' Materials*, Watson-Guptill, New York, 1998, p. 163.

30. ASTM (American Society for Testing and Materials), www.astm.org. A search under the topic of "Artist paints and art materials" yields 67 items, accessed March 2021.

31. E. M. Zafran, *Cavaliers & Cardinals: Nineteenth-century French Anecdotal Paintings*, Taft Museum, Cincinnati, 1992, p. 21.

32. A. Abendschein, *The Secret of the Old Master*, Appleton, New York, 1909, p. 34.

33. J. G. Vibert, *The Science of Painting*, Percy Young, London, 1892, pp. 162–172.

34. R. D. Harley, *The Conservator*, 1987, **11**(1), 45–60.

35. Claude Monet: The Gare St-Lazare, https://commons.wikimedia.org/wiki/File:Claude_Monet,_The_Gare_St-Lazare,_1877.jpg, accessed June 2021.

36. A. Roy, *National Gallery Technical Bulletin*, 1985, vol. 9, pp. 12–20.

37. D. Bomford and A. Roy, *National Gallery Technical Bulletin*, 1983, vol. 7, pp. 13–19.

38. H. Skelton, *Rev. Prog. Color.*, 1999, **29**, 43–64.

39. B. Berrie and S. Q. Lomax, *Stud. Hist. Art*, 1997, **57**, 8–33.

40. N. Heaton, *J. R. Soc. Arts*, 1922, **70**(3631), 552–565.

41. M. Laver, Titanium dioxide whites, in *Artists' Pigments: A Handbook of Their History and Characteristics*, ed. E. W. FitzHugh, National Gallery of Art, Washington, DC, 1997, vol. 3, pp. 295–355.

42. A. Rogers, *Full Spectrum: How the Science of Color Made Us Modern*, Houghton Mifflin Harcourt, New York, 2021, p. 110.

43. S. Q. Lomax and T. Learner, *J. Am. Inst. Conserv.*, 2006, **45**(2), 107–125.

44. S. Q. Lomax, *Stud. Conserv.*, 2005, **50**(Supp. 1), 19–29.

45. N. C. Scherrer, S. Zumbuehl, F. Delavy, A. Fritsch and R. Kuehnen, *Spectrochim. Acta, Part A*, 2009, **73**, 505–524.

46. A. Iqbal, A. C. Rochat, J. Pfenninger and O. Wallquist, *J. Coat. Technol.*, 1988, **60**, 37–45.

47. Ciba-Geigy, *US Pat.*, 4415685, 1983.

48. Vincent van Gogh: Undergrowth with Two Figures, https://commons.wikimedia.org/w/index.php?title=File:Vincent_van_Gogh_-_Undergrowth_with_Two_Figures_(F773).jpg&oldid=391570017, accessed June 2021.

49. J. E. Fieberg, P. Knutås, K. Hostettler and G. D. Smith, *Appl. Spectrosc.*, 2017, **71**(5), 794–808.

50. S. A. Centeno, C. Hale, F. Carò, A. Cesaratto, N. Shibayama and J. Delaney, *et al.*, *Herit. Sci.*, 2017, **5**, 18–29.

51. G. Zhuang, S. Pedetti, Y. Bourlier, P. Jonnard, C. Méthivier and P. Walter, *et al.*, *J. Phys. Chem. C*, 2020, **124**, 12370–12380.

52. F. Vanmeert, G. Van der Snickt and K. Janssens, *Angew. Chem., Int. Ed.*, 2015, **54**(12), 3607–3610.

53. S. Everts, *Chem. Eng. News*, 2016, **94**(5), 32–33.

54. B. Halford, *Chem. Eng. News*, 2020, **98**(19), 7.

55. H. A. L. Standeven, The History and Manufacture of Lithol Red, a Pigment Used by Mark Rothko in his Seagram and Harvard Murals of the 1950s and 1960s, *Tate Papers 10*, Autumn, 2008, https://www.tate.org.uk/research/publications/tate-papers/10, accessed January 2021.

56. E. Ghelardi, I. Degano, M. P. Colombini, J. Mazurek, M. Schilling, H. Khanjian and T. Learner, *Dyes Pigm.*, 2015, **123**, 396–403.

57. R. Frost, *The Road Not Taken*, 1915.

CHAPTER 14

The Forest Primeval: Arboreal Bounty[†]

What was this forest savage, rough, and stern,

Which in the very thought renews the fear....;[1]

Dante

14.1 INTRODUCTION: THE KILLER TREE

How is it that the Forest of Mirkwood struck fear into the hearts of Bilbo Baggins and his fellow Hobbits? We don't have to travel to Middle Earth to find out.[2] We need only look at the fright provoked when people in the Middle Ages encountered one forest denizen, the walnut tree. As one hapless poet (Yours Truly) expressed it:

The presence of the walnut conjures dread.

Look! Round about it all the plants lie dead!

[†]Glossary entries and chapter cross-references are in boldface.

March of the Pigments: Color History, Science and Impact
By Mary Virginia Orna
© Mary Virginia Orna 2022
Published by the Royal Society of Chemistry, www.rsc.org

The folks back then were not dreaming. Although walnut bark and roots gave a better black dye than any other plant, people resisted using it. The concurrence of popular belief and botanical understanding was due to the evidence: not only did the tree kill all the vegetation around it, but it was said that livestock, too, sickened and died if their cowsheds lay too near. People claimed that sleeping under a walnut tree led to fever, headache, evil spirits, and even a visit from the devil himself. In Italian, the word for walnut, "noce," has its roots in "nuocere," to do harm. Modern chemists have enlightened us that the brooding, mortal aura of the walnut was not a medieval superstition. The tree may not have entertained Satan, but it did and does exude a substance that, in quantity, could indeed produce most of the effects observed.[3] Identified in the 1880s, the volatile molecule juglone, $C_{10}H_6O_3$, is the culprit. All members of the walnut family biosynthesize it, but the black walnut produces it in the greatest concentration. A close relative of lawsone, the coloring matter in henna (**Chapter 3.6.2**), it is sometimes used as hair pigmentation and bears a *Colour Index* number of Natural Brown 7.

14.2 CHEMICAL SELF DEFENSE

Why do walnut trees produce juglone? It is simply one of the many chemical self-defense mechanisms that plants and animals have evolved over the eons. We have seen "chemical warfare" at work in the murex shellfish (**Chapter 6.2.2.2**) and the cochineal insect (**Chapter 10.3.1**). This strategy is quite widespread among pigment-producing trees of the forest as well. Since trees, like other large perennials, are easy prey for a host of attackers, they have evolved chemical defenses that are present throughout their lives,[4] as we shall see in the following examples.

14.2.1 Dragon's Blood

Another group of trees associated with menace and peril is the resin- and latex-producing genera *Croton*, *Dracaena*, *Daemonorops* and *Pterocarpus*. They have all been linked with the myth of a dragon-like basilisk and elephant mixing their blood in a death-struggle. The resulting red liquids came to be used as a highly prized set of pigments, all seemingly identical but of very different chemical compositions. The basilisk itself gained literary fame as the denizen that lurked in the chamber of secrets lying beneath the Hogwarts dungeons.[1] The dragon tree itself was regarded as the epitome of exoticism among Renaissance artists. Since, along with the baobab tree, it was thought to be the oldest known representative of the vegetable kingdom,[5] artists have taken it to be a symbol of eternity.[‡] It shows up in Albrecht Dürer's (1471–1528) woodcut of *The Flight into Egypt* (1502)[6,7] and in Hieronymus Bosch's (1450–1516) *The Garden of Earthly Delights* (1504).[8] In the Bosch version, the tree is most certainly the *Dracaena cinnabari*, native to the Canary Islands, doubling in the image as Paradise's Tree of Knowledge of Good and Evil[9] (Figure 14.1).

Figure 14.1   Left: Albrecht Dürer, The Flight into Egypt, Woodcut, 1502, National Gallery of Art, Washington, DC. Reproduced from ref. 7 under the terms of a CC0 1.0 license, https://creativecommons.org/publicdomain/zero/1.0/deed.en; Right: H. Bosch, The Garden of Earthly Delights, left-hand panel, oil on oak, 1504, Museo del Prado, Madrid. Reproduced from ref. 8. The dragon tree is clearly depicted on the left-hand side in both works.

[‡]Since the Renaissance, many older vegetable species have been identified, such as bristlecone pine tree dated to be 5000 years old.

The dragon trees, as a group, secrete resinous materials that go under the names of resins, latexes and saps as forms of the many defense mechanisms that protect plants against a variety of pathogens and insect invasions. Depending on the species, the red exudate is produced by the stem, leaves or fruit, sometimes taking the form of drops or chips. Secretion always takes place as a response to induced trauma, which is essential for its formation. Dragon's blood, sometimes known as Indian cinnabar, is also sticky allowing it to entrap small organisms in its toxic embrace.[10] At least 100 chemical constituents have been isolated from the various species of dragon's blood trees. The major types of compounds comprise sterols, lignans, **terpenoids**, flavonoids, **carotenoids** and **aromatic** compounds.[11] Many of them have antimicrobial activity and have been used as folk medicines since antiquity. The exudates offer huge therapeutic potential for purified compounds; some of them have gone on to clinical trial evaluations at modern pharmaceutical firms. The dragon trees seem to have no other phylogenetic relationship to one another other than belonging to the major plant group, Angiosperms. The four genera are described in Table 1 in ref. 12.[12]

In the ancient world, the crimson red **resin** was coveted not only for its medicinal properties but for its beauty. The major red colorants in *Daemonorops draco* are **dracorhodin** and **dracorubin**, with other minor constituents also contributing to the color. It was used to illuminate manuscripts,[5] varnish violins and furniture, stain glass and marble, dye wool, paint plaster and decorate pottery. Since red in China symbolizes luck, joy and happiness, the crimson resin has been used to color the surfaces of paper and posters especially for weddings and for Chinese New Year. Cosmetically, it finds use in lipsticks, toothpaste, body oils and incense. In modern times, it has been used as a resist in photoengraving.[12]

Art restorers and archaeologists have been intensely interested in the data gleaned from chemical identification of the pigments in dragon's blood. In art restoration work, it is vital to exactly pinpoint the compounds in an artifact or painting in order to find suitable replacements. In addition, it is possible to trace trade routes by tracking the spread of plants through documented use of their resins. **Raman spectroscopy** has turned out to be the instrument of choice in these determinations.[13,14]

14.2.2 Sandarac

The term "dragon's blood" tends to be given to any red resinous substance, and the case of sandarac is no exception. One source, calling it "cinnabar of the Indies,"[15] identifies it as a resin from the fruit of the rotang, or Asian palm, otherwise called *Calamus draco*. Wikipedia includes *Calamus rotang* on its dragon's blood page[16] but does not identify it with sandarac. Laurie[17] says that it is a **resin** from the *Callitris quadrivalvis* native to the north African coast and once exported to Europe in quantity from Berenice (now Benghazi in Libya), mistakenly called juniper resin. Gettens and Stout[18] concur with Laurie, as does Wikipedia, placing sandarac on the *Tetraclinis* page[19] and indicating that *Callitris* is a synonym. To confuse matters further, the tree is the national tree of Malta, where it is called *għargħar*. The one common denominator is that sandarac is a red resin that has its origin in a tree. There is evidence that resin production may be triggered by a traumatic event: the tree's burls, arising always as a defense against mold or insect infestation, have a high resin content.[20]

In 2017, a research group from the Centre nationale de la Recherche scientifique (CNRS), Université de Paris VI,[21] determined the chemical identities in samples of sandarac **resin** purchased from six commercial suppliers. Aware that Leonardo da Vinci had extolled this resin, but cautioned that it must be taken and used fresh from the cypress tree (*Tetraclinis articulata*) in the spring, they had to be satisfied with aged samples that may have polymerized sitting on the shelf. They found that the resin contained six compounds belonging to the diterpenoid family present in all of the samples, but differing widely in quantity from supplier to supplier. Paclitaxel, an anticancer agent, is also a close relative – a member of the same diterpenoid family containing isoprene (C_5H_8) building blocks. In the next section, we will see another example of the chemical ingenuity of trees.

14.3 OAK APPLE INK

Apples play a role in evil intent in the Garden of Eden and the story of Snow White. However, the oak apple is not only benign but, with a little chemical manipulation, it can also produce a superb writing ink. But its origin, like dragon's blood and sandarac, is self-defense. This time, when certain flies, wasps and aphids lay their larvae on the surface of an oak leaf, the chomping worms cause secretion of an irritant which in turn stimulates the tree to form a spherical growth around the site of parasitic development as the larvae mature to adulthood. These growths, variously called oak apples, oak galls, marble galls or Aleppo galls, have distinct sizes and shapes, but one important characteristic in common: they all have a high tannin concentration.[22]

Tannins are ubiquitous in plants, sometimes comprising up to 50% of the dry mass of tree leaves. Tannins in plants evolved as protection against predation: they have a bitter, astringent taste and are hard to digest. Dietary sources are tea and coffee; they are also present in wines, particularly reds, arising from the skins, stems and seeds of the grapes. They are polyphenols made up of building blocks of gallic acid, $C_6H_2(OH)_3COOH$, with molecular weights ranging from 500 to 3000. Historically, **gallic acid** derived from oak galls, became the basis of a virtually ideal permanent writing ink for almost 1000 years.

Ink had to be invented once people moved away from awkward cuneiform clay tablets and began to use papyrus, and later paper, as a writing surface. Tomb records tell us that the ancient Egyptians used a fine black ink made of soot or lampblack. Ditto for the inkwells discovered in Pompeii (**Chapter 7.4.1.5**). Pliny tells us that a black product from gall nuts was known from the 3rd century BCE and, in fact, scientists at the Louvre have identified "encre métallogallique" in papyri dating from as early as 250 BCE.[23] The first Western document containing it is the Vercelli Gospels, from the 4th century CE.[24] The galls had to be imported from Asia; Aleppo galls from Syria were the best and largest. As with all imported commodities, this raised the price enormously. The second ingredient was green vitriol, ferrous sulfate, $FeSO_4 \cdot 7H_2O$, often manufactured by treating iron with sulfuric acid. Recipes for extracting the tannins from the gall nuts varied but there was always a heating and a fermentation step because the desired end product was **gallic acid**. Simple as it may sound, mixing gallic acid with ferrous sulfate produced the iron gall ink (IGI) with a little patience and some gum Arabic to make it more fluid.[25] Here is a poetic "recipe" which may be difficult to reproduce:[26]

To taste the flavor of iron gall ink history thy taketh ...

...one gall wesp on a mediterranian oak tree twig

3000 miles of gum Arabic trade routes

take likewise 5 ounces of mint-green vitriol from a deep dark mine

whereinto put 1 pint of ancient rain water.

Choose a moment in time and a thought to convey.

Mix and stir them every day three or four times thoroughly.

Finish off with a pinch of blotting sand.

Wait a few centuries and discover a testimony of history.

Nowadays, simply buy a pound of wine tannins online and some ferrous sulfate from the garden shop. Stir equal quantities with some water and wait a half-hour or so. A dark precipitate emerges and with a little refinement, you can produce calligraphy-quality permanent black ink (Figure 14.2).[27] The word "permanent" is important: the ink is very hard to remove from any surface, especially from mom's good carpet or stainless steel sink. A simpler recipe is to brew some "real" (non-herbal) tea and drop in an iron supplement tablet from your medicine chest. Teas contain

Wine   Ferrous sulfate
Tannin   heptahydrate

Figure 14.2   Left: Iron gall ink ingredients; Right: Black iron gall ink complex.

tannins and will react immediately with the ferrous ion in the tablet and voilà! History recreated in an instant! The chemical reaction is:

$$2Fe^{2+} + 2C_7O_5H_6 + \tfrac{1}{2}O_2 \rightarrow 2Fe(C_7O_5H_3) + H_2O + 4H^+ \quad (14.1)$$

$$\underbrace{\phantom{2Fe^{2+}}}_{\text{Ferrous ion}} \; \underbrace{}_{\text{Gallic acid}} \qquad \underbrace{}_{\text{IGI complex}}$$

Using **X-ray photoelectron spectroscopy** (XPS), chemists have shown that the insoluble IGI precipitate is composed almost exclusively of Fe(III), intact **gallate ligands**, ($C_7O_5H_3$), and water, with no detectable Fe(II) component. This result shows that the deep blue-black IGI color initially arises from charge transfer bands but is not with the charge transfer type that occurs in Prussian Blue (**Chapter 11.3.2**).[28]

Due to its permanence, intense black color and ease of use, IGI quickly gained acceptance throughout Europe and became the most common writing ink in the Western world for a millennium.[29] It is the ink on some of the most important documents in history such as the Magna Carta, the US Declaration of Independence and the Emancipation Proclamation. It is contained in Ludwig van Beethoven's (1770–1827) original scores and in John Milton's (1608–1674) Bible inscribed in his own hand.[28,30]

While documents written with IGI have lasted for centuries, there is no guarantee that they will last forever. In fact, there are some disconcerting instances of corrosion largely due to any Fe^{2+} present oxidizing to Fe^{3+} and producing harmful peroxides in the process. The peroxides propagate, abstract protons from the parchment or cellulose substrate, and cause discoloration and mechanical weakening (Figure 14.3).[31] The only known effective treatment is with phytic acid (inositol polyphosphate) salts which act as **chelating** (sequestering) agents to the Fe^{2+}, thus blocking further oxidation.[32,33]

Numerous other black inks, obtained mainly from the incomplete combustion of vegetal material, were common in the absence of the necessary ingredients to make IGI.[34,35] In fact, carbon black has been and is the most frequently used pigment to the present day.

The wily and toxic *Garcinia* family is awaiting your visit in the next section.

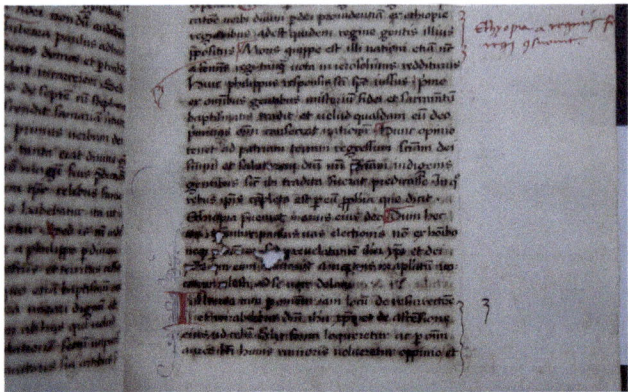

Figure 14.3   This 15th century parchment manuscript exhibits severe burn-through, haloing and lacing. "Haloing" is when a light brown halo spreads out from the inked area. "Burnthrough" or "strike-through" is when the ink appears to sink through the paper and become increasingly visible on the reverse side, and "lacing" is when the inked areas become so weak and brittle that they crack, crumble and fall out. Photo courtesy of Melina Avery, University of Chicago Library. Reproduced from ref. 31.

14.4   GAMBOGE

Gamboge stands alone among the tree-related pigments in that it is an exudate, but one triggered simply by making incisions in the tree trunk. It is also classified as a **resin** but never confused with dragon's blood since it is bright yellow. A native of Southeast Asia produced by trees of the *Garcinia* genus, its name derives from one of its countries of origin, Cambodia, formerly Cambodja. A yellow-orange solid, chemically its most abundant ingredient is gambogic acid, a xanthonoid with the formula $C_{38}H_{44}O_8$. It may have arrived in Europe as early as the 15th century and found use by the early Flemish painters,[36] but was identified much earlier in a 12th century Japanese scroll[37] and tentatively in a 14th century Armenian Gospel book.[38] Its ubiquity is confirmed by its presence in a 1641 Rembrandt[39] and in a J. M. W. Turner paint box.[40,41]

Gamboge has a rather checkered, and deadly, history. Respected by some as the chosen dye for Buddhist monks' robes and discarded by artists as fugitive and unreliable, it has found its way into the courtroom as a murder weapon. Since it is one of the most powerful purgatives and diuretics known, its presence in some British quack medicines resulted in more than a few deaths. The sad tale is told of one Rebecca Cross, aged 15, who took some pills laced with gamboge and died of "mortification of the bowels."[42] It's a great insect repellent as well, albeit an expensive one: no self-respecting cockroach would dare give it a taste. Its poisonous nature identifies it as a passive antipredation agent, always lurking in the *Garcinia* tree, and passed off as a pigment. Other things may lurk in packets of gamboge shipped from Southeast Asia. In the 1970s, Winsor & Newton, a reputable London pigment house, reported finding exploded bullets in their shipment, apparently the remainders of a gun battle in a *Garcinia* grove during the Vietnam war.[43] Gamboge is still on the shelves of pigment dealers today, mainly finding use as a watercolor.

14.5 TWO HEARTWOOD TREES

One of the novel colorants coming out of the New World was a Spanish import called logwood, a tree of the genus *Haematoxylum campechianum*, discovered growing along the Bay of Campeche in Mexico.[44] Its red heartwood contains a colorless dye precursor called hematoxylin, which oxidizes, on exposure in air, to dark red-brown hematein which produces a rich black or navy blue color on textiles, especially when **mordant**ed with iron sulfate.[45] Other metal salts such as copper, chromium or aluminum produce different shades of blue, black and gray.[46] Despite Spanish efforts to maintain a monopoly on logwood, British privateers targeted it as a prize worth fighting for and much of it made its way to England as a result. Mexico forbids its export and thus maintains a monopoly on this product.

As a pigment, logwood was rated by 17th century painters as a cheap colorant good only for watercolor work. Harley[47] says that by about 1730 it had fallen out of use due to its unsuitability for any kind of permanent work. But not soon enough. The applied ink medium in the 18th century *Nova Rhetorica* manuscript held by the Biblioteca Diocesana di Lanciano looked very much like iron gall ink and exhibited typical iron migration across the pages. Spectroscopic analysis showed that the ink contained no gall at all, but logwood, with added carbon black and an iron salt. Addition of iron darkens the logwood and stretches its use, but eventually this complex degrades to reddish-brown smudges due to ink migration.[48] No wonder that logwood fell out of use in that century, but it lives on as a well-regarded stain in modern usage.

Our second heartwood tree, brazilwood, did not fare as well. Though trees have been most successful in warding off predators *via* their chemical expertise, they have been powerless against the most efficient predator of all, *Homo sapiens*. As a result, the Brazil tree (after which the country is named) is an endangered species now confined to a small portion of that great country where it was once ubiquitous. The heartwood of trees of the genus *Caesalpina* contains a colorless precursor, brazilin ($C_{16}H_{14}O_5$) which air oxidizes to deep red-brown brazilein ($C_{16}H_{12}O_5$), closely related to the hematoxylin of the logwood tree.[18] Often disparaged as a fugitive dye good only for watercolor, its Asian cousin, hematoxylin, also of the *Caesalpina* genus, has been identified in 15th century French and Portuguese medieval manuscripts.[49] The wood, called pernambuco, is valued for its extraordinary hardness, most suitable for making violin bows. I recall the great pride with which I first merited such a bow along with a marvelous red Jan Juzek violin. Alas, violinists are very nervous that the pernambuco supply may soon dry up when the genus goes extinct. Brazilwood is a textbook example of "boom and bust."

14.6   QUERCITRON: A HYBRID NAME BY A DOUBLE AGENT

Quercitron is a pigment name coined by its discoverer, Edward Bancroft (1744–1821), from its origin, the Eastern black oak, *Quercus velutina*, prefix "quer", and the yellow color of its inner bark, suffix "citron." While in Britain it retains this name, it has many synonyms elsewhere: yellow lake, brown lake, Italian lake, Dutch pink, yellow madder, and a host of others. It is one of only two naturally occurring colorants whose discoverer is known; the other is cudbear, discovered, named and patented by the Scot, Cuthbert Gordon, in 1766.[50]

Quercitron is harvested from the yellow inner bark of the oak, which is ground down to a powder from which is extracted quercitrin. Hydrolysis of quercitrin yields the pentose, rhamnose, and the coloring principle, quercetin, a flavonol with the empirical formula $C_{21}H_{20}O_{12}$. It was marketed as a dye and Field, in his 1885 treatise on colors and pigments, recommends it as a so–so lake pigment that is "rich, powerful and transparent...but does not resist the action of light and dries rather badly."[51] While it was employed mainly as a dye in England, in the US its use seemed to have been confined to wall painting and carriage decoration, basically because of the demerits mentioned by Field.[52] The life of quercitron's discoverer, described as "perfidious" and "dishonorable" in much of the literature, is more colorful than the colorant itself – a good friend of Benjamin Franklin (1706–1790), Bancroft spied on the British for Franklin, but was also on the payroll of the British Prime Minister, Lord North (1732–1792). In fact, his services were so well appreciated in England that his salary was doubled within a year.[53] He was also a very creditable chemist, an author of a major treatise on dyestuffs, and a friend of Joseph Priestley (1733–1804).[45] His treachery was not discovered until about century after his death – Franklin remained convinced of his integrity until the end.

The final act of this treatise on trees outshines all the others in terms of its universal and ongoing significance in the worlds of botany, color chemistry, chemical instrumentation, ecology, biogeochemistry, energy and Nobel-quality science.

14.7 CHLOROPHYLL AND CHROMATOGRAPHY

Forest green, the pigment that universally clothes photosynthesizing plants, is one of Nature's cleverest evolutionary devices. How it efficiently utilizes the sun's energy is worthy of a PhD in chemical engineering. Green chlorophyll, a **macrocyclic compound**, reflects the entire middle of the visible spectrum (500–600 nm), but absorbs the low energy red-near infrared end (600–800 nm) as well as the high-energy blue-violet-near ultraviolet end (300–500 nm)[54] (**Chapter 1.6**). Both absorbed energy bands[§] power the enzymatic chemical processes necessary for both photosynthesis and respiration. If this pigment did not exist, neither would we. However, green is not the whole story.

One of the great perks of living in New England, despite some terrible winter weather conditions, is the change in tree leaf colors during the fall that draws thousands of "leaf peepers" from all over the world to witness this annual recurring event. The magnificent reds, oranges, and yellows that gradually appear in deciduous trees in September and October is due to the breakdown of the green chlorophyll that masks these colors so that they can appear in all their glory for just a small window of time. But these colors do not have a merely aesthetic purpose. One informative study on plant physiology suggested that the orange and red carotenoid pigments serve as effective light traps (in spite of their relatively low concentrations) able to provide protection against the harmful effects of high-energy blue-light irradiation. The carotenoids absorb in the blue-green, extending the spectral range for photosynthesis through transfer of the absorbed energy to chlorophyll. This activity explains the physiological significance of their retention in leaves up to terminal stages of the aging process[55] before the leaves drop off. And not only that, if these fall leaves appear to glow in the process, it is not your imagination. Another study[56] determined that the leaves really are fluorescing due to the protective action of the carotenoids. By trapping light in aging leaves, carotenoids boost photosynthesis in degraded chlorophyll. They also protect the chlorophyll from harmful UV radiation by dissipating trapped light out of the cell, creating a glow.

The glow of chlorophyll and other green leaf components has suffused the name of Mikhail Tsvet, the hero and inventor of chromatography, down through the years, although he never got credit for his discovery during his lifetime. Tsvet actually utilized the fluorescence phenomenon described above in his own research almost a century before.

[§]There is a classic and very elegant experiment done in 1882 to show that these two bands are the most involved in photosynthesis. The author Th. W. Englemann used aerobic bacteria sensitive to oxygen that was produced by cellular algae. He projected a micro spectrum calibrated at different wavelengths onto a microscope glass slide (chamber) which contained the algae and aerobic bacteria. He knew what wave lengths were projected onto the slide, noting relative activity of the bacteria (movement) when active photosynthesis (oxygen production) occurred *vs.* resting when light was absent or very low.

As its name implies, chromatography literally means "color writing." The principle of the technique has a long history, but at the same time, this history does not necessarily have anything to do with color. The best-known modern household application is the use of filters to remove unpleasant odors and tastes from drinking water. Typically, the liquid water (mobile phase) is passed through a column of finely divided material such as charcoal (stationary phase). Impurities in the water will have an affinity for the particles of charcoal and will adsorb to their surface.

As early as the late 18th century, this method was in use by Carl Wilhelm Scheele (1742–1786) to adsorb gases on charcoal, and by sugar chemists to clarify sugar solutions. In the 19th century, petroleum chemists were using the same technique in a more refined way: they would filter crude petroleum through a column of fuller's earth or other inert adsorbing material. By removing the liquid that filtered through the entire column at different time intervals, or by removing the material from the column at different levels through little portholes, they were able to prepare different fractions of the petroleum that had different physical and chemical properties.[57]

The breakthrough experiment came in 1906, although hardly anyone noticed until decades later – and then, as they say, the rest is history. At that time, the Italian-Russian botanist, Mikhail Tsvet (1872–1919), was trying to separate the plant pigments, chlorophylls and carotenoids, by adsorbing them from a petroleum ether solution onto a solid material. Let us read his own words as he describes his discovery.[58]

If a petroleum ether solution of chlorophyll filters through a column of an adsorption material (I use chiefly calcium carbonate which is firmly pressed into a narrow glass tube), the pigments will separate according to [an] adsorption series from above downward in differently colored zones, and the more strongly adsorbed pigments will displace the more weakly held ones which will move downward. This separation will be practically complete if, after one passage of the pigment solution it is followed, by a stream of pure solvent through the adsorbing column. Like the light rays of the spectrum, the different components of a pigment mixture in the calcium carbonate column will be separated regularly from each other, and can be determined qualitatively and also quantitatively. I call such a preparation a chromatogram and the corresponding method the chromatographic method.

In this remarkable paragraph, Tsvet described the following for a mixture of several components of green leaf material:

- differential adsorption on a stationary (solid) phase
- differential solubility in a liquid phase involving more than one solvent
- displacement of one adsorbed species by another on a solid phase
- an adsorption series
- separation of components on a chromatography column into zones
- qualitative analysis
- quantitative analysis

In naming his preparation a *chromatogram*, and in calling his method *chromatography*, he unknowingly laid the foundations for one of the most powerful analytical tools ever devised – now a multimillion dollar industry that has pushed detection and measurement limits down to parts per trillion, part of the chemistry curriculum in every undergraduate and graduate department of chemistry, and consisting of many more chromatographic methods than simply visualizing a separation of colored analytes on a transparent column. One must wonder whether this method would ever have developed the way it did if chlorophyll were colorless?[59]

Tsvet's life was marked by a series of losses – of family, of educational credentials, and eventually even of teaching posts and research career. His Italian mother died soon after he was born, and his father, a Russian diplomat, left him in the hands of a nursemaid in Switzerland when he was recalled to Russia. Tsvet earned his doctorate at the University of Geneva, but when he moved to Russia, his degree was not recognized and he had to earn a master's degree at the University of Kazan. Later he held a number of teaching posts in Poland, during which he obtained a second doctorate at the University of Warsaw in 1910. After only four tranquil research years, his career ended because of war – a series of evacuations broke his health; he died of heart failure at the age of forty-seven. His great scientific strength lay in the fact that he was both a botanist and a chemist who was deeply interested in the molecular structure of plants, a fact that led to his great interest in chlorophyll. Figure 14.4,[60] a 1972 Russian circulated cover, depicts Tsvet in this dual role by the attribution above his name: "Russian scientist: botanist and chemist."

Chromatography has an interesting aftermath in the annals of the Nobel Prize in Chemistry. The first is the failure of the 1915 Nobel Laureate, Richard Martin Willstätter (1872–1942), to use Tsvet's method successfully because of problems with his choice of adsorbing species. However, he and his students went on to develop many of the separation techniques that made the identification of natural products possible. His abiding interest in chlorophyll eventually led to the Nobel Prize: he was able to show that there were only two types of chlorophyll, a and b, and that magnesium was an integral part of the chlorophyll structure.[61] His work with naturally occurring colored compounds, like **carotenoids**, eventually enabled him to distinguish among and identify orange carotene, yellow xanthophyll, and red lycopene.[62]

The great chromatography success story was the development of liquid–liquid chromatography by Archer John Porter Martin

Figure 14.4 A portrait of Mikhail Semyonovich Tsvet on a 1972 Russian circulated cover. Reproduced from ref. 60 from the collection of Daniel Rabinovich, with permission.

(1910–2002) and Richard Laurence Millington Synge (1914–1994), who shared the 1952 Nobel Prize in Chemistry for this achievement. Martin, assisted by Synge, built a liquid–liquid countercurrent extraction apparatus equivalent to 200 separatory funnels in order to separate acetylamino acids – and failed. They decided then to hold one of the liquid phases (water) stationary on an inert solid support and to move the other immiscible liquid (chloroform) – and thus liquid–liquid partition chromatography was born. And it was even colorful: they used methyl orange as an indicator because it forms bright red bands over an orange background to show the separation of the different acetylamino acids.[63,64] Figure 14.5[65] is a 1977 British stamp commemorating the 25th anniversary of their Nobel Prize along with the centenary of the founding of the Royal Institute of Chemistry.

It is a curious thing that when they first read their paper to the biochemical society announcing this discovery, they were greeted with a resounding silence and no questions. In Martin's own words, "it raised not a flicker of interest." And some years later, when they had begun the first investigations into substituting a vapor phase for the liquid mobile phase – the beginnings of gas chromatography with a colorful twist as well – they experienced

Figure 14.5   Starch chromatography. A. J. P. Martin and R. L. M. Synge. Nobel Prize 1952. Royal Institute of Chemistry 1877–1977. Stamp issued by Great Britain in 1977. Reproduced from ref. 65 from the collection of Daniel Rabinovich with permission.

the same reaction. It is fortunate for the future development of these extraordinary techniques that someone in Stockholm was paying attention.[66,67]

14.8 CONCLUSION

Throughout this volume, we have made reference to pigments in medicine and foods. In fact, in ancient times, colored materials were normal additives to foods and enjoyed the status of medicines in their own right. Over time, the tendency to specialization and compartmentalization separated pigments from the grocery store and the pharmacy. In the chapter that follows, we will see that split rectified as pigments take on a new role as diagnostic tools in medicine and as essential additives to foods.

REFERENCES

1. D. Alighieri, Inferno, Canto 1, in *The Divine Comedy*, 1308–1320.
2. J. R. R. Tolkien, *The Hobbit*, George Allen & Unwin, London, 1937.
3. M. Pastoureau, *Black: The History of a Color*, Princeton University Press, Princeton, NJ, 2009, pp. 90–91.
4. D. Walters, *Fortress Plant: How to Survive when Everything Wants to Eat You*, Oxford University Press, New York, 2017, pp. 130–135.
5. A. Wallert and M. R. van Bommel, *Dyes Hist. Archaeol.*, 2008, **21**, 75–88.
6. R. Tavernier, *Ex Officina*, 1989, vol. 2(6), pp. 76–85.
7. A. Dürer, The Flight into Egypt, https://commons.wikimedia.org/wiki/File:Albrecht_Dürer,_The_Flight_into_Egypt,_c._1504,_NGA_6706.jpg, accessed June 2021.
8. Hieronymus Bosch: The Garden of Earthly Delights, https://commons.wikimedia.org/wiki/File:The_Garden_of_Earthly_Delights_by_Bosch_High_Resolution.jpg, accessed June 2021.
9. R. Mullin, *Chem. Eng. News*, 2016, **94**(42), 40.
10. J. Jura-Morawiec and M. Tulik, *Chemoecology*, 2016, **26**, 101–105.
11. P. Maděra, A. Forrest, P. Hanáček, P. Vahalík, R. Gebauer and R. Plichta, *et al.*, *Forests*, 2020, **11**(2), 236–270.

12. D. Gupta, B. Bleakley and R. K. Gupta, *J. Ethnopharmacol.*, 2008, **115**, 361–380.
13. A. G. King, *J. Chem. Educ.*, 2004, **81**(8), 1086–1088.
14. H. G. M. Edwards, L. F. C. de Oliveira and H. D. V. Prendergast, *Analyst*, 2004, **129**(2), 134–138.
15. M. Pastoureau, *Red: The History of a Color*, Princeton University Press, Princeton, 2017, p. 123.
16. Wikipedia Dragon's Blood, https://en.wikipedia.org/wiki/Dragon%27s_blood, accessed December 2020.
17. A. P. Laurie, *The Materials of the Painter's Craft*, T.N. Foulis, London, 1910, p. 284.
18. R. J. Gettens and G. L. Stout, *Painting Materials: A Short Encyclopedia*, Dover Books, New York, 1966, p. 59.
19. Wikipedia Tetraclinis, https://en.wikipedia.org/wiki/Tetraclinis, accessed December 2020.
20. The Gymnosperm Database, https://www.conifers.org/cu/Tetraclinis.php, accessed December 2020.
21. I. Kononenko, L. de Viguerie, S. Roghut and Ph. Walter, *Environ. Sci. Pollut. Res.*, 2017, **24**, 2160–2165.
22. M. Oliver, *From Lazarus To Theophilus: how manuscript digitization led to the historical, chemical, and technological understanding of iron gall ink and its counterparts*, Honors thesis, University of Mississippi, 2015, pp. 13–15. https://egrove.olemiss.edu/hon_thesis/700, accessed December 2020.
23. E. Delange, M. Grange, B. Kusko and E. Menei, *Revue d'Égyptologie*, 1990, **41**, 213–217.
24. M. Aceto, A. Agostino, E. Boccaleri and A. C. Garlanda, *X-Ray Spectrom.*, 2008, **37**, 286–292.
25. D. N. Carvalho, *Forty Centuries of Ink*, The Banks Law Publishing Co., New York, 1904, p. 83.
26. The Iron Gall Ink Website, https://irongallink.org/, accessed March 2021.
27. Iron gall ink synthesis. Photographs by Mary Virginia Orna.
28. A. Ponce, L. B. Brostoff, S. K. Gibbons, P. Zavalij, C. Viragh and J. Hooper, *et al.*, *Anal. Chem.*, 2016, **88**, 5152–5158.
29. C. Krekel, *Int. J. Forensic Doc. Exam.*, 1999, **5**, 54–58.
30. C. A. Schmitt, *J. Chem. Educ.*, 1944, **21**(8), 413–414.
31. 15th century parchment manuscript. University of Chicago Library.

32. L. Botti, O. Mantovani and D. Ruggiero, *Restaurator*, 2005, **26**, 44–62.
33. A. Bazemore, Chelating soluble iron(II) from iron gall ink using calcium phytate in agar gel, https://www.researchgate.net/publication/329983608, accessed December 2020.
34. M. Leonida, *The Materials and Crafts of Early Iconographers*, Springer, Cham, 2014, pp. 34–40.
35. J. Winter, Preliminary investigations on Chinese ink in far eastern paintings, in *Archaeological Chemistry*, ed. C. W. Beck, American Chemical Society, Washington, DC, 1974, pp. 207–225.
36. A. H. Church, *The Chemistry of Paints and Painting*, Seely Co., 1901, p. 153.
37. K. Yamasaki and H. Nakayama, Studies of the Pigments Used in the Genji Monogatari Scroll Paintings, *Bijutsu Kenkyū*, 1954, **174**, 229–234. In Japanese.
38. M. V. Orna and T. F. Mathews, *Stud. Conserv.*, 1981, **26**, 57–72.
39. H. Kühn, *Maltechnik-Restauro*, 1977, **83**, 223–233.
40. N. W. Hanson, *Stud. Conserv.*, 1954, **1**, 162–173.
41. J. Winter, Gamboge, in *Artists' Pigments: A Handbook of Their History and Characteristics*, ed. E. W. FitzHugh, National Gallery of Art, Washington, DC, vol. 3, 1997, pp. 143–155.
42. Anon., Quacks and Quack Medicines, *The Penny Magazine of the Society for the Diffusion of Useful Knowledge*, 1838, pp. 495–496.
43. V. Finlay, *Color: a Natural History of the Palette*, Random House, New York, 2004, p. 221.
44. F. Brunello, *The Art of Dyeing in the History of Mankind*, Neri Pozza, Vicenza, 1973, p. 197.
45. E. Bancroft, *Experimental Researches Concerning the Philosophy of Permanent Colours*, Thomas Dobson, Philadelphia, vol. II, 1814, p. 297.
46. M. Titford, *Biotech. Histochem.*, 2005, **80**(2), 73–78.
47. R. Harley, *Artists' Pigments c 1600–1835*, Butterworths, London, 1970, p. 60.
48. M. Bicchieri, M. Monti, G. Piantanida and A. Sodo, *J. Raman Spectrosc.*, 2008, **39**, 1074–1078.
49. M. J. Melo, V. Otero, T. Vitorino, R. Araújo, V. S. F. Muralha, A. Lemos and M. Picollo, *Appl. Spectrosc.*, 2014, **68**(4), 434–444.

50. H. N. Hansen, The quest for quercitron: revealing the story of a forgotten dye, MA Thesis, University of Delaware, 2011, p. 71.

51. G. Field, *Field's Chromatography A Treatise on Colours and Pigments for the Use of the Artist*, 2nd edn, Winsor & Newton, London, 1885, p. 132.

52. F. Maire, *Modern Pigments and Their Vehicles*, J. Wiley & Sons, New York, 1908, pp. 90–92.

53. C. A. Browne, *J. Chem. Educ.*, 1937, **14**(3), 103–107.

54. Th. W. Englemann, *Botanische Zeitung*, 1882, **40**, 419–426. Republished in English translation in M. L. Gabriel and S. Fogel, ed. *Great Experiments in Biology*, Prentice Hall, Englewood Cliffs, 1955, pp. 165–170.

55. M. N. Merzlyak and A. Gitelson, *J. Plant Physiol.*, 1995, **145**(3), 315–320.

56. P. Matile, B. M. P. Flach and B. M. Eller, *Bot. Acta*, 1992, **105**(1), 13–17.

57. A. J. Ihde, *The Development of Modern Chemistry*. Dover, New York, 1964, pp. 571–572.

58. M. Tsvet, *Ber. Dtsch. Bot. Ges.*, 1906, **24**, 316–323, as translated and excerpted in H. M. Leicester, *Source Book in Chemistry 1900–1950*, Harvard University Press, Cambridge, 1968.

59. M. Tsvet, *Ber. Dtsch. Bot. Ges.*, 1906, **24**, 384–393.

60. Portrait of Mikhail Semyonovich Tsvet on a 1972 Russian circulated cover. Scanned from the collection of Daniel Rabinovich.

61. Z. Barnes, 1915 Nobel laureate Richard Martin Willstätter (1872–1942), in *Nobel Laureates in Chemistry: 1901-1992*, ed. L. K. James, American Chemical Society, Washington, DC, 1993, pp. 108–113.

62. T. L. Sourkes, *Bull. Hist. Chem.*, 2009, **34**(1), 32–38.

63. P. H. Shetty, 1952 Nobel laureate Archer John Porter Martin, in *Nobel Laureates in Chemistry: 1901–1992*, ed. L. K. James, American Chemical Society, Washington, DC, 1993, pp.352–355.

64. P. H. Shetty, 1952 Nobel laureate Richard Laurence Millington Synge, in *Nobel Laureates in Chemistry: 1901–1992*, ed. L. K. James, American Chemical Society, Washington, DC, 1993, pp. 356–358.

65. Royal Institute of Chemistry, UK, commemorative stamp. Scanned from the collection of Daniel Rabinovich.

66. G. A. Stahl, *J. Chem. Educ.*, 1977, **54**, 80–83.

67. M. V. Orna, *The Chemical History of Color*, Springer, Heidelberg, 2013, pp. 95–99.

CHAPTER 15

Dr. Ehrlich Meets the Poison Squad: Pigments in Food and Medicine†

In food excellent medicine can be found, in food bad medicine can be found; good and bad are relative.[1]

Hippocrates, *De Alimento*

15.1 INTRODUCTION

Pigments have always had an ambiguous relationship to food. As colorful, but toxic, illicit food additives, their presence has plagued consumers for centuries. On the other hand, color trumps the taste buds every time: the market for edible and appetizing pigments is a multi-billion dollar industry. Closely related to food is medicine, a field where pigments, in their role as diagnostic tools, gave rise to some modern medical blessings. This chapter will document the role of pigments in food safety and targeted medical therapies.

---

†Glossary entries and chapter cross-references are in boldface.

March of the Pigments: Color History, Science and Impact
By Mary Virginia Orna
© Mary Virginia Orna 2022
Published by the Royal Society of Chemistry, www.rsc.org

15.2 THERE IS DEATH IN THE POT (AND THE PICKLE BARREL)

Certain pigments' presence in the human diet is both necessary and necessarily limited. Going overboard can lead to serious consequences, as this little limerick describes:

There was a young lady named Mickle

Who enjoyed her luscious green pickle.

She ate three or four more,

Then collapsed to the floor.

Soon Death came and wielded his sickle.

Indeed, our young lady never realized what Frederick Accum (1769–1838) had proclaimed in his treatise of 1820: "Vegetable substances, preserved in a state called pickles, whose sale frequently depends greatly upon a fine lively green color, are sometimes intentionally colored by means of copper. A young lady amused herself by eating pickles impregnated with copper. She soon complained of a pain in the stomach. In nine days after eating the pickle, death relieved her of her suffering"[2] (Figure 15.1).[3]

Copper was a very common food additive until well into the 20th century. Relished for its jewel-like color, canning companies routinely "greened up" their canned peas with a hefty dose of copper salts.[4] Like many other mineral substances, copper is an essential trace element; its levels in the human body can be deficient or toxic. Adolescents and adults need a recommended 900 micrograms (µg) a day to carry on normal metabolic functions. However, intake beyond 10 000 µg, or 10 milligrams (mg) a day render it toxic and potentially lethal. For children, of course, the threshold and upper limits are much less.[5]

This thumbnail sketch of copper can be replicated for many other colorants. Accum, a versatile, polymath chemist, in his campaign against unregulated food adulteration, also identified verdigris (dibasic copper(II) acetate, $Cu(C_2H_3O_2)_2 \cdot 2Cu(OH)_2$) in tea, red lead (minium, $PbO_2 \cdot 2PbO$) in cheese and all three plus vermilion (mercury(II) sulfide, HgS) in confectionery items marketed chiefly to children. All of these compounds contain toxic **heavy metals**.

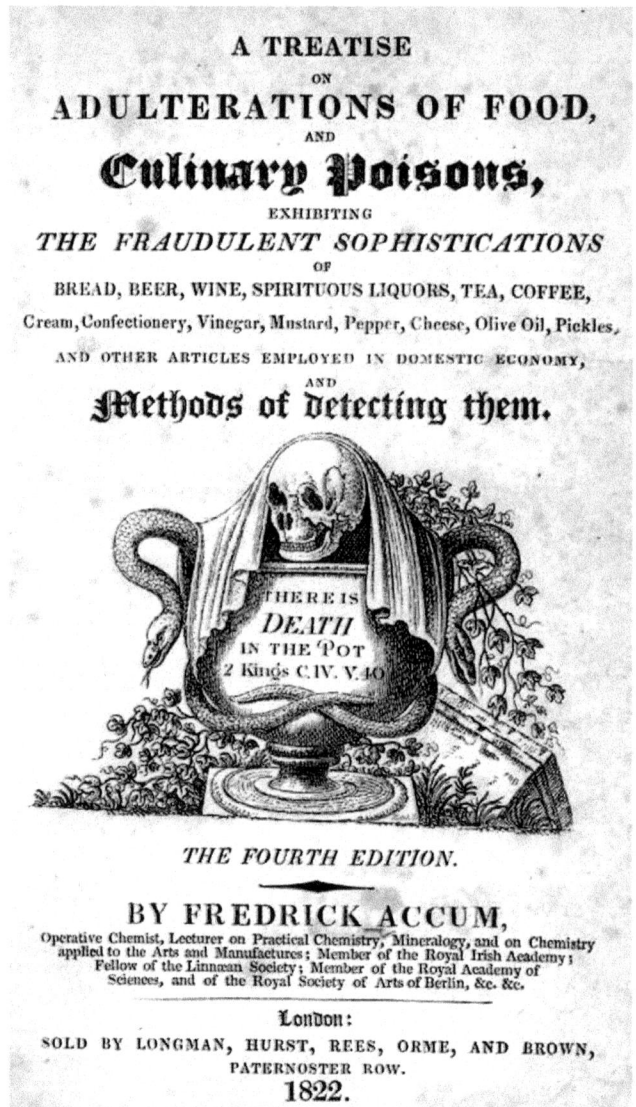

Figure 15.1   Title page of Accum's "Treatise on Adulteration of Food." This work, which passed through many editions in England, the United States, and many European countries, marked the beginning of the modern pure food movement. Reproduced from ref. 3.

Despite his treatise's success,[6] Accum's constant refrain decrying toxic colorants and other substances was bad for business, earning him the enmity of a powerful group of food vendors who vowed to silence him. In 1821, this "cartel" brought him to trial on trumped-up charges, forcing him to decamp to his native Germany, where he continued his campaign unabated.

About ten years later, another "voice in the wilderness" arose in the person of William B. O'Shaughnessy (1809–1889) who drew renewed attention to the practice of adding poisonous colorants to children's cakes and candies. On his own, he secured and analyzed 30 samples to document not only the colors named by Accum, but added substantially to the list: Naples yellow (lead antimonate, $Pb_3(SbO_4)_2$), lead chromate ($PbCrO_4$), massicot (PbO) and copper carbonate ($CuCO_3$).[7] For his pains, the medical profession vilified him to the point where he, too, self-exiled – to India.

Meanwhile a swindled public would continue to consume unsafe, but colorful and attractive-looking food, until truth in labeling and in practice won out through legislation in the early 20th century.

15.3 THE PALETTE OF OUR PALATE

Color in food has been a taste signal from the earliest of times; its desirability in food is estimated to have arisen around 1500 BCE.[8] As a result, artificially colored food has been the norm for well over 3500 years. The presence of these additives had no other purpose other than to impart color; they carried no nutritional value whatsoever. But the perception of flavor based only on sight is powerful enough to override any other sense. A bright yellow pudding might signal lemon, a pale yellow cake, banana. A pale white margarine might call up an unappetizing chunk of lard. These almost visceral perceptions are deeply ingrained in the human psyche, having their roots in habits developed over millennia – color was the most readily accessible clue to avoiding toxic or spoiled substances.[9]

The no–no colors in the Western canon are blue, violet and black, causing loss of appetite in "real foods" but knockout successes in fun foods like Ore-Ida's "Kool Blue Funky Fries."[10] A case in point: One Fourth of July, in this author's experience, a host made up a platter of mashed potatoes to resemble the red, white, and blue of the American flag. When the dish finished its way around the table, the blue part had remained untouched. In a different geographical setting, black burgers are winners in

Tokyo where people are conditioned to accept black in seaweed and squid. Culture is as culture does.[11]

That unscrupulous food purveyors seized on this penchant for color as an invitation to doctor their wares with inexpensive, but toxic, additives only served to point to a larger problem: the whole food industry was rampant with illegal adulteration driven by the profit motive. It took the larger-than-life crusading spirit of Harvey Washington Wiley (1844–1930) to drive through the first food regulatory law by confronting two Presidents, Theodore Roosevelt (1858–1919) and William H. Taft (1857–1930), a hostile Secretary of Agriculture, James Wilson, and the entire gigantic food industry.

15.4   THE POISON SQUAD

Harvey Wiley came to head the federal Bureau of Chemistry in 1881 with an educational background in chemistry and medicine, but the real backdrop to his life and career began at the Indiana farm of his upbringing. Wiley's pre-Civil War childhood days were spent helping his dad, Preston, smuggle runaway slaves from their farm, the southernmost point on the Underground Railroad, to the next stop, eight miles further on. This passion for doing the right thing was to mark his entire career and every major decision of his life.

Realizing that one of the major public health issues his department was dealing with was food quality, he resolved to achieve passage of a comprehensive food safety law. Food manufacturers had *carte blanche* control over what went into their wares. Unregulated and unlabeled food, often laced with toxic preservatives and colorants, victimized an uninformed public, leading to chronic illnesses and sometimes even death. Wiley resolved to build a case for regulation based on solid scientific research in order to combat these practices. For that, he needed some willing human subjects who would document with their very bodies the effects of such additives as borax, formaldehyde and copper salts.

On December 22, 1902, such a body convened – a group of young men sworn to follow Wiley's directives, eat nothing other than what he served up, and live like guinea pigs even through the Christmas holidays. A reporter labeled the group the "Poison

Squad" and the name stuck. The trials lasted over several years and resulted in the report, "Influence of Food Preservatives and Artificial Colors on Digestion and Health"[12] (Figure 15.2).[13] The report documented that the major food preservatives had pronounced deleterious effects on the health of strapping young men.

The Poison Squad experiments were a major turning point in the hard-fought-for passage of the Pure Food and Drugs Act of 1906. Previous advocates for food safety had unsuccessfully introduced as many as 100 bills to Congress over the preceding quarter century. But the context for change was emerging with the growth of progressivism, and a new President, Theodore Roosevelt, was sympathetic to the cause. He signed the act into law on 30 June 1906, giving the chemists at the Bureau of Chemistry some "teeth" with which to carry out mandatory inspections and impose hefty fines. But an uphill battle in this sector became steeper upon the accession of business-oriented W. H. Taft to the Presidency in 1909.[14,15] The Bureau of Chemistry was later renamed the Food and Drug Administration (FDA) in 1930. Dr Wiley was its first Commissioner.

While the new bill was able to reduce the incidences of toxic **inorganic** colors as food additives, a new menace arose in the form of synthetic aniline colorants spawned by the success of

Figure 15.2   Dr Wiley's Poison Squad Dining "in Splendor." Note Wiley himself supervising in the background. Reproduced from ref. 13.

William Henry Perkin's introduction of mauve to the commercial sector in 1856 (**Chapter 12.2.3**). This event provided quasi-unlimited possibilities to color food, but the downside soon became apparent. They provided an opportunity to mask the poor quality of foods and, thus, ultimately to easily mislead consumers.[16] By the early 20th century, more than 800 such colorants had been synthesized and many of them were replacing natural food colors because they were cheap, reproducible, easy to make, and displayed high chromatic intensity. But it soon became evident that many of these colors, completely untested and unregulated, were deleterious to public health: neurotoxicity and carcinogenicity were the major complaints.[17]

15.4.1 Dealing with a New Problem

To counteract this problem, Wiley hired a pharmacist-chemist, Bernhard C. Hesse (1869–1934) who was also an expert on the German dye industry, the major supplier of the colorants in question. Hesse found that of the 80 synthetic colorants that found their way into the American diet, 30 of them had never been tested and therefore were wild cards in the whole food colorant picture. In 1907, he found that only 16 were, more or less, harmless.[18] So Hesse began testing the colorants without negative reports on human and animal models for short-range physiological effects and found only seven additives considered safe.[9] This list did not remain in place for very long. Additional testing between 1918 and 1971 led to colors added to the original list and others delisted. By 1976 only nine colors remained in the inventory; of these nine, Orange B and Citrus Red 2 are effectively unavailable and banned. The remaining seven are Erythrosine (C.I. 45430), Indigotine (C.I. 73015), Tartrazine (C.I. 19140), Fast Green FCF (C.I. 42053), Sunset Yellow FCF (C.I. 15985), Brilliant Blue FCF (C.I. 42090) and Allura Red AC (C.I. 16035). Not all of these are deemed perfectly safe; some are under scrutiny for causing possible allergenic reactions, hyperactivity and learning disabilities in children; they require special labeling and daily consumption limits.

All of these tests and limitations are largely due to a single crusader with a laser-like penetrating, but narrow, vision: Harvey Washington Wiley. His method was science, his psychological weapon was fear, and his legacy was controversy. While we can thank him for a safe, and hopefully unadulterated food supply, Jonathan Rees's recently released book expresses a serious caveat: "When evaluating Harvey Wiley's entire career, it is best not to use the criteria that he did."[19]

The synthetic food dye palette is not as bleak in other parts of the world. Many colorants such as carmoisine, amaranth and quinolone yellow, banned in the US, are alive and well in foods in many parts of Europe and Pacific Rim countries. However, the problem always remains: what synthetic food dye will be banned next? Are there any suitable substitutes on the horizon?[20]

15.4.2   Nature to the Rescue

Realizing the strict limitations regarding usage of synthetic food colorants, manufacturers began to take another look at the attractions of natural colorants. Although not universally safe, and indeed, in some instances downright toxic, natural pigments present a very broad alternative to the synthetics (Figure 15.3).

In Section 15.1, we saw that many **inorganic** pigments crept into the food chain as food colorants even though many were known to be toxic. Formed by geological processes, a limited number of them are acceptable in animal feed (iron oxide), to

Figure 15.3   Nature provides a gamut of colors from plant, animal and microbial sources. Gourds at a vegetable market, Padua, Italy. Photograph: Mary Virginia Orna.

ripen olives (ferrous gluconate and ferrous lactate), or for general use (titanium dioxide).[21]

The other sources of natural food pigments are plant, animal and microbial. All three types are **organic** compounds abundant in nature and biosynthesized in response to the needs of the organism, typically for metabolic processes, energy needs, protection from predators or other functional requirements.[22]

Since almost all plants, being immovable, ready-to-eat food packages, are unable to run away from predators, they have developed sophisticated chemical defense mechanisms instead.[23,24] Many of these defenses are highly colored. For example, **anthraquinone** derivatives, present in rhubarb,[25] as well as others we met in Chapters 10, 12 and 14, can be quite toxic if present in appreciable quantities.

However, large amounts of inexpensive, commercially available natural colorants are safe substitutes for many of the more desirable food color hues. Plant sources for highly favored red, orange and yellow hues are grapes and berries (anthocyanins),[26] beet powder (betalain),[27] and carrots, corn and tomatoes (**carotenoids**).[28] Alfalfa is a good source of the green pigment, chlorophyll,[29] and yellow is derived from the rhizome, *Curcuma longa* L. (the curcuminoid, turmeric).[30] Synthetic titanium dioxide is the only white pigment approved for food use in the US. Its suspected genotoxicity bans it from food use in the European Union.[31,32] Alternative sources of inexpensive and reliable pigments are microorganisms, particularly derived from the *Monascus* spp. yeast. Although not yet approved for food use in the US, commercial exploitation is quite encouraging because of renewability and ability to control growth conditions.[33] They have an excellent track record in Asia where they have a long history of use, providing pigments ranging from yellow, to orange, to red.[34] One great advantage is their applicability to replacing nitrates and nitrites in meats to improve stability and to enhance color.[35]

We have seen that it has been a long, hard road to achieving safe colorants in foods. Humans' perceptions of beauty equated to goodness didn't help matters – only chemical testing, legislative persistence and subsequent vigilance triumphed. Separating harmful pigments from food was a conceptual as well as a practical accomplishment.

From time immemorial, pigments were associated with curative and medicinal properties. Well into the 17th and 18th
centuries, pharmacies were the locus of pigment production
and purchase. In the next section, we will see how pigments
migrated out of the pharmacy and into the medical research
laboratory.

15.5   COLOR IN MEDICINE[‡]

The practice of interweaving colors and cures has a very long and
common history. Initially, they were indistinguishable; they were
one and the same, whether applied topically or taken internally.
We have seen that as late as the 16th century, paints and pre-
scriptions could do double duty (**Chapter 9.6.2**), almost like con-
joined twins. The apothecary shop was the place for artists to buy
their precious colors. All this began to change as Venice took the
upper hand in the pigment trade; it was there that the profession
of *vendecolori*, or colormen, arose by the late 15th century.[36] By
the mid-18th century, the split was complete in most of the rest
of Europe: colors in the hands of artists, medicines in the hands
of physicians.

   Then, in the mid-19th century, a strange thing began to happen: colorants, in the form of new synthetic organic pigments
(SOPs), regrouped with medicine in a rather odd way. With the
advent of magnifying lenses, biologists began to study the structure of animal and plant tissues to understand how organisms
worked. Marcello Malpighi (1628–1694) pioneered this new field
that later came to be called histology (from Greek *histos* = tissue;
logos = study). In the 1700s, histologists evolved to using tech-
nicolor in the form of natural dyes, like hematoxylin, to better
visualize their tissues. And when they became available in 1858,
artists were not the only specialists to get excited about the new
SOPs. Biologists and physicians began to employ them routinely
as well thanks to a seminal treatise on the topic by Joseph von
Gerlach (1820–1896) that kicked off a flurry of experiments in
histological staining.[37,38]

[‡]Sections 15.5 and 15.6 are adapted with permission from ref. 60, M. V. Orna, *The Chemical History of Color*, Springer, Heidelberg, 2013, pp. 113–121.

In 1878, a German anatomy professor, Walther Flemming (1843–1905), began to use SOPs to observe cell division more easily. He noted that the cell nucleus owed the nature of its reactions and, in particular, its affinity for dyes, to a substance that he tentatively called "chromatin," (from Greek *chroma* = color) precisely because it sopped up colors so readily.[39] Ten years later, another German anatomist, Wilhelm von Waldeyer (1841–1923), called the threadlike structures in the nucleus "chromosomes" for the same reason.[40] Thus, the words we use today to describe the components of heredity are forever linked to color and the color reactions of pigments and dyes (Figure 15.4).[41] Within just a few decades, color was to play a role in biology and medicine that was far more in depth and global than just a name.

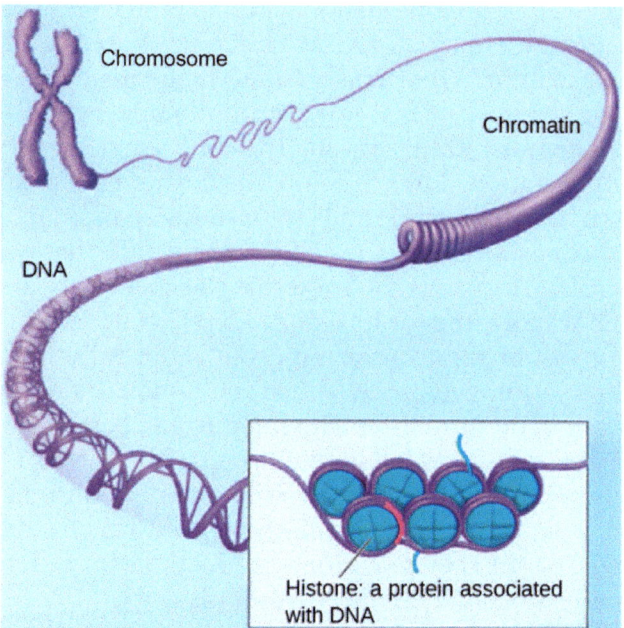

Figure 15.4 DNA is organized into small packets by proteins called histones to form chromatin; the chromatin consolidates further into more highly ordered chromosomes. Reproduced from ref. 41 courtesy of MedlinePlus from the National Library of Medicine.

15.6   THE TOWERING FIGURE OF PAUL EHRLICH[§]

Right around the time that Harvey Wiley was targeting toxic colorants in food, a young German medical student, Paul Ehrlich (1854–1915), was completing his doctoral thesis, "Contribution to the Theory and Practice of Histological Staining."[42] This work constituted the first systematic study of the biological staining properties of the new SOPs.[43] Ehrlich had been captivated by colored compounds and by structural organic chemistry, particularly the work of O. N. Witt. Witt theorized that two types of atomic groups on a colored molecule were necessary for the color: a color-fixing "auxochrome" and a color-rendering "**chromophore**."[44] For colorants, Ehrlich was the beneficiary of the strong ties between German academe and the dye industry. As a result, he always had an ample supply of them for his work.[45] For organic chemistry, he had an instinctive talent for visualizing these molecules in three dimensions. Yet, torn between the influence of the great organic chemist, Adolf von Baeyer (1835–1917), and the great histologist, von Waldeyer, he chose the latter.

Ehrlich achieved renown as an organic chemist, giving him the advantage to excel in histology, immunology, hematology and pharmacology by spinning out the embryonic ideas outlined in his dissertation. First, he noted that the mechanism of the binding of colored compounds to tissues was more than merely physical adhesion but suggested an interaction that was chemical in nature. He also recognized that the chemical group on the molecule responsible for its color was generally not the same as the group responsible for the molecule's affinity for the tissues.[46] This observation eventually led to his famous "side-chain theory" of antibody formation which led to his being awarded the 1908 Nobel Prize in Medicine; it also was the germ of his ideas on chemoreceptors, a concept that eventually led to the theoretical and practical basis of chemotherapy.[47] Ehrlich's biographer observed that the theme of different functions for different parts of the same molecule ran throughout his work.[48]

Another great idea arose from Ehrlich's further experimental work on the affinities of various dyes and pigments for specific tissues. In 1886, doing staining with the dye methylene blue led him to consideration of what he called "localized organ therapy." In

---

[§]Sections 15.5 and 15.6 are adapted with permission from ref. 60, M. V. Orna, *The Chemical History of Color*, Springer, Heidelberg, 2013, pp. 113–121.

other words, he discovered that methylene blue had a special affinity for nerve cells and wondered if similar selectivity on the part of other colorants for other types of cells might exist. In particular, if a dye were to have an affinity for a disease-causing organism, then it might also be possible to target that organism. He took this idea a step further: if a toxin acted in a similar manner as the molecule, then one could deliver a toxin to that organism along with the agent of selectivity. Using this principle, he successfully treated certain experimental infections of the trypanosome parasite with **azo** dyes.[49] The fixing of the dyes could be by either of two mechanisms, the first like **lake** formation – the combination of the dye with a constituent of the fabric to form an insoluble salt-like compound – thus immobilizing or localizing the drug. The second, the formation of solid solutions, where the dye forms a homogeneous mixture with the substance of the fabric. He suggested that certain drugs might be fixed in cells through a similar process.[50] This idea formed the theoretical basis for his work on chemotherapy, and the technique became popularly known as the "magic bullet" technique, the use of a drug that would kill only the agent being targeted. In 1940, Warner Bros. released a popular film, "Dr Ehrlich's Magic Bullet," based on this work starring Edward G. Robinson in the title role.[51] Figure 15.5[52] shows Ehrlich in his study.

Ehrlich was an innately intelligent man who was not afraid to learn from others and not afraid to change his mind. He was also a keen observer who thought about what he observed, asking for the chemical reasons behind the observation, and then devising brilliant and novel experimental methods for verification. He was also a very hard worker, spending almost every waking minute at his work in the laboratory. But working with color had its hazards: colleagues and cleaning ladies alike accused him of leaving traces behind wherever he went.[42] Although his area of work was mainly biochemical/biomedical, he never strayed from the chemical principles of structure and bonding – he could even be called the father of molecular biology. Listed below are some of his accomplishments and qualities:

- Dyes for the most part adhere to animal fibers without the need of a mordant since animal fibers normally contain nitrogen with a lone pair of electrons affording a foothold for the dye – this led Ehrlich to try staining animal tissues using those same dyes.

Figure 15.5   Paul Ehrlich in his "Arbeitszimmer" (workspace). Reproduced from ref. 52.

- Colored molecules were his labeling, measuring, and diagnostic tools.[53]
- He used the products of the dye industry to argue for a new, chemical approach to biomedical research.
- In his doctoral thesis of 1878 he outlined and classified the major synthetic dyes and pigments, resolved problems of commercial names and contaminants, and described how dyes could be used as staining agents to differentiate tissues.
- He classified dyes into three groups: the primary amino dyes, the sulfonic acid derivatives of aniline blue, and the acidic dyes containing nitro and halogen substituents.
- He learned how to exploit the loss of color as dyes were reduced to develop a tool to enable semi-quantitative measurements and provide information about cell surfaces – molecules of dyestuffs in his hands became probes and measuring devices.
- His tight connections with industry enabled him to transform commercial chemical products into analytical tools in the service of biological investigations.
- His early grasp of the versatility and potential use of the new dyestuffs technology came shortly before a time of increasing interest both in public health services, as a means of

ensuring social stability, and in the conquest of tropical diseases.

- He had the ability to think independently, and to pursue unconventional thinking about dye interaction with cells, convinced that nature is unreliable in its adaptive reactions to pathological insults.

One can cite many more reasons for Ehrlich's brilliant success in virtually every field. All of them are necessary, but not sufficient, to account for his monumental achievements. Yet, for the most part, his working conditions were very humble. He said he never needed more than a few test tubes, a Bunsen burner, and blotting paper – and could work in a barn if necessary.[54] But there is one more ingredient that has not been mentioned and is absolutely necessary: Ehrlich was passionate about his subject. He was intrigued, he was curious, he could not delve deeply enough; in short, he was a man in love.

15.7   MONOCLONAL ANTIBODIES

Many years later, the scientific trio of Georges J. F. Köhler (1946–1995), César Milstein (1927–2002), and Niels K. Jerne (1911–1994), still wearing Ehrlich's chemotherapeutical mantle, walked off with the 1984 Nobel Prize in Physiology or Medicine for their work on monoclonal antibodies.[55]

Early in the twentieth century, Ehrlich, from his tissue staining experience, had postulated that if a dye could be made that directly targeted a specific disease-causing organism that perhaps a toxin could be made to accompany the dye and selectively destroy the organism. This idea, dubbed early on as a "magic bullet" mechanism, was later expanded beyond colored molecules to include any type of molecular toxin delivery system. But Ehrlich's dream was far from being realized as no molecular species was selective enough at the cellular level: this had to wait for an understanding of the antibody-antigen relationship, the technique of cell cloning, and the fusing of different types of cells in the hopes of bringing about the needed antibody specificity.

An antigen is an antibody generator, from which the word is derived. An antigen, usually protein in nature, is signaled as a foreign body to a given organism, thus triggering a response on the part of the immune system: the generation of an antibody to combat the invader. A monoclonal antibody is one produced by a single clone of cells, that is, a single cell and its progeny. But what was yet needed was the ability to produce identical antibodies specific to a given antigen, a technique that was finally developed, but not without controversy[56] by Cambridge University's Milstein and Köhler, a postdoctoral fellow working in his laboratory.[57] Jerne, working out of the Basel Institute for Immunology, developed the theory that buttressed Milstein's and Köhler's work.[58] The three shared the Nobel Prize for Physiology or Medicine in 1984 for the discovery.

Although the development of monoclonal antibody technology may seem a long step from the vision of "little Ehrlich," face and hands covered with biological stains, this cornerstone of modern immunochemistry may never have happened if it were not for Ehrlich's dream – a dream now come true.[59,60]

15.8 CONCLUSION

While there are many other applications of pigments and dyes to medicine, none reach the magnitude of the major boons to human health that are Paul Ehrlich's legacy: chemotherapy, immunology, immunochemistry and antibiotics. While there are many colors associated with foods, it took a crusader like Harvey Wiley to assure the public that only safe ones would make their way into their diet. Judiciously applied color, in the form of pigments and dyes, has been transformational in the area of human health and wellness.

In the next, and final, chapter, we will try to envision how far the development of dyes and pigments can take us in terms of new sources, new materials, new uses, new requirements, new ways of seeing, new attitudes and new definitions.

REFERENCES

1. Hippocrates, *Commentarii in librum Hippocratis "De Alimento,"*, Sectio III, Textus XIX.
2. F. Accum, *A Treatise on Adulterations of Food, and Culinary Poisons*, Longman, Hurst, Rees, Orme and Brown, London, 2nd edn, 1820, p. 295.

3. 'A Treatise on Adulterations of Food, and Culinary Poisons' by Friedrich Chritisan Accum, https://commons. wikimedia.org/w/index.php?title=File:%E2%80%98A_ Treatise_on_Adulterations_of_Food,_and_Culinary_ Poisons%E2%80%99_by_Friedrich_Chritisan_Accum. jpg&oldid=240206709, accessed June 2021.
4. D. Blum, *The Poison Squad*, Penguin Press, New York, 2018, p. 187.
5. National Institutes of Health, Copper, https://ods.od.nih. gov/factsheets/Copper-HealthProfessional/, accessed March 2021.
6. M. Barron, Frederick Accum and 'Death in the pot,' https:// blog.nli.org.il/en/accum/, accessed April 2021.
7. W. B. O'Shaughnessy, *Lancet*, 1931, **2**, 1193–1198.
8. H. McKone, *ChemMatters*, 1999, December, 6–7.
9. A. Burrows, *Compr. Rev. Food Sci. Food Saf.*, 2009, **8**, 394–408.
10. J. Hevrdejs, *Chicago Tribune*, 14 July 2000, 1 Tempo.
11. T. Wen, *The Atlantic Daily Idea*, 30 September 2014.
12. *Bureau of Chemistry*, Bulletin 84, Parts I–V, Government Printing Office, Washington, DC, 1904–1908.
13. Harvey Wiley & Poison Squad, https://commons.wiki-media.org/w/index.php?title=File:Harvey_Wiley_%26_ Poison_Squad_(FDA012_%26_FDA004)_(7030346087).jpg& oldid=474507986, accessed June 2021.
14. J. P. Swann, *Chem. Heritage*, 2006, **24**(2), 6–11.
15. J. Rees, Harvey Wiley and the transformation of the American diet, in *Chemistry's Role in Food Production and Sustainability: Past and Present*, ed. M. V. Orna, G. Eggleston and A. F. Bopp, American Chemical Society, Washington, DC, 2019, pp. 75–84.
16. R. M. Schweiggert, *J. Agric. Food Chem.*, 2018, **66**, 3074–3081.
17. C. Cobbold, *A Rainbow Palate: How Chemical Dyes Changed the West's Relationship with Food*, University of Chicago Press, Chicago, 2020.
18. Colorants for foods, drugs and cosmetics, *Kirk-Othmer Encyclopedia of Chemical Technology (Online)*. John Wiley & Sons: New York, 2014. https://pdfslide.net/documents/ kirk-othmer-encyclopedia-of-chemical-technology-colorants-for-foods-drugs.html, accessed March 2021.
19. J. Rees, *The Chemistry of Fear: Harvey Wiley's Fight for Pure Food*, Johns Hopkins University Press, Baltimore, 2021, p. 231.

20. M. V. Orna, Carotenoids, cochineal and copper: food coloring through the ages, in *Chemistry's Role in Food Production and Sustainability: Past and Present*, ed. M. V. Orna, G. Eggleston and A. F. Bopp, American Chemical Society, Washington, DC, 2019, pp. 85–109.
21. G. T. Sigurdson, P. Tang and M. M. Giusti, *Annu. Rev. Food Sci. Technol.*, 2017, **8**, 261–280.
22. A. Aberoumand, *World J. Dairy Food Sci.*, 2011, **6**(22), 71–78.
23. S. Cotton, *Educ. Chem.*, 2005, November, 156–158.
24. D. Walters, *Fortress Plant: How to Survive when Everything Wants to Eat You*, Oxford University Press, London and New York, 2017, pp. 76–98.
25. A. Brunning, *Chem. Eng. News*, 2020, **98**(46), November, 29.
26. J. He and M. M. Giusti, *Annu. Rev. Food Sci. Technol.*, 2010, **1**, 163–187.
27. K. M. Herbach, F. C. Stintzing and R. Carle, *J. Food Sci.*, 2006, **71**(4), R41–R50.
28. S. M. Rivera and R. Canela-Graryoa, *J. Chromatorgr.*, 2012, **A 1224**, 1–10.
29. A. M. Humphrey, *J. Food Sci.*, 2004, **69**(5), C422–C425.
30. DDW, *Turmeric/Curcumin*, DDW Colour House, Louisville, KY, 2014, https://ddwcolor.com/colors/yellow-natural-food-colors/, accessed March 2021.
31. A. Weir, P. Westerhoff, L. Fabricius, K. Hristovski and H. von Goetz, *Environ. Sci. Technol.*, 2012, **46**(4), 2242–2250.
32. B. Erickson, *Chem. Eng. News*, 2021, **99**(18), 13.
33. V. K. Joshi, D. Attri, A. Bala and S. Bhushan, *Indian J. Biotechnol.*, 2003, **2**, 362–369.
34. H. Danuri, *Biosciences*, 2008, **15**(2), 61–66.
35. P. Pattanagul, R. Pinthong and A. Phianmongkhol, *Chiang Mai J. Sci.*, 2007, **34**(3), 319–328.
36. L. C. Matthew, *The Burlington Magazine*, 2002, **144**(1196), 680–686.
37. J. von Gerlach, *Mikroskopische Studien aus dem Gebiete der menschlichen Morphologie*, Ferdinand Enke, Erlangen, 1858, pp. 1–4.
38. K. P. Krafts, E. Hempelmann and B. J. Oleksyn, *Biotech. Histochem.*, 2011, **86**, 7–35.
39. W. Flemming, *Schriften des naturwissenschaftlichen Vereins für Schleswig-Holstein*, 1878, **3**, 23–27.

40. A. Winkelmann, *Clin. Anat.*, 2007, **20**(3), 231–234.
41. NIH, MedlinePlus website, Genetics Home, http://ghr.nlm.nih.gov/handbook/basics/chromosome, accessed March 2021.
42. F. H. Kasten, *Biochem. Histochem.*, 1996, **71**(1), 2–37.
43. J. S. Fruton, *Molecules and Life*, Wiley-Interscience, New York, 1972, p. 195.
44. O. N. Witt, *Ber. Dtsch. Chem. Ges.*, 1876, **8**, 522–527.
45. T. Lenoir, *Minerva*, 1986, **26**, 66–88.
46. P. Ehrlich, Contribution to the Theory and Practice of Histological Staining, in *The Collected Papers of Paul Ehrlich*, ed. H. Himmelweit, Pergamon Press, London, 1956, pp. 73–74.
47. J. Parascandola and R. Jasensky, *Bull. Hist. Med.*, 1974, **48**, 199–220.
48. M. Marquardt, *Paul Ehrlich*, Schuman, New York, 1951, p. 18.
49. B. Witkop, Paul Ehrlich: His ideas and his legacy, in *Science, Technology and Society in the Time of Alfred Nobel*, ed. C. G. Bernhard, E. Crawford and P. Sörbom, Pergamon Press, Oxford, 1982, pp. 146–166.
50. P. Mazumdar, *Bull. Hist. Med.*, 1974, **48**, 1–21.
51. Dr Ehrlich's Magic Bullet, https://en.wikipedia.org/wiki/Dr._Ehrlich%27s_Magic_Bullet, accessed March 2021.
52. Paul Ehrlich Arbeitszimmer, https://commons.wikimedia.org/w/index.php?title=File:Paul_Ehrlich_Arbeitszimmer.jpg&oldid=534474501, accessed June 2021.
53. A. S. Travis, *Sci. Context*, 1989, **3**, 383–408.
54. C. Bäumler, *Paul Ehrlich: Scientist for Life*, Holmes & Meier, New York, 1984, p. 62.
55. The Nobel Prize in Physiology or Medicine for 1984, https://www.nobelprize.org/prizes/medicine/1984/summary/, accessed March 2021.
56. A. Cambrosio and P. Keating, *J. Hist. Biol.*, 1992, **25**, 175–230.
57. G. Köhler and C. Milstein, *Nature*, 1975, **256**, 495–497.
58. N. K. Jerne, The Nobel Prize, https://www.nobelprize.org/prizes/medicine/1984/jerne/biographical/, accessed March 2021.
59. T. A. Waldman, *Science*, 1991, **252**, 1657–1662.
60. M. V. Orna, *The Chemical History of Color*, Springer, Heidelberg, 2013, pp. 113–121.

CHAPTER 16

An Evolving Universe:
The Pigments March On†

...if enough colors are dropped into the magic pot, rainbows or
any other pattern may be brought out. In fact, this is the only
way that improvement can come about – by mutation.[1]

F. Whitaker

16.1 INTRODUCTION

Evolution in the natural world takes place by mutation, so why
not among pigments, those colorful extracts from ores and organisms? Drawing on the past, creative scientists, engineers and artists are overturning our traditional definition of pigments as stuff
to slap on a surface. This chapter examines their emergence.

16.2 THE FORBES PIGMENT COLLECTION

Standing at the nexus of the distant past and an undreamed-of
future lies that Shangri-La of "cultural passion,"[2] the Forbes
Pigment Collection. Situated in the Straus Center for Conservation

---

†Glossary entries and chapter cross-references are in boldface.

March of the Pigments: Color History, Science and Impact
By Mary Virginia Orna
© Mary Virginia Orna 2022
Published by the Royal Society of Chemistry, www.rsc.org

and Technical Studies at Harvard University, it is splayed out in a vast color array presided over by Narayan Khandekar, Director of the Straus Center. The 2700-plus samples hail from Egyptian tombs to slick **nanomaterials** laboratories (Figure 16.1),[3] the brain child of Edward Waldo Forbes (1873–1969), Director of the Fogg Art Museum at Harvard from 1909 to 1944. The collection concretizes Forbes's credo that art and science are so firmly linked that a work of artwork cannot be understood without also understanding the materials from which it is made. With that statement, Forbes kicked off a new discipline: technical art history, thus following in the intellectual tradition of his maternal grandfather, Ralph Waldo Emerson (1803–1882).[4]

Although this lineup of pigments-past and pigments-present is not open to the public, an audio tour[5] introduces them: 27 selected pigments from one of the oldest, charcoal, to a very new arrival, YInMn blue, discovered in 2009. On 18 February 2021, this author arranged to inquire about their future with the person most able to read that crystal ball, Dr Khandekar himself.

Figure 16.1 Dr Narayan Khandekar, director of the Straus Center and senior conservation scientist, and conservation coordinator Alison Cariens handle objects in the Forbes Pigment Collection and the Gettens Collection of Binding Media and Varnishes (arranged side-by-side) inside the Straus Center for Conservation and Technical Studies. © President and Fellows of Harvard College; Photo: Caitlin Cunningham Photography.

16.2.1 Author's (MVO) Interview with Narayan Khandekar (NK), 18 February 2021. (Excerpts)

MVO: "Dr Khandekar, since you are right at the hub of the pigment kingdom – what do you see as evolving in the world of pigments? Where is it going in terms of usage or various impacts that pigments will have on society?"

NK: "I think that there are some things that are coming up like (Mas) Subramanians's discovery of YInMn blue. It is important, not only because of its color but also for its infrared reflective properties. So that means that you've got a paint that actually cools down things from sunlight – it's got a beautiful color – but also has a practical purpose."

ASIDE. YInMn blue, featured in **Chapter 11.6**, is the first new blue pigment to hit the artists' palette in 200 years. Khandekar tracked it down and welcomed it into the Forbes Pigment Collection and Crayola Crayons named a YInMn-inspired new shade "bluetiful" in its honor.[6] Its hefty price tag of $69.00 for just a third of an ounce thrusts it into the same category that natural ultramarine blue occupied centuries ago: worth its weight in gold (almost). Subramanian's team is now in search of a cooler, less expensive blue by using the mineral hibonite $((Ca,Ce)(Al,Ti,Mg)_{12}O_{19})$ as a substrate. The group found that by systematically substituting cobalt (Co^{2+}), nickel (Ni^{2+}) and titanium (Ti^{4+}) ions for the aluminum in the original formula, they could produce a stunning range of intense blue pigments. The new pigments are stable to acids and bases and also are infrared reflective like YInMn blue.[7] The cooling effects of reflective pigments are also a subject of intense research.[8,9]

NK: "I would think that there's lots of work in the nanoworld – there is a race toward nothing – the blackest black and so you've got these people using carbon nanotubes that are totally light absorbent. And there's...something called Morgan white which has a flat reflection across the visible spectrum and so it's different from the other white pigments that you have available: you can mix it with something without adding or subtracting or adjusting the color in any way."

ASIDE. In an attempt to grow carbon nanotubes (structures that resemble rolled up chicken wire, but more than 50 000 times smaller than the diameter of a human hair) onto electrically conducting materials, MIT researchers Brian Wardle and Kehang Cui inadvertently stumbled on the blackest black substance that absorbs 99.995% of incoming light from every angle. They see applications for this ultrablack[10] in optical blinders that reduce unwanted glare and to help space telescopes spot orbiting exoplanets.[11] A new "whitest white," Morgan™White, is a proprietary laboratory-synthesized particulate substance with an exceptionally flat response across the spectrum. Its developer, NanoLab, describes it as a pigment in flake form of nanoscale thickness that can be added as a filler to any substance without adding a hint of tint.[12] It has a competitor in an ultrawhite $BaSO_4$ paint developed by Xiulin Ruan and co-workers at Purdue University.[13]

MVO: "Do you see any developments that you might call a mutation, something that actually changes direction, something that's never been done before?"

NK: "There is something that someone called Natsai Chieza is working with using bacteria to produce dyes; the difference between dyes and pigments is nominal in a way – you can always turn a dye into a pigment – and so she's using bacteria and DNA to produce dyed fabric and that, I think, is interesting. It's something that is not really done much."[14]

MVO: "I've been reading about that in terms of green chemistry. People are trying to get away from the terrible pollution that so many of these pigment manufacturers have done in the past like putting gobs of SO_2 into the atmosphere."

NK: "On the subject of green chemistry and environmental concerns – there are people taking carbon particles out of the air and turning it into ink."[15]

MVO: "Aha. Now that's a real breakthrough. I was wondering, what other influences might have given rise to any new pigment uses. What about polymers and the rise of the different media like **acrylics** and the **alkyds** and their becoming very popular among amateurs?"

NK: "I think that different pigments have different colors in different media and so it is obvious that there is one pigment or one color that may not work so well with oil or acrylic and so you've got these different manufacturers featuring pigments. Some of them work well across the board, but they give different tonalities, different colors and they age differently as well. So the choice of pigments and binding media is crucial."

MVO: "What about the Fine Arts? If you look at the past many pigments are used by artists but also are used as workhorses in industry whereas in industry they don't care how long these pigments will last – whether they are fugitive or not as long as they last the lifetime of the object. But an artist is looking for immortality maybe by using the very same pigments. What would you say is the difference between the short term *versus* the long term usage of pigments?"

NK: "I think that artists will use whatever it is that they need to get the idea of their world across and it's up to us as museums or collectors to work with that – it's not up to us to tell people what's wrong, what works – it's up to the artist to make those decisions."

MVO: "What would you say that probably is the most important project or work that you're doing right now in terms of either the Straus Center or the Pigment Collection?"

NK: "I think that the work we do in the Straus Center is really preserving the collection of the Harvard Art Museum and understanding it. We want to look after it and we also want to find out what it's made of and our library has standards that we use to understand the material and techniques of artists and that is an amazing thing – when you look at a painting – we found a painting that was, incredibly, in original condition from 1605. We have to look at the pigments by that artist and see if they tally with the inventory in the Spanish royal collection from that artist's time and identify the pigments that were being used – in a way to have a conversation with that artist – because every work of art is a series of decisions that were made and so we are able to interrogate the work of art under the same conditions as the artist and I think that's a really important aspect, the core of the work that we do – actually understanding a work of art and its material."

MVO: "How do you use the pigments in the Forbes Collection?"

NK: "We use it daily – when we are analyzing pigments we need standards – if we are analyzing a Canaletto, we're looking at the sky and we can verify that it is Prussian Blue. We use it a lot for teaching – to show students original materials like cochineal beetles or a chunk of ochre."

The interview ended with a discussion on the kinds of students who can take advantage of the collection and the role of art conservation in the curriculum.

As we shall see in the rest of this chapter, the pigments-yet-to-come march on into the more ephemeral realms of nanotechnology and emergence from a biochemical ooze.

16.3 STRUCTURAL COLOR

Mother Nature, by an evolutionary process, invented the structures that display the iridescent colors in butterfly wings (Figure 16.2)[16] long before physicists began studying them in the late 19th century. Now called photonic crystals, and finding use as

Figure 16.2 Structural Color. Left: Natural iridescent colors in the butterfly, *Teinopalpus imperialis*. Reproduced from ref. 16, the terms of the CC BY-SA 4.0 license, https://creativecommons.org/licenses/by-sa/4.0/deed.en. Right: Plasmonic colors. Reproduction of *Monet's Impression, Sunrise* (Musée Marmottan Monet) using the expanded palette by color toning and mixing strategies. (A) Original input image. Reproduced from ref. 21. (B) Reproduction using only the limited palette of "primary plasmonic colors" falls short of the original image. (C) Realistic reproduction of the artwork using an expanded palette of colors, allowing for the subtle variations in tone and color in the original to the replicated. (D) Higher magnification image of dotted box in (C) highlighting the brush strokes that are resolved in the plasmonic painting. (E) Tilted scanning electron microscope image of pixels showing how the colors observed in the plasmonic painting are the manifestation of a predefined structural layout. B, C, D, E Reproduced from ref. 20 with permission from American Chemical Society, Copyright 2014.

light guides in fiber optic devices, they display a range of colors due to light scattering or interference. Due to their unique optical properties, that include iridescence, strong reflectance and resistance to bleaching, new formulations of photonic pigments will be in high demand. Creation of photonic materials libraries can provide low-cost access to a range of possible applications.[17]

The science of **plasmonics** is another subset of structural color. When the interference or scattering is caused by orderly arrays of highly reflective metallic materials such as aluminum **nanoparticles**, called plasmonics, they can reproduce the yellow, magenta and cyan color space in ultra-high-resolution (**Chapter 1.5**).[18] This allows for their formulation as plasmonic pigments for incorporation into paints, inks, cosmetics and color displays[19,20] (**Figure 16.2** right).[21] Ironically, the imaging principle that inspired this work was the earliest semi-empirical photographic method, daguerreotyping, that formed images consisting of silver-mercury nanoparticles. The method, introduced in 1839, was able to reproduce images of extraordinary resolution and quality – it is indeed the first example of plasmonic color printing, an art accomplished long before the science was understood and developed.[22]

16.4   HIGH PERFORMANCE PIGMENTS

Traditional pigments are defined as insoluble particulate colorants that are stable during processing and during their useful lifetime. This definition applies most especially to a workhorse group technically called complex inorganic colored pigments (CICPs). What makes members of this group "high performance" are qualities that are highly desirable in today's economy: sustainability in terms of manufacture, environmental friendliness, economy of scale, increased functionality (*e.g.*, corrosion inhibition, optical variability, heat-reflecting characteristics) and better technical performance (*e.g.*, ease of application, higher tinting strength). Chemically, as a class, they can be called "mixed metal oxides," namely oxides containing more than one metallic element in their formulas.

Because the color range of these pigments is broad, the aesthetics are vivid, and the utility is high, these pigments will continue to be in great demand. Though artistic use is limited, industrial and architectural use will burgeon with their use as protective coatings, preventing the damaging effects of ultraviolet light, or averting heat buildup by reflecting infrared light.[23] It is also quite possible that some really ancient pigments, like ultramarine blue, could be revitalized by improved manufacturing technology, and that they too could join the ranks of high performance pigments with their enhanced physical properties.[24]

16.5   MICROBIAL PIGMENTS

Can you imagine painting with *Penicillium* or dipping your dashiki in a golden yellow bath of *Staphylococcus*? Imagine no more. In response to the public's growing concern over the harmful environmental impact of petroleum-derived organic pigments, colorant manufacturers are giving more attention to the desirability of natural substitutes, and especially of microorganisms, for use in the food, textile and cosmetic industries. Development of new pigments for these uses will also affect artists who are always on the lookout for novel materials to express their own visions of reality.

In the world of textiles, Dr Khandekar already alluded to the use of bacteria to dye textiles. While we have been using lichen and mushrooms for that purpose for centuries, bacteria, especially genetically engineered ones, are new to the game. Chief scientific officer and co-founder of Tinctorium Tammy Hsu's approach to "greener" blue jeans is a case in point. She overcame the insolubility of the desired dye, indigo, by genetically tricking *Escherichia coli* into synthesizing water-soluble indican instead. Subsequent hydrolysis to indigo right on the fabric avoids the use of corrosive and eco-unfriendly chemicals.[25,26] New gene splicing techniques like CRISPR (clustered regularly interspaced short palindromic repeats) can enable a talented biochemist to program various bacterial species to produce a desired dye with the snip of a genetic scissor. A Korean team succeeded in synthesizing blue-pigmented actinorhodin from *Streptomyces coelicolor* by this method.[27]

Lots of microorganisms can gussy up our veggies and burgers with a spectacular show of pigments. Yeasts have stepped up to the plate with some great pinks and orange-reds. Algae specialize in reds, yellows and oranges. But bacteria are the champions with blues, greens and purples as well as a good range of reds, oranges and yellows. An added value is that many of these colors have health benefits as antioxidants, anti-inflammatories, antimicrobials and immunosuppressants.[28] Some key food pigments are the **carotenoids** canthaxanthin, zeaxanthin, lycopene and β-carotene.[29] Bacterial pigments are a safe, but untapped, resource for the classroom as well. Using bacteria-derived pigments, students can create living art by painting on agar medium or by mixing them with conventional acrylic binders.[30]

Lest our fungal friends feel left out, let be it known that they have attracted notice as sources of an extraordinary range of colors (Figure 16.3).[31] A major challenge to their use on an industrial scale is development of a cheap and efficient extraction technology. Due to their tendency to form mycotoxins along with pigments, fungal sources are currently banned for use in foods and pharmaceuticals. This is another major difficulty likely to be solved by genetic engineering.

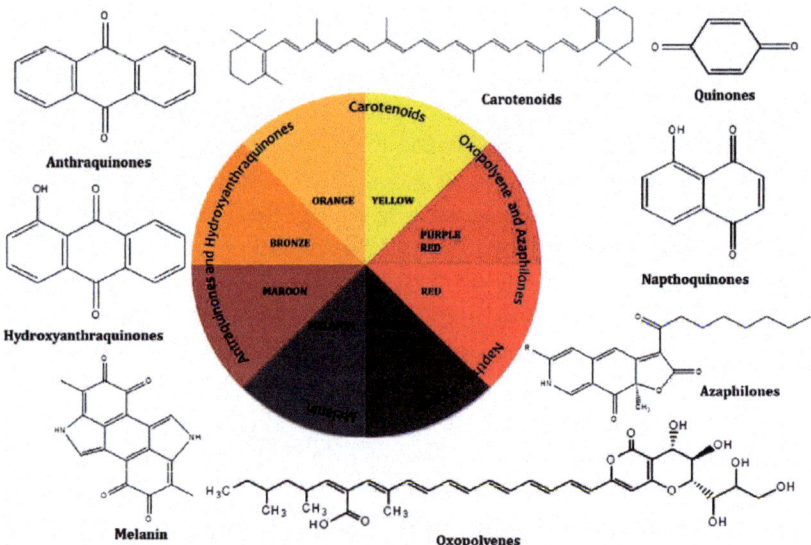

Figure 16.3   Color range of fungal pigments surrounded by the formulas of the different classes of compounds that produce these colors. Reproduced from ref. 31, https://doi.org/10.3389/fchem. 2020.00369, under the terms of the CC BY 4.0 license, https://creativecommons.org/licenses/by/4.0/.

16.6 SEMICONDUCTING PIGMENTS

There was a day when the transistor radio was the hottest new device in the teenage ambit. No one ever dreamed that its circuitry might be replaced by – of all things – pigments from the lowly shellfish (**Chapter 6**).

We saw in **Chapter 1.6** that many pigments exhibit color due to electrons undergoing transitions which correspond to discrete energy levels from 400 to 700 nm in the pigment molecule. However, some pigments are colored due to an entirely different mechanism that is common to transistors: semiconduction. Semiconductors have an electrical conductivity midway between that of metals and of insulators such as glass. This confers on them a range of useful properties such as passing current more easily in one direction than another or showing increased sensitivity to light or heat. Controlled introduction of impurities into the semiconductor crystal can modify these properties, making them useful as electronic switches, amplifiers and detectors. The most commonly used semiconducting materials have been silicon- and germanium-based. Modern electronics relies almost entirely on semiconductor technology.

Even though the semiconducting properties of some organic compounds were discovered as early as the 1960s,[32,33] semiconducting pigments‡ were largely overlooked.§,[34] It took over 40 years for them to be recognized as inexpensive, sustainable, flexible, biodegradable and biocompatible substitutes for traditional semiconductors. They show promise for such applications as 'smart' radio-frequency identification (RF-ID) packaging tags, incorporation into textiles and as wearable electronic devices. **Carotenoids**,[35] **indigo**, dibromoindigo (Tyrian purple)[36] and isoindigo (a geometric isomer of indigo)[37] derivatives have also garnered considerable attention for their potential applications in the biomedical field as resorbable implants, in regenerative medicine and for *in vivo* imaging. Additional

---

‡Actually, semiconducting pigments like aniline black were successfully used commercially as fabric dyes dating back to 1860, although their semiconducting nature was not known at that time.

§Note: This groundbreaking paper was the basis of the awarding of the Nobel Prize in Chemistry in 2000 to three of the co-authors, Shirakawa, MacDiarmid and Heeger. It spawned a tsunami of innovative electronic devices including organic light emitting diodes (OLEDs) that are ubiquitous in smartphones and high-definition TV.

applications inspired by natural indigo are in the areas of transistors, photovoltaics, photoacoustic imaging, photodetectors and organic lasers.[38] A start-up company, Natron Energy, is designing safe and reliable batteries using a traditional blue pigment, Prussian Blue, as both the anode and the cathode. While these batteries may not have the power density to replace lithium-ion systems, they are reliable substitutes in applications that need quick power discharge like small electrical grids.[39]

16.7 "ERGOCHROMIC" PIGMENTS: THERMO-, PHOTO-, MECHANO-, ELECTROCHROMISM

Perhaps some of you remember the "mood rings" enclosed as "prizes" in some cereal boxes which, when worn, exhibited a range of colors. This late-stage exploitation of thermochromism, or color change on application of heat, was well-known to the Paleolithic artists who changed yellow ochre into red ochre in their hearths. Today we know that virtually any stimulus can effect a color change in certain pigments and these have been the subject of intense study in search of new applications. I'd like to limit my remarks to pigments that I've called "ergochromic," *i.e.*, pigments that change color when energy in the form of heat, light, mechanical stress or electricity impinges upon them and that the color change can be reversible or irreversible, depending upon the application.

16.7.1 Thermochromism

A wide range of organic and inorganic pigments exhibit thermochromism, which may be reversible or irreversible depending on the material. For example, some copper compounds undergo a change in geometry accompanied by a color change from green to yellow on the application of heat. Likewise, some organic compounds, like Schiff bases and conjugated polymers, can display a range of colors due to molecular rearrangement when heated. Applications involve temperature indicators in the food industry, thermal printing in the textile industry, and environmental control of building temperature using "smart" windows.

16.7.2 Photochromism

Photochromism is familiar to us in light-sensitive sunglasses and color changes on toys and T-shirts. They also find significant use as security markers for documents, clothing and packages. A typical mechanism for the effect is that UV light cleaves a chemical bond causing a color change in the molecule, which reverts to its original colorless form when the light is removed. Some main classes of compounds that show this behavior are spiropyrans and naphthopyrans.[40]

16.7.3 Mechanochromism

At 10:22 PM on Monday, 3 May 2021, a subway overpass in Mexico City collapsed, plunging a passenger train to the ground and killing dozens of people. Could this accident have been prevented? What if the bridge had been covered with a pigment which, on slight stretching, changed color and revealed the points of stress on the structure beforehand? Such a material exists in a fast-acting "mechanophore" – a molecule that changes color in response to physical stress. The compound, oxazine, changes color reversibly upon repeated stretching and releasing.[41] Experts predict that the impact of mechanoforce probes will continue to grow.[42]

16.7.4 Electrochromism

Electrochromism (EC) is an oxidation/reduction process that involves the passage of an electric current to provoke a color change that is reversible. Essentially color-switching by loss or gain of electrons, it was first reported in 1815 by Jöns Jacob Berzelius (1779–1848) when he noticed that pale-yellow tungsten oxide (WO_3) changed, on reduction, into a deep blue form.[43] In 1969, S.K. Deb documented the first EC device.[44] Among its first commercial applications were voltage-regulated windows and switchable rear-view mirrors, but lack of electrochromics exhibiting a full range of colors over the visible spectrum limited application development. A breakthrough came in 2010, from the work of a team headed by John R. Reynolds, then working at the University of Florida. Following a 10 year systematic structure–property study of electrochromic polymers, his group reported on the first set of soluble materials that exhibited the

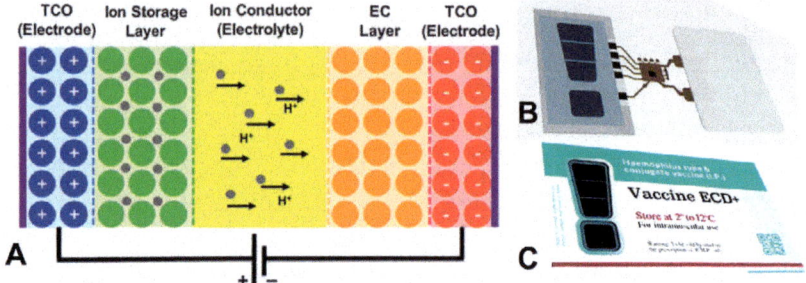

Figure 16.4 (A) Bare bones of an EC device. Reproduced from ref. 46 with permission from the Royal Society of Chemistry. (B) EC device connected for an application and (C) Cover for the application. Image courtesy of Ynvisible Interactive Inc.

whole gamut of colors possible.[45] Figure 16.4A[46] is a schematic to show how an EC device works. The two electrodes consist of transparent conducting oxides (TCO), the EC layer is the electrochromic material and the ion storage layer protects the TCO from ion insertion or from irreversible color change. The conducting layer is mainly responsible for carrying the charge from a power source to the EC layer. (B) and (C) show how the device is connected for an application.[47]

This advance opened the floodgates in the development of laboratory-based color-changing electrochromic devices with a wide range of applications such as for textiles and energy-efficient buildings, displays, self-dimming mirrors for automobiles, electrochromic e-skin, textiles, and smart windows for energy-efficient buildings, logistics monitoring, consumer electronics and medical technology.[48]

16.8 CONCLUSION

Our colorful journey has taken us from the vast reaches of unrecorded time up to a vibrant and vigorous present. Throughout this span of time, the pigments have marched to different drummers in an unbroken succession of new developments. Little by little, the pigment palette expanded, sometimes empirically, sometimes by accident, sometimes by a systematic scientific study. At times, there were quantum leaps, mutations, to entirely new materials; at others, there was a gradual, steady and often deliberate evolution of a technology that was handed down for centuries. But in every period, we have witnessed the indomitable resolve and dazzling genius of the human spirit at work. Pigments once thought to be good only for coloring a surface have now been put to work in energy-efficient and life-saving devices. It seems that their applications are endless but, since this book will not be endless, the reader is invited to continue observing future developments as they unfold in other sources.

REFERENCES

1. F. Whitaker, *The Artist and the Real World*, North Light Publishers, Westport, CT, 1980, p. 35.
2. S. Sharma, *The New Yorker*, 2018 (September 3), 28–32.
3. Forbes Pigment Collection, https://harvardartmuseums.org/article/pigment-collection-colors-all-aspects-of-the-museums, accessed June 2021.
4. M. N. Bolotnikova, Up Close (Virtually) with the Forbes Pigment Collection, *Harvard Magazine*, 6 November 2020, https://harvardmagazine.com/2020/11/up-close-virtually-with-the-forbes-pigment-collection, accessed May 2021.
5. A History of Color: An Audio Tour of the Forbes Pigment Collection, https://harvardartmuseums.org/tour/a-history-of-color-an-audio-tour-of-the-forbes-pigment-collection, accessed May 2021.
6. E. N. Brown, It's not every day we get a new blue, *New York Times*, 8 February 2021.
7. B. A. Duell, J. Li and M. A. Subramanian, *ACS Omega*, 2019, **4**(26), 22114–22118.
8. J. Mandal, Y. Fu, A. C. Overvig, M. Jia, K. Sun and N. N. Shi, *et al.*, *Science*, 2018, **362**, 315–319.
9. Y. Chen, J. Mandal, W. Li, A. Smith-Washington, C.-C. Tsai and W. Huang, *et al.*, *Sci. Adv.*, 2020, **6**, eaaz5413.
10. The MIT black is blacker than Surrey NanoSystems Vantablack, https://www.surreynanosystems.com/about/vantablack, accessed April 2021.
11. K. Cui and B. L. Wardle, *ACS Appl. Mater. Interfaces*, 2019, **11**(38), 35212–35220.
12. Morgan-white, NanoLab, https://www.nano-lab.com/morgan-white.html, accessed April 2021.

13. X. Li, J. Peoples, P. Yao and X. Ruan, *ACS Appl. Mater. Interfaces*, 2021, 21733–21739.

14. A. Peters, These gorgeous colors come from dye made by bacteria, not chemicals, https://www.fastcompany.com/90257662/these-gorgeous-colors-come-from-dye-made-by-bacteria-not-chemicals, accessed April 2021.

15. C. Voon, Converting air pollution into inks and pigments for artists, *Hyperallergic*, https://hyperallergic.com/319765/converting-air-pollution-into-inks-and-pigments-for-artists/, accessed April 2021.

16. Teinopalpus imperialis imperatrix, https://commons.wikimedia.org/w/index.php?title=File:Teinopalpus_imperialis_imperatrix.jpg&oldid=491427896, accessed June 2021.

17. Y.-L. Li, X. Chen, H.-K. Geng, Y. Dong, B. Wang and Z. Ma, *et al.*, *Angew. Chem., Int. Ed.*, 2021, **60**, 3647–3653.

18. D. Franklin, Z. He, P. M. Ortega, A. Safaei, P. Cencillo-Abad, S.-T. Wu and D. Chanda, *Proc. Natl. Acad. Sci. U. S. A.*, 2020, **117**(24), 13350–13358.

19. R. Brazil, *ACS Cent. Sci.*, 2020, **6**(3), 332–335.

20. S. J. Tan, L. Zhang, D. Zhu, X. M. Goh, Y. M. Wang and K. Kumar, *et al.*, *Nano Lett.*, 2014, **14**, 4023–4029.

21. Claude Monet: Impression, Sunrise, https://commons.wikimedia.org/w/index.php?title=File:Monet_-_impression-sunrise.jpg&oldid=540611988, accessed June 2021.

22. A. Schlather, P. Gieri, M. Robinson, S. A. Centeno and A. Manjavacas, *Proc. Natl. Acad. Sci. U. S. A.*, 2019, **116**(28), 13791–13798.

23. M. C. Comstock, *J. Surf. Coat. Aust.*, 2016, December, 10–29.

24. G. Buxbaum, Introduction to inorganic high performance pigments, in *High Performance Pigments*, ed. E. B. Faulkner and R. J. Schwartz, Wiley, New York, 2nd edn, 2009, pp. 3–6.

25. E. Landhuis, *Chem. Eng. News*, 2019, **97**(44), 22–25.

26. T. M. Hsu, D. M. Welner, Z. N. Russ, B. Cervantes, R. L. Prathuri, P. D. Adams and J. E. Dueber, *Nat. Chem. Biol.*, 2018, **14**, 256–261.

27. S. Cho, J. Shin and B.-K. Cho, *Int. J. Mol. Sci.*, 2018, **19**(4), 1089.

28. T. Sen, C. J. Barrow and S. K. Deshmukh, *Front. Nutr.*, 2019, **6**, 7.

29. K. Heer and S. Sharma, *Int. J. Pharm. Sci. Res.*, 2017, **8**(5), 1913–1922.

30. L. K. Charkoudian, J. T. Fitzgerald, C. Khosla and A. Champlin, *PLoS Biol.*, 2010, **8**(10), e1000510.
31. R. Kalra, X. A. Conlan and M. Goel, *Front. Chem.*, 2020, **8**, 369. https://pubmed.ncbi.nlm.nih.gov/32457874/.
32. *Organic Semiconductors. Proceedings of an Inter-industry Conference*, ed. J. J. Brophy and J. W. Buttrey, Macmillan, New York, 1962.
33. K. Okamoto and W. Brenner, *Organic Semiconductors*, Reinhold, New York, 1964.
34. H. Shirakawa, E. J. Louis, A. G. MacDiarmid, C. K. Chiang and A. J. Heeger, *J. Chem. Soc., Chem. Commun.*, 1977, **16**, 578–580.
35. K. S. Sattiraju and H. S. Potapragada, *Adv. Res. Elect. Electron. Eng.*, 2015, **2**(10), 97–101.
36. E. D. Glowacki, G. Voss, L. Leonat, M. Irimia-Vladu, S. Bauer and N. S. Sariciftci, *Isr. J. Chem.*, 2012, **52**, 540–551.
37. Y. Wang, Y. Yu, H. Liao, Y. Zhou, I. McCulloch and W. Yue, *Acc. Chem. Res.*, 2020, **53**, 2855–2868.
38. K. J. Fallon and H. Bronstein, *Acc. Chem. Res.*, 2021, **54**(1), 182–193.
39. M.-A. Watson, *Chem. Eng. News*, 2020, **98**(44), 40–41.
40. P. Bamfield and M. G. Hutchings, *Chromic Phenomena: Technological Applications of Color Chemistry*, RSC Publishing, Cambridge, 2010, pp. 9–59.
41. H. Qian, N. S. Purwanto, D. G. Ivanoff, A. J. Halmes, N. R. Sottos and J. S. Moore, *Chem*, 2021, **7**(4), 1080–1091.
42. B. Halford, *Chem. Eng. News*, 2021, **99**(9), 5.
43. R. J. Mortimer, *Am. Sci.*, 2013, **101**(1), 38–55.
44. S. K. Deb, *Appl. Opt.*, 1969, **8**, 192–195.
45. A. L. Dyer, E. J. Thompson and J. R. Reynolds, *ACS Appl. Mater. Interfaces*, 2011, **3**(6), 1787–1795.
46. M. H. Chua, T. Tang, K. H. Ong, W. T. Neo and J. W. Xu, Introduction to Electrochromism, in *Electrochromic Smart Materials: Fabrication and Applications*, ed. J. W. Xu, M. H. Chua and K. W. Shah, Royal Society of Chemistry, Cambridge, 2019, pp. 1–21, Figure 1.1, p. 3.
47. Ynvisible Interactive, Inc. https://www.ynvisible.com/products/segment-display, accessed June 2021.
48. J. W. Xu, M. H. Chua and K. W. Shah, *Electrochromic Smart Materials: Fabrication and Applications*, RSC Publishing, Cambridge, 2019.

Glossary

Acrylics. An acrylic polymer is polymethylmethacrylate formed by the co-polymerization of vinyl acetate and methyl methacrylate. Scheme 1 shows the formation of the monomer, vinyl acetate, from acetic acid and ethylene, and its subsequent addition to methyl methacrylate to form the polymer, polymethylmethacrylate (not shown).

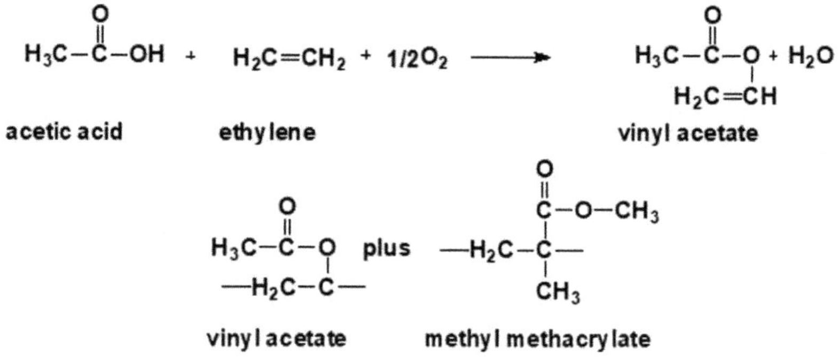

$$H_3C-\overset{\overset{O}{\|}}{C}-OH \; + \; H_2C{=}CH_2 \; + \; 1/2\,O_2 \longrightarrow H_3C-\overset{\overset{O}{\|}}{C}-O \; + \; H_2O$$

$$H_2C{=}CH$$

acetic acid   ethylene   vinyl acetate

$$H_3C-\overset{\overset{O}{\|}}{C}-O \quad \text{plus} \quad -H_2C-\overset{\overset{\overset{O}{\|}}{C}-O-CH_3}{\underset{CH_3}{C}}-$$

$$-H_2C-C-$$

vinyl acetate   methyl methacrylate

March of the Pigments: Color History, Science and Impact
By Mary Virginia Orna
© Mary Virginia Orna 2022
Published by the Royal Society of Chemistry, www.rsc.org

The most popular acrylics are water-based emulsions which are quick drying, flexible, impervious to cracking, mix well with other media, and are inexpensive. With an almost limitless spectrum of vivid colors embedded into an acrylic binder, many artists looked no further for their perfect medium.

Alkyd resin. An alkyd resin forms by reaction of a monoglyceride ester of glycerol (an alcohol) and a fatty acid with an acid anhydride such as phthalic anhydride to form an alkyd polymer according to Scheme 2 (*"al"* + *"cid"* = "alcid" modified to "alkyd").

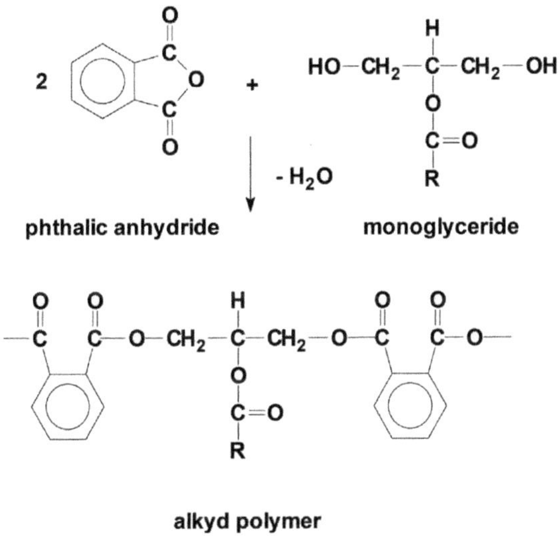

phthalic anhydride   monoglyceride

alkyd polymer

Anthraquinoids, Anthraquinone. Anthraquinoids are colored derivatives of the colorless parent molecule, anthraquinone. Anthraquinone has eight substitution points on which groups such as –OH and –COOH, called auxochromes, can be placed. These groups change the internal electronic energy levels of the molecule, enabling electronic transitions in the visible region of the spectrum, yielding a range of red to yellow colors. Each of the colorants shown is a unique variant of anthraquinone. Table 1 shows the structures of anthraquinone and some of its major colored derivatives.

Table 1 The structures of anthraquinone and its colored derivatives.[a]

Position on anthraquinone	1	2	3	4	5	6	7	8
Anthraquinone	H	H	H	H	H	H	H	H
Alizarin	OH	OH	H	H	H	H	H	H
Purpurin	OH	OH	H	OH	H	H	H	H
Pseudopurpurin	OH	OH	COOH	OH	H	H	H	H
Carminic acid	CH_3	COOH	OH	H	OH	OH	Gp	OH
Kermesic acid	CH_3	COOH	OH	H	OH	OH	H	OH
Flavokermesic acid	CH_3	COOH	OH	H	H	OH	H	OH
Laccaic acids	COOH	COOH	OH	H	OH	OH	V	OH

[a]Gp = glucopyranose $(C_6H_{11}O_5)$; V = variable sidechain.

Aromatic, aromatic amines. Historically, compounds described as "aromatic" had pleasant odors. It was later found that many of these compounds contained a benzene ring, C_6H_6, bonded to other substituents. As chemists use the term today, an aromatic compound generally contains at least one benzene ring, although not all aromatic compounds have pleasant odors, and occasionally "aromaticity" can include other compounds with a "benzene-like" electronic arrangement. A benzene ring bonded to one or more amino groups, $-NH_2$, is an aromatic amine. Aniline, $C_6H_5NH_2$, is the simplest of this type; its structure is shown in the next entry (A).

Azo dyes; azo pigments are characterized by the presence of the azo linkage, $-N=N-$. Aniline (A), an aromatic amine, is commonly the starting material to synthesize azo compounds. Azobenzene (B) is the simplest azo compound and the starting material for many modern dyes and pigments. C is an azo pigment, Pigment Yellow 111 (aka Hansa Brilliant Yellow). (Color code: Gray = carbon; red = oxygen; blue = nitrogen; white = hydrogen; green = chlorine, Scheme 3)

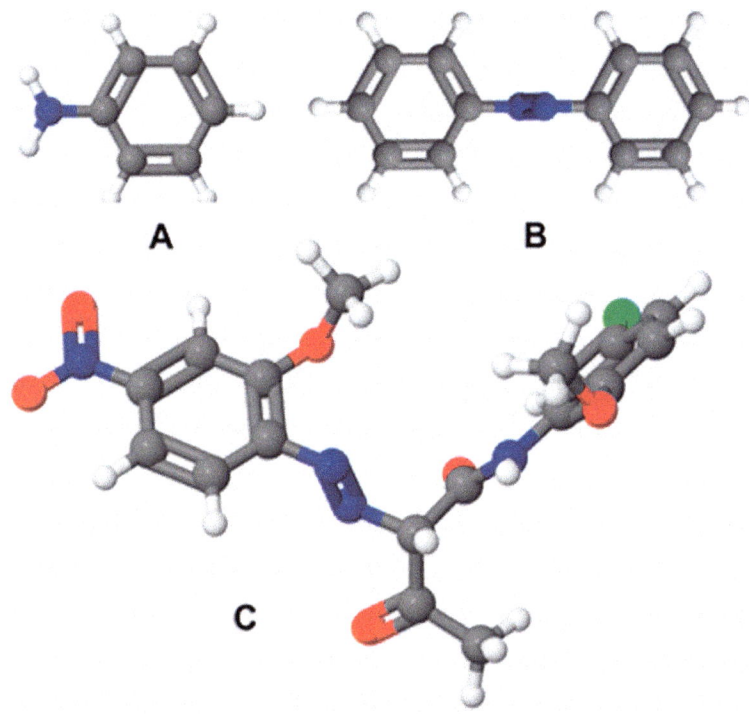

A **B**

C

Carotenoids. Carotenoids are the pigments that produce the bright yellows, reds and oranges in plants, algae and some bacteria. There are over 600 naturally-occurring carotenoids, of which two major classes, the carotenes (hydrocarbons) and the xanthophylls (alcohols), occur in common fruits and vegetables. Among the carotenes, β-carotene is the chief colorant in carrots and lycopene confers the red color on tomatoes, both with the molecular formula $C_{40}H_{56}$. Lutein and zeaxanthin are two important xanthophylls that impart the yellow color to egg yolks and sweet corn, respectively. Both have the molecular formula, $C_{40}H_{56}O_2$. Lutein has been associated with the prevention of age-related vision loss.

Chelate, chelating. A chelate complex is so-called because of the claw-like structure (chela = claw in Greek) formed by some colorants with metal ions. For example, an aluminum atom can coordinate with two carminic acid molecules through the 5-hydroxy oxygen and the adjacent carbonyl oxygen to form a six-membered chelate ring structure.

The color of this chelate complex is dark red in alkaline solution. If tin, Sn^{2+}, substitutes for the aluminum, the complex becomes a bright scarlet, sometimes called Drebbel's scarlet (**Chapter 11**).

Chromophore. A chromophore is an atom or a group of atoms whose presence on a molecule is responsible for the color exhibited by the molecule.

Conjugated system. When a molecular framework, usually either in chains or rings of carbon atoms, is linked by alternating single and double bonds, the system is said to be conjugated. This "joining" takes place through available "p" electrons that can participate in the bonding throughout the dimensions of the molecule. This delocalization of the electrons lowers the overall energy of the molecule, conferring greater stability on the system.

Dracorhodin $(C_{17}H_{14}O_3)$; **dracorubin** $(C_{32}H_{24}O_5)$ are the principal components of the crimson red resin from dragon's blood trees. Both compounds belong to the bioactive flavonoid family.

Energy dispersive X-ray spectroscopy (EDX) involves focusing an electron beam on a sample and recording the energy and number of the X-rays emitted, yielding information on the identity and relative abundance of each element present.

Enzyme. An enzyme is a protein biosynthesized by a living organism for the purpose of catalyzing a specific reaction necessary for the proper functioning of that organism.

Fiber optic reflectance spectroscopy (FORS). A technique that utilizes a fiber optic probe to measure and analyze the reflectance spectrum of a surface for identification purposes.

Fluorescence spectroscopy. A method whereby appropriate ultraviolet radiation excites the molecules in a sample causing emission of (usually) visible radiation which, when measured, can identify and quantify the specific material being examined.

Fourier transform infrared spectrophotometry (FTIR). A digital technique used to collect high-resolution spectral data over a wide frequency range to produce an infrared spectrum of a sample, allowing for identification of the component or components in the sample.

Functional group. A specific group of atoms in a molecule with its own characteristic properties that affects the physical and chemical properties of the parent molecule. Some important functional groups are alcohols (C–OH), ketones (C=O), carboxylic acids (COOH) and amines (C–NH$_2$).

Gallate ligands; Gallic acid. Gallic acid is a polyphenol, *i.e.*, a compound with several alcohol (OH) groups bonded to a benzene ring. A major component of tannins, it can chemically react with iron in solution to produce black products. The final product is formed stepwise according to the reaction (Scheme 5) where gallate ligands (anything that binds, from Latin "ligare," to bind) bind to newly-oxidized Fe(III) to form bis(gallato) diiron(III), commonly called iron gall ink (IGI):

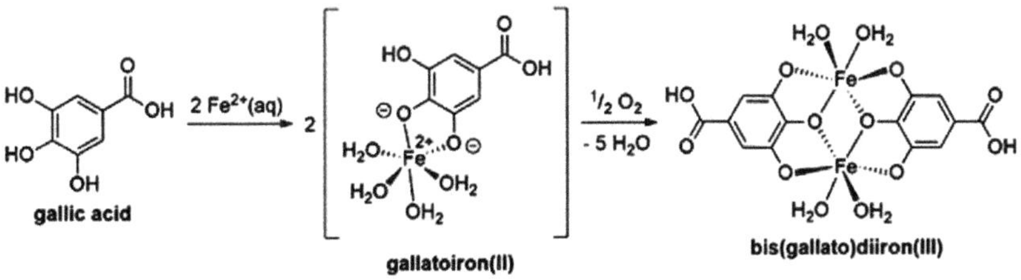

Gas chromatography/mass spectrometry (GC/MS); (pyr/GC/ MS). This tandem instrumental method first separates the components in a mixture (by GC) and as they flow out of the chromatographic column, they enter the MS portion of the instrument where they are identified according to their molar masses. GC/ MS is one of the most powerful and accurate tools for analyzing organic compounds such as synthetic organic pigments (SOPs). Coupled with pyrolysis (pyr) by which samples are heated to the decomposition point before entering the GC, this becomes the method of choice for identifying polymeric materials such as pigment binders.

Half-life. The half-life is the time it takes for a given amount of a radioactive isotope to decay to half of its original activity. The time ranges are from billionths of a second to billions of years. Since half-life is an atomic property, no amount of physical or chemical stimulus has any effect on its duration. Because for each half-life duration the quantity remaining is halved, the quantity can approach, but never reach, zero. On average, it takes about ten half-lives for a radioactive substance's activity to reach a virtual (undetectable) "zero" indistinguishable from the background radiation which is part of the environment.

Heavy metals. Heavy metals are metallic elements with fairly high densities and toxicities at low concentrations. Included are arsenic (As), cadmium (Cd), chromium (Cr), lead (Pb), mercury (Hg) and thallium (Tl).

Hydrogen bonding. Hydrogen bonding is a force of attraction between molecules that is weaker than a true covalent or electrostatic bond but stronger than van der Waals forces. The bond is typically formed between a hydrogen covalently bonded to a highly electronegative atom (such as F, O or N) and another electronegative atom on an adjacent molecule.

Indigo. Indigo is an economically important colorant that exists in two forms that can interchange reversibly by an oxidation–reduction reaction. The left-hand structure is blue, oxidized indigo, insoluble in water; the right-hand structure is the near-colorless, leuco reduced form, which is water soluble. For indigo, X = H. When both X's = Br, the compound is purple and is called 6,6'-dibromoindigo (DBI). When only one of the X's = Br, the compound is bluish-purple and is called 6-bromoindigo or monobromoindigo (MBI).

Induced luminescence (VIL and UVL). Induced luminescence is a technique whereby electromagnetic radiation impinges on a sample, usually a pigment, which then absorbs and re-emits the radiation at a longer wavelength. If visible light (visible induced luminescence, VIL) is used, the emission is usually in the infrared region; if ultraviolet light (ultraviolet induced luminescence, UVL) is used, the emission is usually in the visible region. The emitted light can be recorded photographically or visually. This method has been able to detect the presence of pigments invisible to the naked eye or even to any visual detection equipment.

Inorganic. Inorganic chemistry has traditionally been defined as the chemistry of all elements and their compounds excluding carbon. This is tricky since some carbon compounds, such as carbonates, are considered inorganic, so the definition is not hard and fast. Furthermore, some animals secrete shells that consist largely of calcium carbonate, an "inorganic compound" made by a living creature. It is no wonder that some textbooks decline to define "inorganic."

Isotopes. Isotopes are atoms having the same atomic number (number of protons) but different atomic masses (due to different numbers of neutrons). For example, carbon-12 is the isotope of carbon containing six protons and six neutrons; carbon-14 contains six protons and eight neutrons.

Lake. Pigments soluble in water are called dyes. They normally cannot be applied directly to surfaces with a binder because of their inherent solubility and tendency to bleed, as well as their poor covering power: they are largely transparent. To transform them into artists' pigments, it is necessary to render them insoluble. This is often done by dissolving the dye in an alkaline medium by adding alkali (often calcium or sodium salts as carbonates) and alum (potassium aluminum sulfate) to precipitate the colorant onto a substrate of aluminum oxide, forming the insoluble lake pigment.

Macrocyclic compounds are molecules or ions containing twelve or more members in a ring. Some important pigments and other colored compounds belong to this class of organic compounds, including phthalocyanine pigments, heme, a major blood component, and chlorophyll, found in green plants.

Phthalocyanine   **Heme**   **Chlorophyll**

Mohs hardness scale is a measure of a substance's resistance to scratching based on the comparative softness-hardness of 10 common minerals. Talc (number 1) is the softest and diamond (number 10) is the hardest.

Mordant. Commonly, metal salts serve as bridging agents between a textile and a dye. The metal acts as a mordant (from Latin, "mordere," to bite) by being "bitten" by functional groups on the dye molecule to form a metal–organic complex. Mordanting is accomplished by adding the mordanting agent to the dyebath itself, by treating the textile substrate with the mordant prior to dyeing, or by treating the already dyed fibers with the mordant. If the metal ion is fixed in the fiber through ionic bonding or other means, then the dye molecules are also fixed. However, since this is a chemical reaction, then changing the metal ion will change the identity of the complex, and hence, the color.

Nanomaterial, nanoparticles. Nanomaterials are defined by their particle size which measure, by definition, between one and 100 nm (nanometer = a billionth of a meter). Nanotechnology, a very active research area, is the control and manipulation of nanoscale materials, nanoparticles, *i.e.*, substances on an atomic or molecular scale.

Organic. Organic chemistry has traditionally been defined as the chemistry of carbon compounds in which one or more carbon atoms are covalently bonded and to other elements, commonly hydrogen, oxygen and nitrogen. Initially thought to have been exclusively produced by living organisms, and hence called "organic," these compounds can arise from any source. The boundary between organic and inorganic compounds is fluid since some inorganic compounds contain carbon and some organic compounds contain metals.

Oxidation is the chemical process by which an atom, ion or molecule loses one or more electrons. It can also more narrowly be defined as the gain of oxygen by a substance.

Plasmonics. Plasmons are surface waves of oscillating electrons at the interface between a noble metal and a dielectric (insulating material) that can capture and process light to produce vivid colors, offering a new and materially simpler route to color printing technologies.

Polymer, polymerization. A polymer is any natural or synthetic molecular structure composed of multiples of similar smaller units called monomers. Some examples are plastics and resins. Polymerization is the process of the chemical linking of monomers to form a polymer.

Purple dye formation. The simplified schema in Scheme 8 follows one dye precursor to the formation of the purple dye, 6,6'-dibromoindigo, realizing that other precursors would follow parallel paths. Once a murex snail's hypobranchial gland is ruptured, the enzyme purpurase comes into contact with tyrindoxyl sulfate and hydrolyzes it to tyrindoxyl, the enol form. Tyrindoxyl undergoes a molecular rearrangement, called tautomerism, to the keto form, tyrindolinenone, which dimerizes on oxidation to form tyriverdin. The final step, exposure to light and air, yields the purple colorant, 6,6'-dibromoindigo (DBI).

tyrindoxyl sulfate   tyrindoxyl   tyrindoleninone

dibromoindigo   tyriverdin

Quinacridones. A class of diketo synthetic organic pigments first prepared in 1915 and commercialized in 1958. About 150 quinacridone derivatives have been made from the parent structure, quinacridone (Pigment Violet 19). Quinacridone colors range from bluish-red to magenta-yellow; they are used extensively in artists' materials.

Radioactivity, Radioactive decay. The spontaneous disintegration of an unstable atomic nucleus by emission of subatomic particles and/or electromagnetic radiation. The most common types of emission are alpha (helium nuclei), beta (electrons) and gamma (high energy electromagnetic radiation).

Raman spectroscopy is an analytical tool whereby scattered visible light is used to determine the vibrational energy transitions of a sample, enabling sample identification.

Reduction; reducing agent. Reduction is the chemical process by which an atom, ion or molecule gains one or more electrons. A reducing agent is a substance that causes reduction by, in turn, becoming oxidized.

Refractive index (RI). RI is the ratio of the velocity of light in a vacuum to the velocity of light in a given medium, such as oil or water. It is the difference between the refractive indices of a paint medium and a pigment that determines the opacity or transparency of a paint. The larger the refractive index difference, the greater is the opacity of the paint.

Resins; terpenoids. Natural resins, called terpenoids, are viscous exudates of plants, principally of conifers. They are hydrocarbons, compounds containing only carbon and hydrogen, that are classified according to the number of carbon atoms in their formula. Monoterpenoids contain 10 carbons, sesquiterpenoids, 15, and diterpenoids, 20. Some examples are α-pinene ($C_{10}H_{16}$), cadinene ($C_{15}H_{24}$) and halimane ($C_{20}H_{38}$).

Vat dyeing. Vat dyeing, or vatting, is a process in which a dye is reduced in alkaline solution to render it soluble (usually near colorless), followed by impregnation of the dye material on a fabric, and subsequent oxidation to the insoluble colored form. Vat dyes have their own category in *The Colour Index*, and some colorants carry more than one usage number as a result. For example, indigo is listed both as Pigment Blue 66 and Vat Blue 1.

Wavelength. The distance between two successive points on a periodic wave such as light waves and other forms of electromagnetic radiation. As the wavelength increases, the frequency of occurrence decreases, and since frequency is the measure of the energy of a wave, increased wavelength leads to decreased energy. For example, red light has longer wavelengths than blue light, and is therefore less energetic.

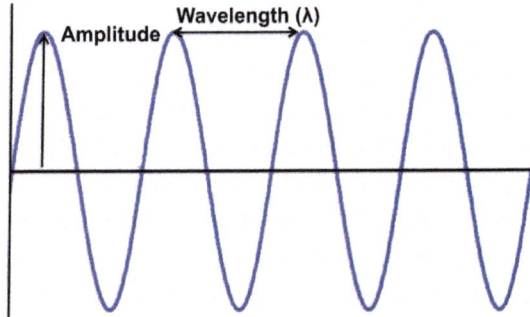

X-ray diffraction (XRD). When X-rays of a single wavelength, called monochromatic, strike a crystalline substance, they are diffracted at angles that can be correlated to the lattice spacings in the crystal. Each crystalline material has a unique set of these spacings, enabling its identification by comparison with standard reference values.

X-ray fluorescence (XRF). When a sample is irradiated with high energy X-rays, electrons in the inner shells of the atoms in the sample are dislodged. Electrons from the higher energy levels of the atoms drop to the lower energy state to fill the vacancies, thereby emitting fluorescent X-rays which are characteristic of each element.

X-Ray photoelectron spectroscopy. When a beam of X-rays irradiates a sample, the number and kinetic energies of ejected electrons can be measured. This technique, based on the photoelectric effect, can not only determine which elements are present but also their chemical environment.

Further Reading

1. R. Adunka and M. V. Orna, *Carl Auer von Welsbach: Chemist, Inventor, Entrepreneur*, Springer, Cham, 2018.
2. G. Agricola, *De Re Metallica*, trans. from the first Latin edition of 1556 by H. C. Hoover, and L. H. Hoover, , Salisbury House, London, 1912.
3. L. B. Alberti, *On Painting (1435–1436)*, trans. J. R. Spencer, Yale University Press, New Haven, 1970.
4. M. A. Angeloglou, *A History of Make-up*, Macmillan, New York, 1971.
5. *Archaeological Chemistry VIII*, ed. Armitage, R. A. and Burton, J. H., American Chemical Society, Washington, DC, 2013.
6. S. Augusti, *I Colori Pompeiani*, De Luca, Rome, 1967.
7. J. Ayres, *The Artist's Craft*, Phaidon, London, 1985.
8. G. H. Bachhoffner, *Chemistry as Applied to the Fine Arts*, J. Carpenter, London 1837.
9. J. Balfour-Paul, *Indigo – Egyptian Mummies to Blue Jeans*, Firefly Books, London, 2011.
10. P. Ball, *Bright Earth: Art and the Invention of Color*, Farrar, Straus and Giroux, New York, 2001.
11. P. Ball, *The Beauty of Chemistry: Art, Wonder and Science*, MIT Press, Cambridge, 2021.

March of the Pigments: Color History, Science and Impact
By Mary Virginia Orna
© Mary Virginia Orna 2022
Published by the Royal Society of Chemistry, www.rsc.org

12. *Chromic Phenomena: Technological Applications of Color Chemistry*, ed. Bamfield, P. and Hutchings, M. G., Royal Society of Chemistry, Cambridge, 2010.
13. P. B. Barcilon and P. C. Marani, *Leonardo, the Last Supper*, trans. H. Tighe, University of Chicago Press, Chicago, 2001.
14. J. Barnes, *Keeping an Eye Open*, Knopf, New York, 2015.
15. M. Bar-Zohar, *Bitter Scent: The Case of L'Oréal, the Nazis and the Arab Boycott*, Dutton Books, London, 1996.
16. F. E. Basten, *Max Factor: The Man Who Changed the Faces of the World*, Arcade Publishing, New York, 2008.
17. C. Baudelaire *The Mirror of Art*, Doubleday Anchor Books, New York, 1956.
18. C. Bäumler, *Paul Ehrlich: Scientist for Life*, Holmes & Meier, New York, 1984.
19. J. J. Beer, *The Emergence of the German Dye Industry*, University of Illinois Press, Urbana, 1981.
20. R. S. Berns, *Billmeyer and Saltzman's Principles of Color Technology*, 4th edn, Wiley, New York, 2019.
21. *Artists' Pigments: A Handbook of Their History and Characteristics*, ed. Berrie, B., National Gallery of Art, Washington, DC, 2007, vol. 4.
22. V. Biringuccio, *Pirotechnia: The Classic 16th Century Treatise on Metals and Metallurgy*, pub. 1540, trans. C. S. Smith and M. T. Gnudi, Dover Publications, Mineola, NY, 1990.
23. D. Blum, *The Poison Squad*, Penguin Press, New York, 2018.
24. D. Bomford, J. Leighton, J. Kirby and A. Roy, *Impressionism: Art in the Making*, National Gallery Publications, London, 1991.
25. D. Bomford, J. Dunkerton, D. Gordon and A. Roy, *Italian Painting before 1400: Art in the Making*, National Gallery Publications, London, 1989.
26. *The Cradle of Language*, ed. Botha, R. And Knight, C., Oxford University Press, New York, 2009.
27. *Cennino Cennini's Il Libro dell'Arte, a New English Translation and Commentary with Italian Transcription*, ed. Broecke, L., Archetype Publications, London, 2015.
28. F. Brunello, *The Art of Dyeing in the History of Mankind*, Neri Pozza, Vicenza, 1973.
29. F. Caron, *Dynamics of Innovation: The Expansion of Technology in Modern Times*, Berghahn Books, New York, 2013.

30. R. M. Christie, *Colour Chemistry*, Royal Society of Chemistry, Cambridge, 2015.

31. A. H. Church, *The Chemistry of Paints and Painting*, Seeley, Service, Co., London, 3rd edn, 1901.

32. *La Grotte Chauvet. L'art des origines*, ed. Clottes, J., Editions du Seuil, Paris, 2010.

33. C. Cobbold, *A Rainbow Palate: How Chemical Dyes Changed the West's Relationship with Food*, University of Chicago Press, Chicago, 2020.

34. M. Combs, The Polychromy of Greek and Roman Art; An Integration of Museum Practices, MA Dissertation, City University of New York, New York, 2012.

35. M. Corballis, *The Recursive Mind: The Origins of Human Language, Thought and Civilization*, Princeton University Press, Princeton, 2011.

36. *Physical Techniques in the Study of Art, Archaeology and Cultural Heritage*, ed. Creagh, D. and Bradley, D., Elsevier, New York, 2006, vol. 1.

37. M. P. Crosland, *Historical Studies in the Language of Chemistry*, Harvard University Press, Cambridge, MA, 1962.

38. G. Curtis, *The Cave Painters: Probing the Mysteries of the World's First Artists*, Knopf, New York, 2006.

39. W. V. Davies, *Colour and Painting in Ancient Egypt*, British Museum Press, London, 2001.

40. *Dermatological Complications with Body Art: Tattoos, Piercings and Permanent Make-Up*, ed. De Cuyper, C. and Pérez-Cotapos, S. M. L., Springer, Cham, 2nd edn, 2018.

41. F. Delamare, *Blue Pigments: 5000 Years of Art and Industry*, Archetype Publications, London, 2013.

42. C. Desdemaines-Hugon, *Stepping Stones*, Yale University Press, New Haven, 2010.

43. *Painting the Skin: Pigments on Bodies and Codices in Pre-columbian Mesoamerica*, ed. Dupey García, E. and Vásquez de Ágredos Pascual, M. L., University of Arizona Press, Tucson, 2018.

44. *The Pigment Compendium: A Dictionary of Historical Pigments*, ed. Eastaugh, N., Walsh, V., Chaplin, T. and Siddall, R., Elsevier, New York, 2004.

45. J. Edmonds. *The History of Woad and the Medieval Woad Vat*, 1st edn, 3rd reprint, 2006. Published by John Edmonds. Available from: Lulu.com.

46. L. Eldridge, *Face Paint: The Story of Make-Up*, Abrams, New York, 2015.

47. *Episodes from the History of the Rare Earth Elements*, ed. Evans, C. H., Springer, Heidelberg, 1996.

48. *High Performance Pigments*, ed. Faulkner, E. B. and Schwartz, R. J., Wiley-VCH, Weinheim, 2009.

49. G. Field, *Field's Chromatography or Treatise on Colours and Pigments as Used by Artists*, ed. Taylor, J. S., Winsor & Newton, London, 1885.

50. *Artists' Pigments: A Handbook of Their History and Characteristics*, ed. Feller, R. L., National Gallery of Art, Washington, DC, 1986, vol. 1.

51. V. Finlay, *The Brilliant History of Color in Art*, Getty Publications, Los Angeles, 2014.

52. V. Finlay, *Color. A Natural History of the Palette*, Random House, New York, 2004.

53. *Artists' Pigments: A Handbook of Their History and Characteristics*, ed. FitzHugh, E. W., Washington, DC: National Gallery of Art, 1997, vol. 3.

54. W. M. Flinders Petrie, *Prehistoric Egypt*, British School of Archaeology in Egypt, Bernard Quaritch, London, 1920.

55. M. Fontani, M. Costa and M. V. Orna, *The Lost Elements: The Periodic Table's Shadow Side*, Oxford University Press, New York, 2015.

56. M. R. Fox, *Dye-Makers of Great Britain 1856–1976*, Imperial Chemical Industries, Manchester, 1987.

57. M. R. Fox, *Vat Dyestuffs and Vat Dyeing*, Chapman and Hall, London, 1948.

58. J. N. Friend, *An Introduction to the Chemistry of Paints*, Longmans, Green, London, 1910.

59. R. H. Fritze, *Egyptomania: A History of Fascination, Obsession and Fantasy*, Reaktion Books, London, 2016.

60. J. S. Fruton, *Molecules and Life*, Wiley-Interscience, New York, 1972.

61. S. Garfield, *Mauve*, W.W. Norton, New York, 2001.

62. R. J. Gettens and G. L. Stout, *Painting Materials: A Short Encyclopaedia*, Dover Publications, New York, 1966.

63. *Essays in Global Color History: Interpreting the Spectrum*, ed. Goldman, R. B., Gorgias Press, Piscataway, 2016.

64. M. D. Gottsegen, *The Painter's Handbook*, Watson-Guptill, New York, 2006.

65. A. B. Greenfield, *A Perfect Red*, Harper Collins, New York, 2005.
66. F. Gunn, *The Artificial Face: A History of Cosmetics*. David and Charles, Newton Abbott, 1973.
67. R. D. Guthrie, *The Nature of Paleolithic Art*, University of Chicago Press, Chicago, 2006.
68. W. S. Haine, *The History of France*, ABC-CLIO, Santa Barbara, CA, 2nd edn, 2019.
69. J. Hamblin, *Clean: The New Science of Skin*, Riverhead, New York, 2020.
70. R. Harley, *Artists' Pigments. c. 1600-1835*, Butterworths, London, 1970.
71. M. K. Hartwig, *A Companion to Ancient Egyptian Art*, Wiley-Blackwell, London, 2014.
72. *The Collected Papers of Paul Ehrlich*, ed. Himmelweit, H., Pergamon Press, London, 1956, vol. 1.
73. J. B. Hurry, *The Woad Plant and its Dye*, Oxford University Press, London, 1930.
74. A. J. Ihde, *The Development of Modern Chemistry*, Dover Publications, New York 1984.
75. *Nobel Laureates in Chemistry: 1901-1992*, ed. James, L. K., American Chemical Society, Washington, DC, 1993.
76. H. W. Janson and D. J. Janson, *The Picture History of Painting*, Harry N. Abrams, New York, 1957.
77. I. Kakoulli, *Greek Painting Techniques and Materials from the Fourth to the First Century BC*, Archetype Publications, London, 2009.
78. F. Karim-Cooper, *Cosmetics in Shakespearean and Renaissance Drama*, Edinburgh University Press, Edinburgh, 2012.
79. V. J.-R. Kehoe, *The Technique of Film and Television Make-up*, Hastings House Publishers, New York, 1958.
80. K. L. Kelly, *Color Universal Language and Dictionary of Names*, National Bureau of Standards SP440, Washington, DC, 1976.
81. R. King, *Il Papa e il suo Pittore: Michelangelo e la Nascita avventurosa della Cappella Sistina*, Rizzoli, Milano, 2003.
82. K. Krawczynski, *Daily Life in the Colonial City*, Greenwood Press, Denver, 2013.
83. R. G. Kuehni and A. Schwarz, *Color Ordered: A Survey of Color Order Systems from Antiquity to the Present*, Oxford University Press, New York, 2008.

84. *Archaeological Chemistry III*, ed. Lambert, J. B., American Chemical Society, Washington, DC, 1984.
85. *Collaborative Endeavors in the Chemical Analysis of Art and Cultural Heritage Materials*, ed. Lang, P. and Armitage, R. A., American Chemical Society, Washington, DC, 2012.
86. A. P. Laurie, *The Materials of the Painter's Craft*, T.N. Foulis, London, 1910.
87. A. P. Laurie, *The Pigments and Mediums of the Old Masters*, Macmillan, London, 1914.
88. V. Lautman, *The New Tattoo*, Abbeville Press, New York, 1994.
89. M. D. Leakey, *Olduvai Gorge Volume III: Excavations in Beds I and II, 1960-1963*, Cambridge University Press, Cambridge, 1971.
90. W. F. Leggett, *Ancient and Medieval Dyes*, Chemical Pub. Co. Brooklyn, 1944.
91. M. D. Leonida, *The Materials and Craft of Early Iconographers*, Springer, Cham, 2014.
92. P. Levi, *The Periodic Table*, Schocken Books, New York, 1984.
93. *The Pigment Handbook*, ed. Lewis, P. A., Wiley, New York, 2nd edn, 1988, vol. 3.
94. K. T. Lillios, *The Archaeology of the Iberian Peninsula: From the Paleolithic to the Bronze Age*, Cambridge University Press, Cambridge, 2020.
95. C. Lloyd and R. Thomson, *Impressionist Drawings: From British Public and Private Collections*, Phaidon Press, Oxford, 1986.
96. A. Loeb, *Extraterrestrial: the First Sign of Intelligent Life beyond Earth*, Houghton Mifflin Harcourt, New York, 2021.
97. A. Lucas, *Ancient Egyptian Materials and Industries*, E. Arnold, London, 1962.
98. M. Marquardt, *Paul Ehrlich*, Schuman, New York, 1951.
99. R. Mayer, *The Artist's Handbook of Materials and Techniques*, Viking Press, New York, 3rd edn, 1970.
100. B. McFarland, *A World from Dust*, Oxford University Press, New York, 2016.
101. S. B. McGrayne, *Prometheans in the Lab: Chemistry and the Making of the Modern World*, McGraw-Hill, New York, 2001.
102. M. P. Merrifield, *Original Treatises on the Arts of Painting*, John Murray, London, 1849, vol. 2.

103. M. A. Miczak, *Henna's Secret History: The History, Mystery and Folklore of Henna*, Writers Club Press, New York, 2001.

104. M. Minnaert, *The Nature of Light and Color in the Open Air*, Dover Publications, New York, 1954.

105. *Color for Science, Art and Technology*, ed. Nassau, K., Elsevier, New York, 1998.

106. I. Newton, *Opticks*, Dover Publications, New York, 1952.

107. C. Nicholl, *Leonardo da Vinci: Flights of the Mind*, Viking, New York, 2004.

108. *Ancient Egyptian Materials and Technology*, ed. Nicholson, P. T. and Shaw I., Cambridge University Press, Cambridge, 2000.

109. R. Nietzki *Chemistry of the Organic Dyestuffs*, trans. A. Collin and W. Richardson, Gurney and Jackson, London, 1892.

110. K. Okamoto and W. Brenner, *Organic Semiconductors*, Reinhold, New York, 1964.

111. *Archaeological Chemistry: A Multidisciplinary Analysis of the Past*, ed. Orna, M. V. and Rasmussen, S. C., Cambridge Scholars Publishing, Newcastle upon Tyne, 2021.

112. M. V. Orna, *The Chemical History of Color*, Springer, Heidelberg, 2013.

113. *Chemistry's Role in Food Production and Sustainability: Past and Present*, ed. Orna, M. V., Eggleston, G. and Bopp, A. F., American Chemical Society, Washington, DC, 2019.

114. *Science History: A Traveler's Guide*, ed. Orna, M. V., American Chemical Society, Washington, DC, 2014.

115. S. Panayotova and P. Ricciardi, *Masters' Secrets in Colour – the Art and Science of Illuminated Manuscripts*, Harvey Miller/Brepols, London and Turnhout, 2016.

116. M. Pastoureau, *Black: The History of a Color*, Princeton University Press, Princeton, 2009.

117. M. Pastoureau, *Blue: The History of a Color*, Princeton University Press, Princeton, 2018.

118. M. Pastoureau, *Green: The History of a Color*, Princeton University Press, Princeton, 2014.

119. M. Pastoureau, *Red: The History of a Color*, Princeton University Press, Princeton, 2017.

120. M. Pastoureau, *Yellow: The History of a Color*, Princeton University Press, Princeton, 2019.

121. R. H. Peters, *Textile Chemistry, vol. III: The Physical Chemistry of Dyeing*, Elsevier, New York, 1975.

122. J. Pipe, *Egyptian Mummies: A Very Peculiar History*, The Salariya Book Co., Brighton, 2012.
123. Pliny the Elder, *The Elder Pliny's Chapters on the History of Art*, trans. K. Jex-Blake and ed. Sellers, E., Macmillan, London, 1896.
124. Pliny the Elder, *Natural History*, trans. J. Bostock and H. T. Riley, Henry G. Bohn, London, 1855–1857.
125. *A Companion to the Archaeology of the Ancient Near East*, ed. Potts, D. T., Blackwell Publishing, Oxford, 2012.
126. D. Rankine, *Heka – The Practices of Ancient Egyptian Ritual and Magic*, Avalonia, London, 2006.
127. G. Rapp, *Archaeomineralogy*, Springer, Heidelberg, 2009.
128. *Chemical Technology in Antiquity*, ed. Rasmussen, S. C., American Chemical Society, Washington, DC, 2015.
129. J. Ray, *The Rosetta Stone: And the Rebirth of Ancient Egypt*, Profile Books, London, 2008.
130. M. Rayner-Canham and G. Rayner-Canham, *Women in Chemistry: Their Changing Roles from Alchemical Times to the Mid-twentieth Century*, American Chemical Society, Washington, DC, 1998.
131. J. Renoir, *My Father*, Collins, London, 1962.
132. A. Rogers, *Full Spectrum: How the Science of Color Made Us Modern*, Houghton Mifflin Harcourt, New York, 2021.
133. O. N. Rood, *Students' Text-Book of Color: Or, Modern Chromatics, with Applications to Art and Industry*, D. Appleton, New York, 1881.
134. *Artists' Pigments: A Handbook of Their History and Characteristics*, ed. Roy, A., National Gallery of Art, Washington, DC, 1993, vol. 2.
135. F. Bernardino de Sahagún, *Historia general de las Cosas de la Nueva España*, A. Valdés, México, 1829–1830.
136. *The Conservation of Subterranean Cultural Heritage*, ed. Saiz-Jimenez C., CRC Press, New York, 2014.
137. M. Sanz de Sautuola, *Breves apuntes sobre algunos objetos prehistóricos de la provincia de Santander*, Telesforo Martínez Blanca, Santander, 1880.
138. B. Schaefer *Natural Products in the Chemical Industry*, Springer, New York and Heidelberg, 2014.
139. *Science and Art: The Contemporary Painted Surface*, ed. Sgamellotti, A., Brunetti, B. G. and Miliani, C., Royal Society of Chemistry, Cambridge, 2020.

140. *Science and Art: The Painted Surface*, ed. Sgamellotti, A., Brunetti, B. G. and Miliani, C., Royal Society of Chemistry, Cambridge, 2014.

141. V. Sherrow, *For Appearance' Sake: The Historical Encyclopedia of Good Looks, Beauty and Grooming*, Oryx Press, Westport, 2001.

142. M. Sommer, *Bones and Ochre: The Curious Afterlife of the Red Lady of Paviland*, Harvard University Press, Cambridge, 2007.

143. *The Royal Purple and the Biblical Blue*, ed. Spanier, E., Keter Publishing House, Jerusalem, 1987.

144. R. Sugg, *Mummies, Cannibals and Vampires: The History of Corpse Medicine from the Renaissance to the Victorians*, Routledge, New York, 2nd edn, 2016.

145. *The Value of Colour. Material and Economic Aspects in the Ancient World, Berlin Studies of the Ancient World 70*, ed. Thavapalan, I. S. and and Warburton, D. A., Edition Topoi, Berlin, 2020.

146. Theophilus, *On Divers Arts*, trans. J. G. Hawthorne and C. S. Smith, Dover, New York, 1979.

147. D. V. Thompson, *The Materials and Techniques of Medieval Painting*, Dover Books, New York, 1956.

148. G. Tierie, *Cornelis Drebbel, 1572–1633*, H. J. Paris, Amsterdam, 1932.

149. A. S. Travis, *The Rainbow Makers: The Origins of the Synthetic Dyestuff Industry in Western Europe*, Lehigh University Press, Bethlehem, 1993.

150. G. Vail, *A History of Cosmetics in America*, Toilet Goods Association, New York, 1947.

151. G. Vasari, *The Lives of the Artists*, trans. J. C. P. Bondanella, Oxford University Press, Oxford, 1991.

152. W. L. Voegtlin, *The Stone Age Diet*, Vantage Press, New York, 1975.

153. D. Walters, *Fortress Plant: How to Survive When Everything Wants to Eat You*, Oxford University Press, New York, 2017.

154. M. E. Weeks, *Discovery of the Elements*, 7th edn, Journal of Chemical Education, Easton, PA, 1968.

155. R. White, *L'Art préhistorique dans le Monde*, La Martinière, Paris, 2003.

156. T. Wilkinson, *A World beneath the Sands*, W.W. Norton, New York, 2020.

157. T. A. H. Wilkinson, *Writings from Ancient Egypt*, Penguin Books, London, 2016.

158. *Chemistry and Technology of the Cosmetics and Toiletries Industry*, ed. Williams, D. F. and Schmitt, W. H., Blackie Academic and Professional, London, 2nd edn, 1996.

159. R. J. P. Williams and R. Rickaby, *Evolution's Destiny: Co-Evolving Chemistry of the Environment and Life*, RSC Publishing, London, 2012.

160. *The Heaviest Metals: Science and Technology of the Actinides and Beyond*, ed. Williams, W. J. and Hanusa, T. P., John Wiley and Sons, New York, 2019.

161. J. J. Winckelmann, *Reflections of the Painting and Sculpture of the Greeks: With Instructions for the Connoisseur, and an Essay On Grace in Works of Art*, trans. H. Fusseli, A. Millar and T. Cadell, Andrew Millar, London, 1767.

162. J. Wood, *Personal Narrative of a Journey to the Source of the River Oxus*, John Murray, London, 1841.

163. L. Woodhead, *War Paint: Madame Helena Rubinstein and Miss Elizabeth Arden: Their Lives, Their Times, Their Rivalry*, Turner Publishing, Nashville, 2004.

164. *The Travels of Marco Polo, the Venetian*, ed. Wright, T., George Bell & Sons, London, 1886.

165. F. Wynne, *I Was Vermeer: The Rise and Fall of the Twentieth Century's Greatest Forger*, Bloomsbury, New York, 2006.

166. *Electrochromic Smart Materials: Fabrication and Applications*, ed. Xu, J. W., Chua, M. H., and Shah, K. W., Royal Society of Chemistry, Cambridge, 2019.

167. H. Zollinger, *Color: A Multidisciplinary Approach*, Wiley VCH, New York, 1999.

Subject Index